Biomedical Photonics for Diabetes Research

In 2021, over 537 million people worldwide were diagnosed with diabetes, according to the International Diabetes Federation, and so the diagnosis, care, and treatment of patients with diabetes mellitus have become one of the highest healthcare priorities. Biomedical photonics methods have been found to significantly improve and assist in the diagnosis of various disorders and complications arising from diabetes. These methods have also been widely used in various studies in the field of diabetes, including in the assessment of biochemical characteristics, metabolic processes, and microcirculation that are impaired in this disease.

This book provides an introduction to methods of biomedical photonics. The chapters, written by world-leading experts, cover a wide range of issues, including the theoretical basis of different biophotonics methods and practical issues concerning the conduction of experimental studies to diagnose disorders associated with diabetes. It provides a comprehensive summary of the recent advances in biomedical optics and photonics in the study of diabetes and related complications.

This book will be of interest to biomedical physicists and researchers, in addition to practicing doctors and endocrinologists looking to explore new instrumental methods for monitoring the effectiveness of patient treatment.

Features

- The first collective book combining accumulated knowledge and experience in the field of diabetes research using biophotonics.
- Contributions from leading experts in the field.
- Combines the theoretical base of the described methods and approaches, as well as providing valuable practical guidance and the latest research from experimental studies.

Series in Medical Physics and Biomedical Engineering

Series Editors

Kwan-Hoong Ng, E. Russell Ritenour, and Slavik Tabakov

RECENT BOOKS IN THE SERIES:

Auto-Segmentation for Radiation Oncology: State of the Art
Jinzhong Yang, Gregory C. Sharp, and Mark Gooding

Clinical Nuclear Medicine Physics with MATLAB: A Problem Solving Approach
Maria Lyra Georgosopoulou (Ed)

Handbook of Nuclear Medicine and Molecular Imaging for Physicists – Three Volume Set
Volume I: Instrumentation and Imaging Procedures
Michael Ljungberg (Ed)

Practical Biomedical Signal Analysis Using MATLAB®
Katarzyna J. Blinowska and Jaroslaw Zygierewicz

Handbook of Nuclear Medicine and Molecular Imaging for Physicists – Three Volume Set
Volume II: Modelling, Dosimetry and Radiation Protection
Michael Ljungberg (Ed)

Handbook of Nuclear Medicine and Molecular Imaging for Physicists – Three Volume Set
Volume III: Radiopharmaceuticals and Clinical Applications
Michael Ljungberg (Ed)

Electrical Impedance Tomography: Methods, History and Applications, Second Edition
David Holder and Andy Adler (Eds)

Introduction to Medical Physics
Cornelius Lewis, Stephen Keevil, Anthony Greener, Slavik Tabakov, and Renato Padovani

Calculating X-ray Tube Spectra: Analytical and Monte Carlo Approaches
Gavin Poludniowski, Artur Omar, and Pedro Andreo

Problems and Solutions in Medical Physics: Radiotherapy Physics
Kwan-Hoong Ng, Ngie Min Ung, and Robin Hill

Biomedical Photonics for Diabetes Research
Andrey V. Dunaev and Valery V. Tuchin (Eds)

For more information about this series, please visit: https://www.routledge.com/Series-in-Medical-Physics-and-Biomedical-Engineering/book-series/CHMEPHBIOENG

Biomedical Photonics for Diabetes Research

Edited by
Andrey V. Dunaev and Valery V. Tuchin

CRC Press
Taylor & Francis Group
Boca Raton London New York

CRC Press is an imprint of the
Taylor & Francis Group, an **informa** business

First edition published 2023
by CRC Press
4 Park Square, Milton Park, Abingdon, Oxon, OX14 4RN

and by CRC Press
6000 Broken Sound Parkway NW, Suite 300, Boca Raton, FL 33487-2742

CRC Press is an imprint of Taylor & Francis Group, LLC

ISBN: 978-0-367-62830-7 (hbk)
ISBN: 978-0-367-63091-1 (pbk)
ISBN: 978-1-003-11209-9 (ebk)

DOI: 10.1201/9781003112099

Typeset in Times
by codeMantra

Contents

Editors

Andrey V. Dunaev is a Professor and a Leading Researcher at the Research and Development Center of Biomedical Photonics at Orel State University (Orel, Russia). His research interests are devoted to multimodal optical diagnostics of microcirculatory-tissue systems, including methodological and metrological support of diagnostic systems. He is a senior member of SPIE. Professor Dunaev is an Honorary Worker of Science and High Technologies of the Russian Federation. He is an author of more than 100 papers in peer-reviewed journals, 5 monographs and has over 10 patents.

Valery V. Tuchin is a Corresponding Member of the Russian Academy of Sciences, Professor, Head of the Department of Optics and Biophotonics and Director of the Science Medical Center of Saratov State University. He is also Head of the Laboratory for Laser Diagnostics of Technical and Living Systems at the Institute of Precision Mechanics and Control, FRC "Saratov Scientific Center of the Russian Academy of Sciences." His research interests include biophotonics, biomedical optics, tissue optics, laser medicine, tissue optical clearing, and nanobiophotonics. He is a member of SPIE, OSA, and IEEE, Visiting Professor at HUST (Wuhan) and Tianjin Universities in China, and an Adjunct Professor at the University of Limerick (Ireland) and the National University of Ireland (Galway).

Professor Tuchin is elected Fellow SPIE and OSA (OPTICA), he was awarded many titles and awards, including Honored Scientist of the Russian Federation, Honored Professor of SSU, Honored Professor of Finland (FiDiPro), SPIE Educational Award, Chime Bell Award of Hubei province (China), Joseph Goodman Award (OSA / SPIE) for Outstanding Monograph (2015), Michael Feld Award (OSA) for Pioneering Research in Biophotonics (2019), the Medal of the D.S. Rozhdestvensky Optical Society (2018) and the A.M. Prokhorov medal of the Academy of Engineering Sciences named after A.M. Prokhorov (2021). He is the author of over 1000 articles, 30 monographs, and textbooks, has over 60 patents, his works have been cited over 34,500 times.

Contributors

Robert Bartlett
Swift Medical Inc.
Toronto, Canada

Alla B. Bucharskaya
Science Medical Center
Saratov State University
Saratov, Russia
and
Laboratory of Laser Molecular Imaging and Machine Learning
Tomsk State University
Tomsk, Russia
and
Saratov State Medical University
Saratov, Russia

Alexander V. Bykov
Opto-Electronics and Measurement Techniques Unit
University of Oulu
Oulu, Finland

Olga P. Cherkasova
Biophysics laboratory
Institute of Laser Physics of SB RAS
Novosibirsk, Russia

Petr S. Demchenko
Terahertz Biomedicine Laboratory
ITMO University
St. Petersburg, Russia

Nataliya I. Dikht
Saratov State Medical University
Saratov, Russia

Alexandre Yu. Douplik
Department of Physics
Ryerson University
Toronto, Canada
and
iBest, Keenan Research Centre of the LKS Knowledge Institute
St. Michael's Hospital
Toronto, Canada

Viktor V. Dremin
College of Engineering and Physical Sciences
Aston University
Birmingham, UK
and
Research and Development Center of Biomedical Photonics
Orel State University
Orel, Russia

Andrey V. Dunaev
Research and Development Center of Biomedical Photonics
Orel State University
Orel, Russia

Polina A. Dyachenko (Timoshina)
Science Medical Center
Saratov State University
Saratov, Russia
and
Laboratory of Laser Molecular Imaging and Machine Learning
Tomsk State University
Tomsk, Russia

Petr B. Ermolinskiy
Physics Department
Lomonosov Moscow State University
Moscow, Russia

Anastasia A. Fabrichnova
Faculty of Fundamental Medicine
Lomonosov Moscow State University
Moscow, Russia

Wei Feng
Britton Chance Center for Biomedical Photonics, Wuhan National Laboratory for Optoelectronics
Huazhong University of Science and Technology
Wuhan, China
and
MoE Key Laboratory for Biomedical Photonics, Collaborative Innovation Center for Biomedical Engineering, School of Engineering Sciences
Huazhong University of Science and Technology
Wuhan, China

Yury I. Gurfinkel
Medical Research and Educational Center
Lomonosov Moscow State University
Moscow, Russia

Sviatoslav I. Gusev
Terahertz Biomedicine Laboratory
ITMO University
St. Petersburg, Russia

Mikhail K. Khodzitsky
Terahertz Biomedicine Laboratory
ITMO University
St. Petersburg, Russia

Natalya G. Kiseleva
Department of Pediatrics
Prof. V.F. Voino-Yasenetsky Krasnoyarsk State Medical University
Krasnoyarsk, Russia

Alexander I. Krupatkin
National Medical Research Center of Traumatology and Orthopaedics
Moscow, Russia

Dongyu Li
Britton Chance Center for Biomedical Photonics, Wuhan National Laboratory for Optoelectronics
Huazhong University of Science and Technology
Wuhan, China
and
MoE Key Laboratory for Biomedical Photonics, Collaborative Innovation Center for Biomedical Engineering, School of Engineering Sciences
Huazhong University of Science and Technology
Wuhan, China

Yulia I. Loktionova
Research and Development Center of Biomedical Photonics
Orel State University
Orel, Russia

Andrei E. Lugovtsov
Physics Department
Lomonosov Moscow State University
Moscow, Russia

Igor V. Meglinski
College of Engineering and Physical Sciences
Aston University
Birmingham, UK
and
Opto-Electronics and Measurement Techniques Unit
University of Oulu
Oulu, Finland

Irina A. Mizeva
Institute of Continuous Media Mechanics
Ural Branch of Russian Academy of Science
Perm, Russia

Ravshanjon Kh. Nazarov
Terahertz Biomedicine Laboratory
ITMO University
St. Petersburg, Russia

Nisan Ozana
Faculty of Engineering
Bar-Ilan University
Ramat-Gan, Israel

Alexey P. Popov
VTT Technical Research Centre of Finland
Oulu, Finland

Elena V. Potapova
Research and Development Center of Biomedical Photonics
Orel State University
Orel, Russia

Alexander V. Priezzhev
Physics Department
Lomonosov Moscow State University
Moscow, Russia

Gennadi Saiko
Swift Medical Inc.
Toronto, Canada
and
Department of Physics
Ryerson University
Toronto, Canada

Vladimir V. Salmin
Department of Medical and Biological Physics
Prof. V.F. Voino-Yasenetsky Krasnoyarsk State Medical University
Krasnoyarsk, Russia

Alla B. Salmina
Research Institute of Molecular Medicine and Pathobiochemistry
Krasnoyarsk, Russia
and
Department of Brain Research
Research Center of Neurology
Moscow, Russia

Rui Shi
Britton Chance Center for Biomedical Photonics, Wuhan National Laboratory for Optoelectronics
Huazhong University of Science and Technology
Wuhan, China
and
MoE Key Laboratory for Biomedical Photonics, Collaborative Innovation Center for Biomedical Engineering, School of Engineering Sciences
Huazhong University of Science and Technology
Wuhan, China

Viktor V. Sidorov
SPE "LAZMA" Ltd.
Moscow, Russia

Tatyana E. Taranushenko
Department of Pediatrics
Prof. V.F. Voino-Yasenetsky Krasnoyarsk State Medical University
Krasnoyarsk, Russia

Georgy S. Terentyuk
Science Medical Center
Saratov State University
Saratov, Russia

Valery V. Tuchin
Science Medical Center
Saratov State University
Saratov, Russia
and
Laboratory of Laser Molecular Imaging and Machine Learning
Tomsk State University
Tomsk, Russia
and
A.N. Bach Institute of Biochemistry
Research Center of Biotechnology of the Russian Academy of Sciences
Moscow, Russia
and
Laboratory of Laser Diagnostics of Technical and Living Systems
Institute of Precision Mechanics and Control
Saratov, Russia

Daria K. Tuchina
Science Medical Center
Saratov State University
Saratov, Russia
and
Laboratory of Laser Molecular Imaging and Machine Learning
Tomsk State University
Tomsk, Russia
and
A.N. Bach Institute of Biochemistry
Research Center of Biotechnology of the Russian Academy of Sciences
Moscow, Russia

Tingting Yu
Britton Chance Center for Biomedical Photonics, Wuhan National Laboratory for Optoelectronics
Huazhong University of Science and Technology
Wuhan, China
and
MoE Key Laboratory for Biomedical Photonics, Collaborative Innovation Center for Biomedical Engineering, School of Engineering Sciences
Huazhong University of Science and Technology
Wuhan, China

Zeev Zalevsky
Faculty of Engineering
Bar-Ilan University
Ramat-Gan, Israel

Tianmiao Zhang
Terahertz Biomedicine Laboratory
ITMO University
St. Petersburg, Russia

Elena V. Zharkikh
Research and Development Center of Biomedical Photonics
Orel State University
Orel, Russia

Evgenii A. Zherebtsov
Optoelectronics and Measurement Techniques
University of Oulu
Oulu, Finland
and
Research and Development Center of Biomedical Photonics
Orel State University
Orel, Russia

Angelina I. Zherebtsova
Optoelectronics and Measurement Techniques
University of Oulu
Oulu, Finland
and
Research and Development Center of Biomedical Photonics
Orel State University
Orel, Russia

Dan Zhu
Britton Chance Center for Biomedical Photonics, Wuhan National Laboratory for Optoelectronics
Huazhong University of Science and Technology
Wuhan, China
and
MoE Key Laboratory for Biomedical Photonics, Collaborative Innovation Center for Biomedical Engineering, School of Engineering Sciences
Huazhong University of Science and Technology
Wuhan, China

Jingtan Zhu
Britton Chance Center for Biomedical Photonics, Wuhan National Laboratory for Optoelectronics
Huazhong University of Science and Technology
Wuhan, China
and
MoE Key Laboratory for Biomedical Photonics, Collaborative Innovation Center for Biomedical Engineering, School of Engineering Sciences
Huazhong University of Science and Technology
Wuhan, China

Introduction

Andrey V. Dunaev and Valery V. Tuchin

In 2021, over 537 million people worldwide were diagnosed with diabetes, according to the International Diabetes Federation and so the diagnosis, care, and treatment of patients with diabetes mellitus have become one of the highest healthcare priorities. Biomedical photonics methods have been found to significantly improve and assist in the diagnosis of various disorders and complications arising from diabetes. These methods have also been widely used in various studies in the field of diabetes, including in the assessment of biochemical characteristics, metabolic processes, and microcirculation that are impaired in this disease.

This book provides an introduction to methods of biomedical photonics. The chapters, written by world-leading experts, cover a wide range of issues, including the theoretical basis of different biophotonics methods and practical issues concerning the conduction of experimental studies to diagnose disorders associated with diabetes. It provides a comprehensive summary of the recent advances in biomedical optics and photonics in the study of diabetes and related complications.

This book will be of interest to biomedical physicists and researchers, in addition to practicing doctors and endocrinologists looking to explore new instrumental methods for monitoring the effectiveness of patient treatment.

Features

- The first collective book combining accumulated knowledge and experience in the field of diabetes research using biophotonics.
- Contributions from leading experts in the field.
- Combines the theoretical base of the described methods and approaches, as well as providing valuable practical guidance and the latest research from experimental studies.

Diabetes mellitus (DM) is a chronic disease that not only significantly reduces the quality of life of patients, but also leads to significant social and economic problems around the world. DM is a group of chronic long-term diseases characterized by various metabolic disorders. There are two main forms of DM – type 1 and type 2 diabetes; however, diabetes can also occur during pregnancy and under the influence of other conditions.

Over the past few years, the prevalence of DM has reached alarmingly high levels. The number of diagnosed patients with this disease already amounts to approximately more than half a billion worldwide, and this number will only grow in the near future, according to forecasts. Owing to the increasing prevalence of the disease, complications of diabetes are now considered one of the most important problems of modern healthcare. According to experts, almost every patient with DM will develop at least one or more complications throughout his or her life. Complications of diabetes range from acute, life-threatening conditions (such as severe hypoglycemia or ketoacidosis) to chronic, debilitating complications affecting many organs and organ systems (such as retinopathy, nephropathy, neuropathy, and cardiovascular disease). Chronic complications of diabetes are formed under the influence of prolonged exposure to high levels of glucose in the body. They are associated with disorders of the cardiovascular and nervous systems. Ultimately, these complications can lead to severe vision loss and blindness, end-stage of renal disease and need for hemodialysis or transplantation, development and infection of diabetic ulcers, amputations, heart failure, stroke, and so on. Currently, it has been demonstrated that the length and health-related quality of life of diabetic patients are determined by the presence and severity of chronic complications of this disease, as well as the quality of their treatment. In this regard, there is an urgent need for the early detection and prevention of the development of diabetic complications.

Biophotonics methods represent a viable solution to this problem. Various spectroscopy and imaging technologies can provide information on the optical properties of skin, which are directly related to its blood supply, degree of oxygenation, and the presence of chromophores. Thus, biophotonics gives unique possibilities for both structural and functional analyses of biological tissues, as well as early and noninvasive diagnosis and monitoring of the effectiveness of the therapy in various diseases. Recently, there has been a surge of interest in the use of optical technologies in DM. More and more articles on related topics are published every year in highly rated journals. To date, considerable experience has been accumulated in research in the application of biophotonics methods for the diagnosis of disorders associated with DM. However, to date, these studies are scattered in separate journal articles and book chapters (for e.g., [1–3]). To date, the only book on the optical measurement of free glucose is known, which was published in 2009 [4]. The current book presents new data since 2009 on the development of biophotonics methods for detecting free glucose, but its main difference lies in the discussion of optical methods for monitoring the degree of tissue glycation and associated disorders and complications in the development of the disease. In this book, the knowledge accumulated over last years of research in this area and the prospects for further research in this direction are summarized.

This book contains 11 chapters. Chapter 1 provides a brief analysis of the vital complications that occur in DM, which are largely due to the glycation of protein and lipid components of tissues with an elevation of free glucose in the bloodstream. The two most commonly used animal models of DM based on its chemical induction by alloxan or streptozotocin are described. Noninvasive optical methods that make it possible to determine the rate of molecular diffusion in tissues are presented. Spectroscopic techniques have been used in *ex vivo* as Optical Coherence Tomography (OCT) in *in vivo* studies. The diffusion coefficients of glycerol and glucose in the skin, myocardial, and kidney samples are measured. Noninvasive optical methods for microcirculation monitoring, including laser Doppler flowmetry (LDF), Doppler OCT (DOCT), and laser speckle contrast imaging (LSCI), have been tested in animal models of DM.

Chapter 2 reviews the optical methods in DM foot ulcer screening. This chapter focuses on the overview of the pathophysiology of diabetic foot, including microbial burden, current "gold standard" diagnostics, prediction of development: small and large fiber tests, Doppler ultrasound, skin perfusion pressures (SPP), indocyanine green (ICG) angiography (exogenous fluorescent imaging), and transcutaneous oxygen pressures ($TcPO_2$). This chapter also concisely presents the novel screening optical modalities, which have been translated into the clinic but have not yet received widespread clinical adoption: LDF, laser Doppler (perfusion) imaging (LDPI), laser speckle (perfusion or contrast) imaging (LSPI or LSCI), near-infrared spectroscopy (NIRS), hyperspectral (HSI) and multispectral imaging (MSI), spatial frequency domain imaging (SFDI), orthogonal polarization spectral imaging (OPS), and thermography. Thus, this chapter details the diagnostic methods that are already being used or with the potential to be used by primary care practitioners for screening purposes. The application of efficient screening modalities allows earlier interventions in a patient population that already presents clinically with late-stage complications, significant morbidity, and mortality risk.

Chapter 3 presents the review of the principles and capabilities of optical techniques that are based on the interaction of light with whole blood or red blood cell (RBC) suspensions to assess the parameters of blood microcirculation and microrheology during DM. The capillaroscopy and aggregometry methods providing *in vitro* measurements of microrheologic parameters and *in vivo* monitoring of blood microcirculation are discussed. The former includes aggregation and deformability properties of RBC, forces of their pair interaction, and aggregation rate that are measured by diffuse light scattering, laser diffractometry, laser trapping, and manipulation techniques. In this chapter, it was demonstrated that the alterations of the parameters measured *in vivo* and *in vitro* for patients suffering from DM are interrelated. Good agreement between the results obtained with different techniques, and their applicability for the diagnostics of microrheological properties of blood were demonstrated. The possibility to use the reviewed methods in clinical practice as appropriate techniques for estimating blood microcirculation and microrheological disorders in the case of DM was substantiated.

Chapter 4 is devoted to the diagnostics of functional abnormalities in the microcirculation system using the LDF method. This chapter reviews the anatomical, morphological, and functional features of cutaneous microcirculation, as well as the microcirculatory disorders caused by DM. The possibilities of the LDF technique in the diagnosis of microcirculation disorders are illustrated using the results of a clinical study conducted to evaluate skin blood flow regulation abnormalities in patients with DM and obtained via the wavelet analysis of the LDF perfusion records. A stepwise protocol for the thermal test, the original data processing methods, and specific data interpretation techniques are presented.

In Chapter 5, recent advances in the development and clinical applications of a new class of wearable devices for blood perfusion monitoring and diagnosis of microvascular complications in DM by using LDF method are reviewed. This chapter shows the possibilities of a fine analysis of capillary blood flow structure and its rhythmic oscillations in the time and frequency domains, coupled with a new possibility of round-the-clock monitoring, that provide a valuable diagnostic information about the state of microvascular blood flow.

Chapter 6 is devoted to the optical angiography that offers essential tools for an accurate evaluation of the structural and functional changes for DM-targeted organs and tissues, which provide valuable information for the early diagnosis and therapy of DM. This chapter introduces the structural and functional changes of vasculatures induced by DM within kidneys, skin, and cerebrum, as well as how novel optical imaging techniques *in vitro* and *in vivo* can be used to study these changes induced by DM.

Chapter 7 is devoted to noninvasive detection methods of carbohydrate metabolism-related parameters with UV-induced fluorescence spectroscopy. Approaches are given for assessing such promising diagnostic and prognostic biomarkers as advanced glycation end products (AGEs) and their receptors expressed either in tissues (RAGEs) or in the peripheral blood (sRAGEs).

Chapter 8 describes a diagnostic approach, based on photonics-based technology and innovative solutions in machine learning, which is capable of evaluating the skin complications of DM at an early stage. In this chapter, it is shown that the technique of polarization-based hyperspectral imaging developed, accomplished by implementing an artificial neural network, provides new horizons in the study and diagnosis of age-related diseases.

Chapter 9 provides an overview of research conducted in the field of noninvasive assessment of cutaneous fluorescence for the detection and assessment of the severity of complications of DM. A detailed review of studies assessing the accumulation of AGEs as well as other intrinsic fluorophores of human tissues using the fluorescence method is presented. Multimodal studies involving the combination of other biophotonics methods with fluorescence spectroscopy for the analysis of diabetic complications are also considered. This chapter shows the developed method of diagnosing complications of DM, combining the analysis of skin fluorescence and the LDF signal registered from patients and experimental studies on its validation.

The last two chapters are devoted to research on the possibility of using biophotonics methods to measure blood glucose levels. In particular, Chapter 10 demonstrates the possibility of the application of THz time-domain spectroscopy for the assessment of diabetic complications. In this chapter, the noninvasive glucose measuring technique that utilizes the reflection of the THz pulse from the nail plate/nail bed interface is proposed. The possibility of the development of metafilm-based method for detecting glucose levels in blood using THz time-domain spectroscopy was demonstrated. Also, the influence of concentration of bilirubin, creatine, uric acid, and triglycerides on the refractive index of blood is shown.

Chapter 11 reviews several photonic methods, which are based on temporal-spatial analysis of the collected speckle field, for sensing glucose in the bloodstream. This chapter presents remote optical approach for measuring nanovibrations generated in the skin due to the changes in the bloodstream parameters as a function of the glucose concentration in the blood. It was shown that these temporal-spatial vibration-related changes can be analyzed with different machine learning algorithms (e.g., random forest) to enhance the sensitivity of the measurements. Furthermore, shown an experimental approach in which excited acoustic waves are generated and change the refractive index of the glucose solution and these changes are monitored via remote sensing of temporal speckle size variations. Finally, in this chapter, very promising remote clinical approach for continuous monitoring of diabetic foot patients via laser speckle contrast analysis is presented.

This book is aimed for researchers, postgraduate and undergraduate students, biomedical engineers, and physicians who are interested in designing and applying biophotonics methods and instruments for the diagnosis and treatment of DM complications, and the general application of optical methods in medicine and the medical device industry. Because of the large amount of fundamental and basic research on optical methods presented in this book, it should be useful for a broad audience, including students and physicians. Physicians and biomedical engineers will be particularly interested in the chapters covering clinical applications and instrumentation. Optical engineers will also find many critical applications to stimulate novel ideas of laser and optical design systems.

Finally, the editors would like to thank all the authors who devoted their precious time to contribute very interesting and knowledgeable chapters, all who helped us in the preparation of this book, authors and publishers for their permission of reproducing their figures in this book, and the editorial staff of the publisher.

REFERENCES

1. Geddes, Chris D., and Joseph R. Lakowicz, eds. *Glucose Sensing*, Vol. 11. Springer Science & Business Media, New York, 2007.
2. Cunningham, David D., and Julie A. Stenken, eds. *In Vivo Glucose Sensing*. Vol. 174. John Wiley & Sons, Hoboken, NJ, 2009.
3. Vashist, Sandeep Kumar, and John HT Luong. *Point-of-Care Glucose Detection for Diabetic Monitoring and Management*. CRC Press, Boca Raton, FL, 2017.
4. V.V. Tuchin (ed.), *Handbook of Optical Sensing of Glucose in Biological Fluids and Tissues*, CRC Press, Taylor & Francis Group, London, 2009, 709 p.

1 Optical and Structural Properties of Biological Tissues under Simulated Diabetes Mellitus Conditions

Daria K. Tuchina, Alla B. Bucharskaya,
Polina A. Dyachenko (Timoshina), Nataliya I. Dikht,
Georgy S. Terentyuk, and Valery V. Tuchin

CONTENTS

1.1 INTRODUCTION

Diabetes mellitus (DM) is a first-order priority of national health systems worldwide. In 2017, there were more than 450 million people with DM (9% of the world's population), and about 50% of people with diabetes belong to the most active working age of 40–59 years [1]. The World Diabetes Association predicts that by 2045, there could be as many as 693 million people with diabetes [1,2]. Despite the high quality of drugs used in the treatment of diabetes and development of new technologies for its treatment and monitoring, the levels of disability and mortality of patients are not significantly reduced [1–4].

The high risk of vascular complications in patients with type 2 DM (T2DM) led the American Heart Association to classify DM as a cardiovascular disease [5,6]. It is presumed that the negative prognostic value of any traditional risk factor for cardiovascular diseases in combination with DM increases significantly, compared with individuals without DM [6]. For example, according to the data of epidemiological studies on a combination of DM and arterial hypertension, the probability of developing fatal coronary heart disease increases by 3–5 times, stroke by 3–4 times, complete loss of vision by 10–20 times, and gangrene of the lower extremities by 20 times [6]. Macro- and microvascular complications causing cardiovascular disease, diabetic kidney disease, diabetic retinopathy (DR), and neuropathy lead to increased mortality, blindness, kidney failure, and dementia. To predict the development of vascular complications, along with monitoring clinical risk factors

DOI: 10.1201/9781003112099-1

and glycemic control, it is necessary to analyze the genetic component of both DM and its complications [3,4].

Severe complications of DM are functionally associated with an increase in the concentration of free glucose in the blood plasma, clinically defined as hyperglycemia. Pathophysiological manifestations of long-term hyperglycemia, in turn, can be associated with the development of insulin-dependent type 1 DM (T1DM) and non-insulin-dependent T2DM [7–20].

Biochemical mechanisms of hyperglycemia in T2DM are discussed in Ref. [8]. These include insulin resistance in the muscle tissue and liver and impaired insulin secretion by pancreatic β-cells, whose resistance to glucagon-like peptide 1 (GLP1) contributes to progressive β-cell dysfunction. However, an increased level of glucagon and hypersensitivity of the liver to it contribute to excess production of glucose in the liver. Insulin resistance in adipocytes leads to accelerated lipolysis and an increase in plasma free fatty acids (FFAs), which exacerbates insulin resistance in the muscle tissue and liver and contributes to further β-cell dysfunction. Increased renal glucose reabsorption by sodium/glucose co-transporter 2 (SGLT2) and increased urinary glucose excretion threshold also contribute to hyperglycemia. Resistance to appetite-suppressing effects of insulin, leptin, GLP1, amylin, and YY peptide, as well as low dopamine and high brain serotonin levels, contribute to weight gain, which exacerbates the resistance. In addition, vascular insulin resistance and concomitant inflammation are associated with the expression of adenosine monophosphate (AMP)-activated protein kinase, dipeptidyl peptidase 4, inhibitors of NF-κB, mitogen-activated protein kinase, nuclear transcription factor-κB (NF-κB), receptor agonists, reactive oxygen species (ROS), Toll-like receptor 4, tumor necrosis factor (TNF), and thiazolidinediones.

A critical consequence of hyperglycemia is a spontaneous nonenzymatic reaction between carbonyl groups of reducing sugars and amine groups of proteins, lipids, or nucleic acids, leading to the irreversible formation of advanced glycation end products (AGEs) [9,10,20]. Figure 1.1 shows schematically all AGEs formed in the body due to glycation and four other metabolic pathways [20].

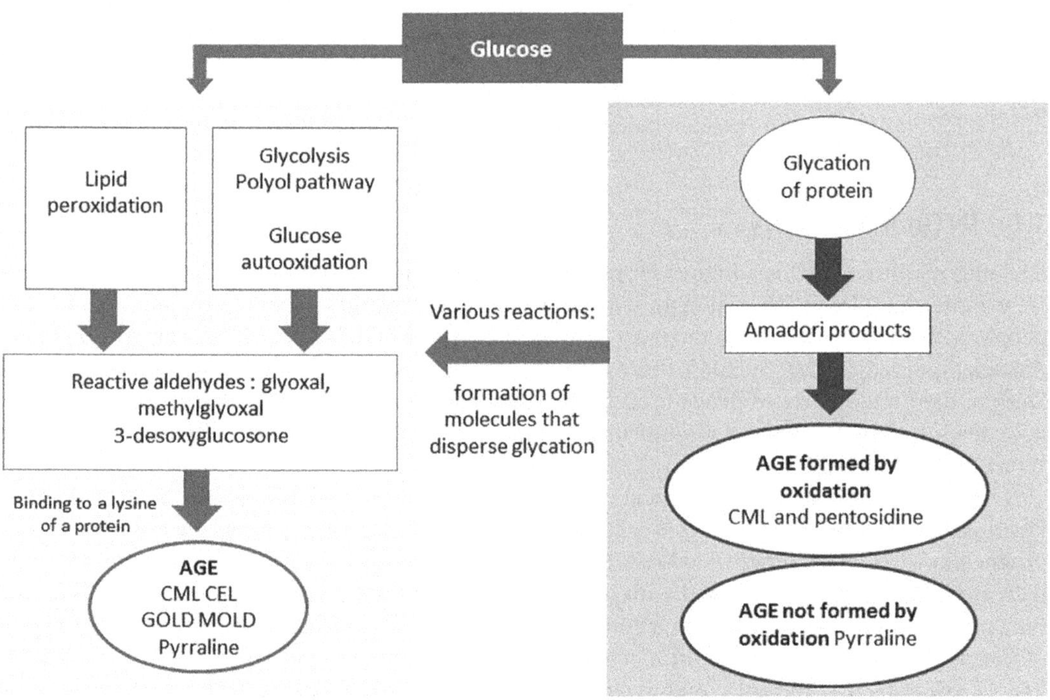

FIGURE 1.1 Schematic representation of all AGEs formed in the body due to glycation and four other metabolic pathways. (Reprinted by CC BY license from Ref. [20].)

Various metabolic pathways including polyol production, glycolysis, glucose autoxidation, and lipid peroxidation produce the same mediators as in glycation, so the term "AGE" refers to all of these resulting products that are expressed in DM, tissue aging, and kidney failure. Aminophospholipids found *in vivo* in cell and mitochondrial membranes, phosphatidylethanolamines, and serines are in the glycated form, which ultimately alters membrane plasticity, membrane potential, and conductivity. The pro-oxidant role of glycated forms is also associated with the activation of NF-κB and dysregulation of autophagy and mitochondrial cellular bioenergy production [20].

Figure 1.2 presents the approximate time scale for the formation of protein AGEs, and Figure 1.3 shows the major side effects associated with AGEs [9]. Prolonged hyperglycemia causes vascular damage in various organs and tissues, which leads to the occurrence and progression of various complications in patients with DM, including retinopathy, neuropathy, infertility-related diseases, diabetic dermopathy, periodontal disease, foot ulcers, and delayed healing wounds [9–18]. In addition, DM, as previously mentioned, can not only promote and accelerate the development of cardiovascular diseases but also increase the likelihood of many neurovascular diseases, such as stroke, atherosclerosis, vascular dementia, and Alzheimer's disease [9,19,20].

Optical properties and permeability of biological tissues for chemical agents depend on the structure of the biological tissues, which can be changed due to pathological processes, such as protein and lipid glycation [9,10,21]. AGE crosslinks are the source of both mechanical and biological consequences of glycation of extracellular tissue matrix collagens *in vivo* [1–10]. In addition, it has

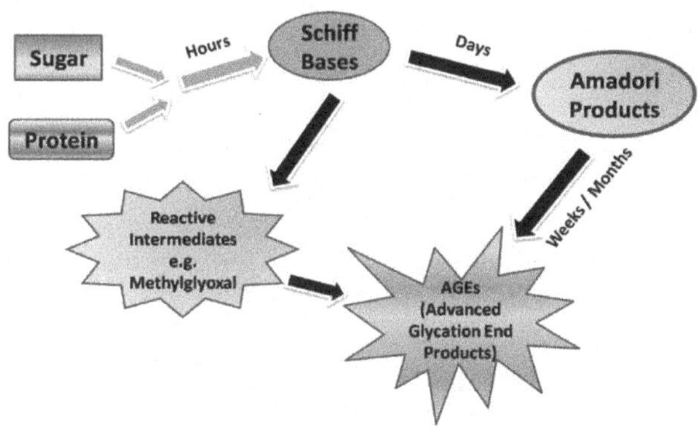

FIGURE 1.2 Approximate time scale for the formation of protein AGEs. (Reprinted by CC BY license from Ref. [9].)

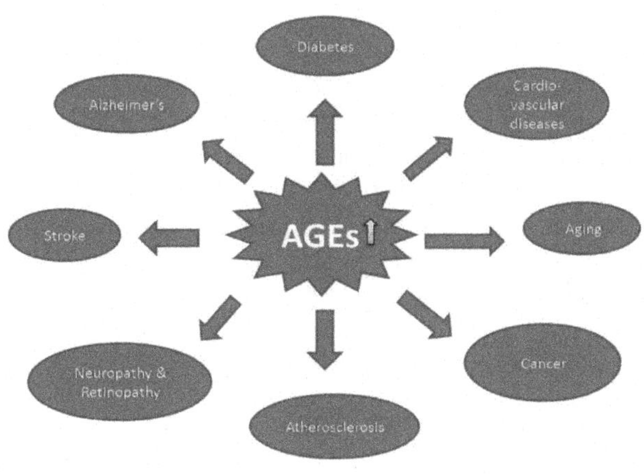

FIGURE 1.3 Side effects associated with AGEs. (Reprinted by CC BY license from Ref. [9].)

recently been shown that when collagen is glycated *in vivo* with glucose, not only AGE crosslinks but also adducts of monovalent lysine sugars play an important role in fibrillary structure stiffness [10]. Consequently, the rate of diffusion of metabolic endogenous or exogenous agents in biological tissues will reflect changes in their molecular structure, which can be used as a DM biomarker based on the monitoring of tissue stiffness.

Currently, to assess protein glycation in the body as a stable and accurate biomarker, glycated hemoglobin (HbA1c) is clinically used to diagnose diabetes with a threshold of 6.5% of total hemoglobin. Since the lifespan of erythrocytes is about 120 days, HbA1c reflects the average glucose concentration increase over the previous 8–12 weeks. However, time-resolved absorption spectroscopy can differentiate the HbA1c fraction within a single erythrocyte and quantify it by phasor analysis and thus can detect the effect of erythrocyte lifespan on its glycation [22]. Plasma AGEs are also proposed as markers of glycemia, and their possible role in the development of complications of DM is studied [23]. However, to assess the pathological complications of glycemia, assessing the degree of protein glycation directly in biological tissues and organs is recommended. Thus, assessing the degree of glycation of tissues of vital organs should provide information about the course of the disease throughout a person's life, not just at an interval of 120 days.

Noninvasive optical methods for determining the degree of glycation of protein components of biological tissues and their impact on vascular dysfunction allow the assessment of tissue and organ structural and functional properties and are rapidly developing, such as fluorescence spectroscopy and imaging, Raman spectroscopy, diffuse reflectance spectroscopy and imaging, hyperspectral imaging, multiphoton microscopy, optoacoustic imaging, optical coherence tomography (OCT), confocal microscopy, terahertz imaging, capillaroscopy, laser Doppler flowmetry (LDF), and laser speckle contrast imaging (LSCI) [13,15,22,24–43].

As DM affects many vital organs, having a broad understanding of the state of the whole organism during the development of DM is necessary. With known correlations between glycation of proteins and that of lipids in different tissues and organs, it is possible to determine the state of the patient's internal vital organs using valuable information about glycation of surficial tissues available for optical noninvasive measurements, such as skin and oral mucosa. All these results dictate the need for a comprehensive study of the mechanisms and complications of DM and their effect on individual organs/tissues.

The vascular endothelium can respond to chemical, physical, and humoral changes in the environment by producing vasodilators and vasoconstrictors, the balance of which determines the tone of vascular smooth muscle cells [14]. With prolonged exposure to damaging factors, such as hyperglycemia, on the vessels, an imbalance occurs between endothelial mediators, called endothelial dysfunction, which is characterized by impaired endothelium-dependent vasodilation, hemodynamic regulation, and fibrinolytic capacity; increased turnover and overproduction of growth factors; overexpression of adhesion molecules, inflammatory genes, and ROS; and increased oxidative stress and permeability [44]. These DM-associated microvascular dysfunctions may eventually lead to more serious complications, which calls for the need for earlier detection of vascular dysfunctions. Blood flow (blood volume per unit time in the vasculature) and perfusion (blood volume per tissue volume per unit time) are important clinical parameters of microcirculation that are closely related to the velocity of blood flow through the morphology of the vasculature (e.g., vessel diameters and vessel density). At the same time, cutaneous microvascular perfusion at rest is not impaired in DM, regardless of the progression of the disease [14,43–45]. The response of microvessels to external influences can be a sensitive indicator of disease progression and complications caused by DM [14,43]. In this regard, imaging of the microcirculatory blood supply to tissues under external stimulus is of paramount importance for monitoring the complications of DM at the early stage.

Nanotechnology is rapidly extending to many fields of medicine. The unique features of gold nanoparticles (GNPs) open up broad prospects for the creation of new nanoparticle-based drugs that can be used in the monitoring and treatment of various diseases, including DM. GNPs are actively used in biomedical applications due to their promising properties, such as biocompatibility, high surface reactivity, oxidation resistance, and plasmon resonance [46]. The few available literature data on the antioxidant and hypoglycemic effects of GNPs suggest that nanoparticles and their conjugates could become a cost-effective diagnostic and therapeutic tool in the treatment of DM and its complications [47].

To combat DM, the development of safe and effective diagnostic and therapeutic tools is necessary, which requires the use of suitable animal models to draw the right conclusions. The review [48] provides key information about the currently available animal models of obesity and T2DM, highlighting the benefits, limitations, and important caveats of each of these models.

For T1DM, the most well-known and widely used method is chemical diabetes induction, which is implemented by using either alloxan or streptozotocin [40,49–56], both of which provoke the destruction of pancreatic β-cells and induce insulin-dependent DM in rodents. Alloxan, like streptozotocin, is a structural analog of glucose. Upon administration, it is transported to β-cells of the pancreas by the GLUT-2 transporter. The destructive effect of alloxan on β-cells is realized by the generation of ROS in a cyclic reaction with 5-hydroxybarbituric acid [54]. Streptozotocin causes alkylation of DNA and subsequent activation of poly(ADP-ribose)synthase in β-cells, which leads to the depletion of nicotinamide adenine dinucleotide (NAD+) and decrease in cellular levels of adenosine triphosphoric acid, both of which cause cell necrosis and development of insulin resistance [56].

This chapter provides a brief analysis of the vital complications that occur in DM, which are largely due to the glycation of protein and lipid components of biological tissues with an increase in the free glucose level in the bloodstream. The two most commonly used animal models of T1DM are presented in this chapter. It has been shown that tissue permeability to foreign (test) molecules is largely determined by the structure of the tissue, which changes during pathological processes, such as glycation of tissue, so the rate of diffusion of test molecules can serve as a biomarker of the degree of tissue glycation. Noninvasive optical methods are presented that make the determination of the rate of diffusion of test molecules in tissues possible. Spectroscopic methods have been used in *ex vivo* studies and OCT in *in vivo* studies. The measured diffusion coefficients of glycerol and glucose in skin, myocardial, and kidney samples are presented. It has been convincingly shown that glycation significantly reduces the diffusion rate of test molecules both in the skin and in the myocardium, which opens the way for noninvasive monitoring of the degree of glycation of internal organs using the measured diffusion rate of the skin and accessible mucous membranes.

Microvascular dysfunctions caused by DM increase the sensitivity of the vascular system to external influences, which may be an indicator of disease progression and complications. Therefore, the response of the microcirculation rate of the skin and accessible mucous membranes to external disturbances can serve as a marker for the development of pathology. This chapter describes noninvasive optical methods for microcirculation monitoring, including LDF, Doppler OCT (DOCT), and LSCI, which have been tested in animal models of T1DM.

1.2 ALLOXAN ANIMAL MODEL OF T1DM

Following a previous study [57], alloxan T1DM in male white outbred rats was modeled in this study. Animal studies were conducted using 24 healthy mature rats according to the University's Animal Ethics Committee and the corresponding national agency regulating experiments on animals. Animals were randomly divided into two groups of 12 animals per group: the experimental group with alloxan T1DM and the control group. Prior to any experiment, the rats were anesthetized with Zoletil 50 (Virbac, France). Experimental T1DM was induced by an intraperitoneal injection of a single dose of alloxan monohydrate 100 μg/ kg body weight. After administration of alloxan, the blood glucose level was monitored every day using Accu-Chek Performa Roche (Switzerland).

In this study, golden nanorods (GNRs) of length 41 ± 8 nm and diameter 10.2 ± 2.0 nm were synthesized using a well-established method [58]. To prevent nanoparticle aggregation in biological tissues and enhance biocompatibility, these GNRs were functionalized with thiolated polyethylene glycol (*Mw*=5000, Nektar, USA). On the 15th day after diabetes induction, 2 ml of GNRs with 400 μg/ml gold concentration was injected intravenously (IV) into six rats of each group. Twenty-four hours after GNR injection, the experiment was stopped, and blood serum and tissue samples of internal organs were collected for biochemical investigation and morphological studies, respectively. The standard histological techniques with hematoxylin and eosin and periodic acid–Schiff (PAS) staining were used. Morphometric analysis of histological preparations was carried out using the digital image analysis system Mikrovizor medical μVizo-101 (LOMO, Russia).

The activity of lipid peroxidation processes in serum was investigated using standard spectrophotometric methods on a RF-5301 PC spectrofluorophotometer (Shimadzu Corporation, Japan), and the content of intermediate products of lipoperoxidation—malondialdehyde (MDA) and lipid hydroperoxides (LPO)—was also assessed. The following molecular markers were studied in blood serum using enzyme immunoassay: TNF-α (Rat TNF-α Platinum Elisa Kit, eBioscience Inc.) and insulin-like growth factor IGF-1 (Rat/Mouse IGF-1 Elisa Kit, Novozymes Inc.).

After administration of alloxan, significant increases in the serum glucose level were observed as follows: up to 12.34±2.86 mmol/l 10 days after administration and up to 18.39±3.49 mmol/l 15 days after administration, which was significantly higher than the control values ($p<0.05$).

The animals showed typical signs of DM clinical manifestations, such as polydipsia and polyuria. On the 15th day after alloxan injection, morphological changes of varying degrees of severity were observed in histological sections of rats with simulated diabetes compared with the control group.

In the liver, the architectonic structure persisted, dystrophic changes in hepatocytes with varying degrees of severity were developed, and the plethora was detected in the central venous and sinusoids. After PAS staining, a decrease in the glycogen content in the cytoplasm of hepatocytes was observed in comparison with the control group of animals. This is due to a decrease in tissue glycogen stores in the liver after damage to the insulin-producing β-cells of the pancreatic islets and the development of alloxan T1DM (Figure1.4a and b). An increase in the number of hepatocytes in necrosis of up to 49 was noted in the group of animals with T1DM, which is three times higher than in the control group (Table 1.1).

Characteristic structural and functional changes in the kidney under T1DM conditions associated with glycosuria were observed (Figure 1.4c). In glomeruli, plethora was seen, and in some cases, thickening of the Bowman-Shymlansky capsule and the plasmatic permeation in small arterioles were noted (Figure 1.4c); in addition, vacuolar dystrophy of renal tubules was revealed in a few instances (Figure 1.4d).

In the pancreas, plethora and focal perivascular sclerosis were marked, and pancreatic acini were closely adjacent to each other (Figure 1.4e). In animals with alloxan T1DM, the number and size of pancreatic islets were reduced, while the islets showed many signs of degeneration (Figure 1.4f).

TABLE 1.1
Results of Morphometric Study of the Liver [57]

	Morphometric Indices in the Field of View (×774) Median [Q1;Q3]			
Animal group	**Hepatocytes with dystrophic changes**	**Necrotic hepatocytes**	**Nonparenchymatous elements in the liver**	**Binuclear hepatocytes**
Control	36[30;43]	16[13;19]	7[5;8]	5 [3;6]
Control after IV GNRs	43[37;47]	17[13;19]	4[3;5]	5[3;7]
Alloxan T1DM	41[40;59]	49[31;57]*	13[10;17]*	1[1;2]*
T1DM after IV GNRs	32[25;35]	37[33;39]*	21[18;24]*	1[0;2]*

* $-p < 0.05$ reliability of the difference with the control group.

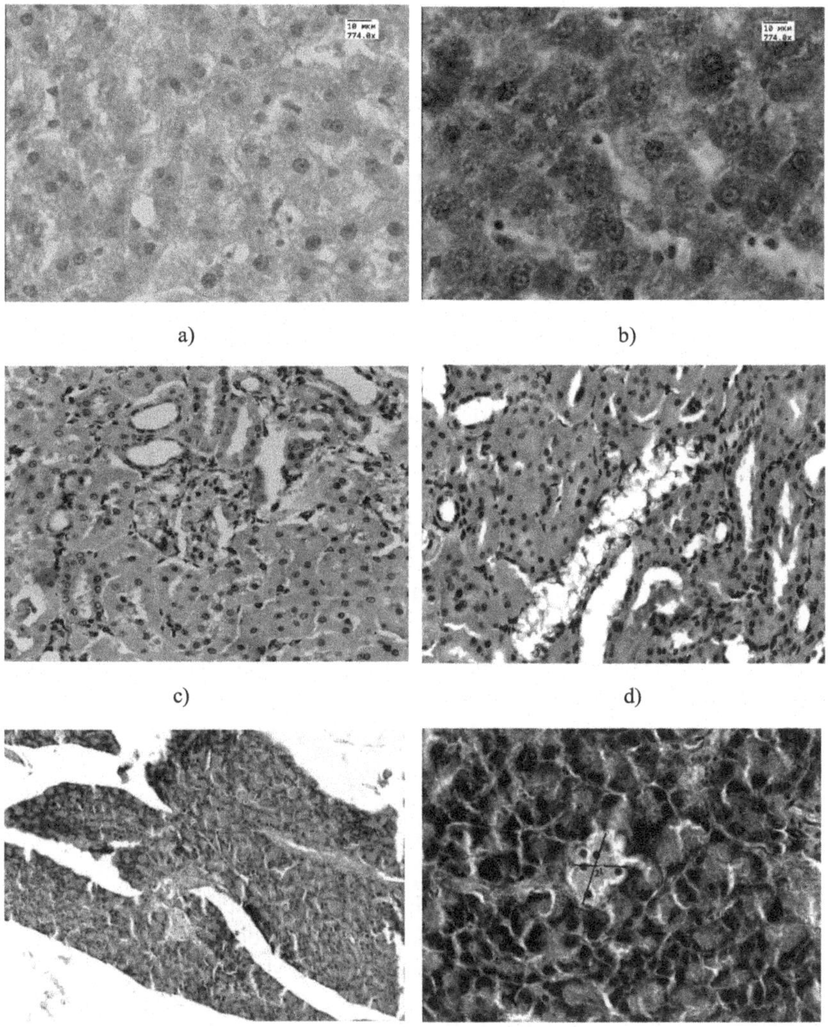

FIGURE 1.4 Microphotographs of histological preparation of tissues at alloxan T1DM in rats: dystrophic changes in liver hepatocytes, H&E, magnification ×774 (a); PAS-positive reaction in hepatocyte cytoplasm, PAS staining, magnification ×774 (b); kidney, thickening of Bowman-Shymlansky capsule, H&E, magnification ×246.4 (c); PAS staining, magnification ×246.4 (d); pancreas, plethora, and focal perivascular sclerosis, H&E, magnification ×64.4 (e); PAS staining, magnification ×246.4 (f). (Reprinted by CC BY license from Ref. [57].)

The atrophy of endocrine islets was associated with the selective destruction of β-cells. During the development of alloxan T1DM, changes in the vascular wall such as a plasmatic permeation associated with hyperglycemia took place as the first stage of changes in blood vessels, leading eventually to the development of diabetic microangiopathy.

Morphological examination in the control group of rats showed no significant morphological changes in the internal organs 1 day after IV injection of GNRs. In the kidney, moderate plethora was observed in the stroma, and in the glomerulus, tubules in some places were edematous and weak dystrophy of epitheliocytes was noted (Figure 1.5a). In the spleen and in the liver, the structure was preserved, and moderate plethora was observed (Figure 1.4b).

Twenty-four hours after the IV administration of GNRs, significant changes were not found in the internal organs in the experimental group of rats with alloxan T1DM (Figure 1.5a and b).

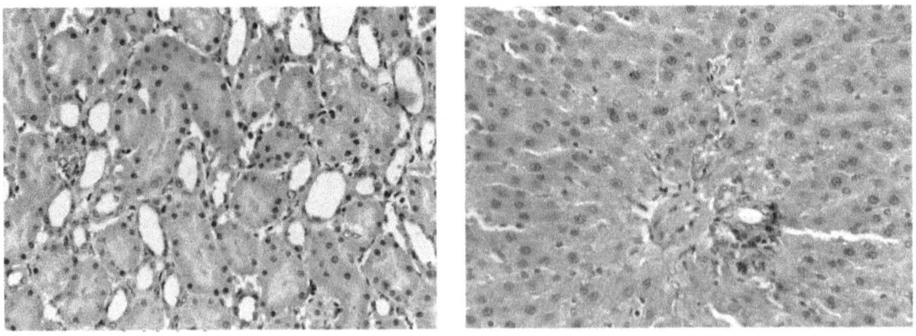

a) b)

FIGURE 1.5 Microphotographs of histological preparation of rat internal organs of the control group after IV injection of GNRs: kidney (a), liver (b); H&E, magnification ×246.4. (Reprinted by CC BY license from Ref. [57].)

TABLE 1.2
Indices of Blood Serum in Rats with Alloxan T1DM and GNR Injection [59]

Animal Group	Control Median [Q1;Q3]	Control after IV GNRs Median [Q1;Q3]	Alloxan T1DM Median [Q1;Q3]	T1DM after IV GNRs Median [Q1;Q3]
TNF-α (pg/mL)	4.4 [4;5]	5[3;13.6]	2.6[2;13.8]	13.4[2.2;60.2]*
IGF-1(ng/L)	1260[722.6;1676]	1045[423.3;1359]	504.4[112;972.7]*	472.2[72.2;1305]

* $-p < 0.05$ reliability of the difference with the control group.

As a response to the IV administration of GNRs, an increase in the total number of macrophage and lymphoid cell elements of the liver was noted. A significant decrease in insulin-like growth factor (IGF) was observed in the serum of rats with alloxan T1DM compared with the control group of animals ($p<0.01$). After GNR administration in animals with simulated alloxan T1DM, the level of TNF-α slightly increased, and the level of IGF-1 did not change in this group of animals.

After GNR administration in the control group of animals, no significant changes in the serum content of intermediate products of lipoperoxidation were observed (Figure 1.6). In rats simulated with alloxan T1DM, activation of free radical oxidation and accumulation of lipoperoxidation products MDA and LPO in blood serum were observed, compared with the control group. However, with the IV injection of GNRs, the serum levels of MDA and LPO did not differ from those in the group of diabetic animals without any exposure.

The results obtained in experiments with alloxan T1DM simulation in animals showed significant systemic metabolic changes and accumulation of malondialdehyde ($p < 0.001$) and lipid hydroperoxides ($p<0.001$) in serum of diabetic animals compared with the control group.

The absence of toxic effects of PEG-coated GNRs in healthy animals during a single IV injection was confirmed by the absence of changes in MDA and LPO indices and molecular markers of intercellular interaction. These results agree with those of Hainfeld et al. [60] regarding the absence of the toxic effect of GNPs on the biochemical parameters of blood and internal organs of animals.

Biochemical investigation in the animal group with simulated T1DM via injection of GNRs revealed a slight decrease in the lipid hydroperoxide level only ($p<0.001$), but the serum content of malondialdehyde did not change significantly.

The presented results are consistent with those of Ref. [61], who established the antioxidant effect of spherical GNPs of about 50 nm in diameter in mice with streptozotocin-induced diabetes.

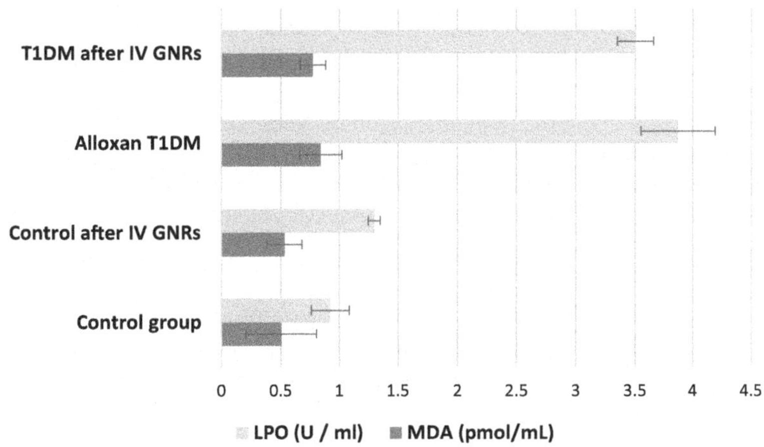

FIGURE 1.6 Lipid peroxidation indices in the serum of rats in control and diabetes groups before and after the IV injection of GNRs. (Reprinted by CC BY-NC license from Ref. [59].)

In this work, GNPs exhibited an insistent control over the blood glucose level and lipid and serum biochemical profiles in diabetic mice to levels similar to those in the control mice, provoking their effective role in controlling and increasing the organ functions for better utilization of blood glucose. Histopathological and hematological studies revealed the nontoxic and protective effect of GNPs over the vital organs when administered at a dosage of 2.5 mg/kg.body.weight/day.

This is due to the significant decrease in lipid peroxidation and ROS generation accomplished in diabetic mice with GNPs compared with diabetic controls, suggesting that GNPs prevent disruption of organs by protecting lipids from peroxidation by ROS under hyperglycemic conditions. It is worth mentioning here that oxidative stress plays an important role in the etiology of several diabetic complications.

In Ref. [62], an increase in IL-1, IL-6, and TNF-α gene expression was noted using the polymerase chain reaction (PCR) method during acute (after 24 hours) exposure to 10- and 50-nm GNPs in the liver of rats. After 5 days, a decrease in the TNF-α expression was observed, suggesting that these changes are reversible.

These results confirmed GNPs' potential as antioxidants due to their ability to reduce the production of ROS, which do not elicit the secretion of proinflammatory cytokines.

1.3 SPECTROSCOPIC *EX VIVO* MEASUREMENTS

Since proteins are the main components of many tissues, protein glycation leads to significant changes in tissue structure [28,63]. As tissue permeability for chemical agents is largely determined by tissue structure and its changes caused by pathological processes, such as glycation, the changes in the molecule diffusion rate in tissues during a certain time interval can indicate the changes in the tissue structure and thus can be used as a biomarker of the degree of tissue glycation [31,64,65]. Studying the permeability of tissues for different molecules is aimed at obtaining information about the mechanisms of tissue interaction with different endogenous (metabolic) and exogenous agents, the drug transport in tissues, and the impact of agents on morphological, optical, functional, and diffusion properties of tissues [32,66–74]. These data are necessary for understanding the pathological mechanisms of DM, effective application of different pharmacological preparations to treat DM, and developing noninvasive optical methods of disease diagnostics and monitoring [66,75], since manifestation of tissue glycation, efficiency of treatment, and diagnostics are determined by the drug (agent) diffusion rate, i.e., the time needed for drug (agent) molecules to reach the target part of the organism.

The 2-week alloxan T1DM regimen in rats, as described in Section 1.2, was used for *ex vivo* investigation of glycerol diffusion in skin and myocardium [31]. Experiments were conducted on myocardium and skin samples obtained from white outbred male laboratory rats weighing 500±20 g. Hair was removed from the skin using the depilatory cream "Veet" (Reckitt Benckiser, France), and skin was washed with saline to remove the residual cream. The skin samples were extracted by surgical scissors, and then, subcutaneous fat was removed using tweezers. Myocardium samples were cut using a scalpel after freezing the whole rat heart and washed with saline to remove the excess blood from the samples. For each series of measurements, ten samples with an area of nearly $10 \times 15\,mm^2$ were used.

The rats were divided into the control and diabetic groups. T1DM in the rats was induced via a single intramuscular injection of alloxan–saline solution as 10 mg of alloxan (Acros Organic, Belgium) per 100 g of the rat body mass [52,54,57,76]. Mean values of free glucose in the blood were measured after fasting from the caudal vein using the "Accu-Chek Performa" glucometer (Roche Diagnostics, Germany), which increased from 128±18 mg/dl (7±1 mmol/l) to 350±147 mg/dl (19±8 mmol/l) on the 15th day (in 2 weeks) after the alloxan injection [31,77].

The aqueous 70% glycerol solution prepared by mixing dehydrated glycerol (Trading Depot No.1 of Chemicals, Russia) and distilled water was used as an immersion agent. The refractive index of the solution (1.427±0.001) was measured at 589 nm by the Abbe refractometer IRF-454B2M (LOMO, Russia). The choice of concentration for glycerol solution was determined by its relatively low viscosity (22.5 cP) in comparison with dehydrated glycerol (1410.0 cP) yet sufficiently high refractive index [78]. Moreover, studies of the efficiency of skin optical clearing using aqueous glycerol solutions of different concentrations [79–81] have shown that 70% concentration of the solution is close to optimal.

To calculate diffusion coefficients of glycerol and the efficiency of their optical clearing, collimated transmittance spectra of the samples were recorded using the multichannel spectrometer USB4000-Vis-NIR (Ocean Optics, USA). Each sample was kept in a plastic plate of area $3.5 \times 1.5\,cm^2$ with a hole in the center of area of $8 \times 8\,mm^2$ and then placed in a glass cuvette with 5 ml of glycerol solution. The cuvette was fixed between two fiber-optical cables QP400-1-VIS-NIR (Ocean Optics, USA) with an inner diameter of 400 µm, and 74-ACR collimators (Ocean Optics, USA) were fixed at the ends of the fibers using SMA-905 connectors. The halogen lamp HL-2000 (Ocean Optics, USA) was used as the light source. Measurements were taken at room temperature (about 20°C). Weight and thickness of the samples were measured before and after their immersion in glycerol solution. The samples were placed between two slides, and thickness of the samples was measured using a micrometer with an accuracy of±5 µm. The samples' weight was measured using an electronic balance (Scientech, USA) with an accuracy of±1 mg.

The method used for estimating the diffusion coefficients of chemical agents is described in detail in Refs. [30,82,83]. In this case, the diffusion coefficient is the mean rate of the exchange flow of the agent into the tissue and water from the tissue, i.e., a relative or bimodal diffusion coefficient. Glycerol diffused into the skin mainly from the dermis side, whereas it penetrated into the myocardium sample from both sides. The skin permeability coefficient P for glycerol (one-side diffusion) was estimated using the following expression [84]:

$$P = \frac{D}{l}, \tag{1.1}$$

where D is the glycerol diffusion coefficient in the tissue and l is the thickness of the tissue sample.

The myocardium permeability coefficient (two-side diffusion) was calculated using the following expression [84]:

$$P = \frac{\pi^2 D}{l}. \tag{1.2}$$

The optical clearing efficiency of the tissue sample was estimated at different wavelengths as follows:

$$\mathrm{OC_{eff}} = \frac{\mu_{s0} - \mu_{s\,\mathrm{min}}}{\mu_{s0}}, \tag{1.3}$$

where μ_{s0} and $\mu_{s\,\mathrm{min}}$ are the initial and the minimal values of the scattering coefficient of the tissue sample, determined from the values of the attenuation coefficient $(\mu_t = \mu_a + \mu_s)$. The attenuation coefficient was calculated using the Bouguer–Lambert law, $\mu_t = -\ln(T_c / l)$, where T_c is the collimated transmittance of the tissue sample and μ_a is the tissue absorption coefficient. For calculations, the absorption coefficient values were taken from Ref. [85]. The possible change in the absorption coefficient due to the osmotic dehydration of the tissue was not considered in the calculations since the absorption coefficient is much smaller than the scattering coefficient for both skin and muscle tissues [85].

Also, 2 weeks of streptozotocin-induced diabetes in rats was studied for the investigation of glycerol diffusion in skin *ex vivo* [74]. Rats of the streptozotocin group obtained an injection of streptozotocin mixed with saline at a dosage of 5 mg of streptozotocin (Alfa Aesar, USA) per 100 g of body weight. The free glucose level in rat blood from the caudal vein was measured after fasting using the "Satellit Express" glucometer (Elta, Russia) before injection and 2 weeks after injection, which was 87 ± 14 mg/dL (5 ± 1 mmol/l) and 390 ± 153 mg/dL (22 ± 8 mmol/l), respectively. Optical measurements were performed in rats 2 weeks after streptozotocin injections.

One month of alloxan-induced diabetes was also studied on rat skin and myocardium *ex vivo* by spectroscopy [70]. The same alloxan model as described above was used. The free glucose level in rat blood before alloxan injection, on the third week, and 1 month after the injection (on the day of the experiment) was 70 ± 5, 290 ± 159, and 420 ± 86 mg/dl (4 ± 1, 16 ± 9, 23 ± 5 mmol/l), respectively. The aqueous 70% glycerol solution was prepared as described above, and its refractive index was measured at 589 nm as 1.4270 and at 930 nm as 1.4188 using the multiwavelength Abbe refractometer DR-M2/1550 (Atago, Japan).

Figure 1.7 presents the typical kinetics of collimated transmittance spectra of rat skin samples taken from control rats (Figure 1.7a) and rats with 4 weeks of alloxan T1DM (Figure 1.7b) during immersion samples in the 70% glycerol solution. The initial thickness of the control and diabetic samples was 0.79 ± 0.04 mm and 0.81 ± 0.01 mm, respectively. For both groups of samples, the collimated transmittance increased and reached saturation. It is observed that collimated transmittance of diabetic skin started to increase slowly, where the "slow" phase took about 60 minutes before the "fast" grow phase was started. However, only about 30 minutes was needed for the control group to complete the "slow" phase. Approximately 145 minutes was needed to complete the "fast" grow

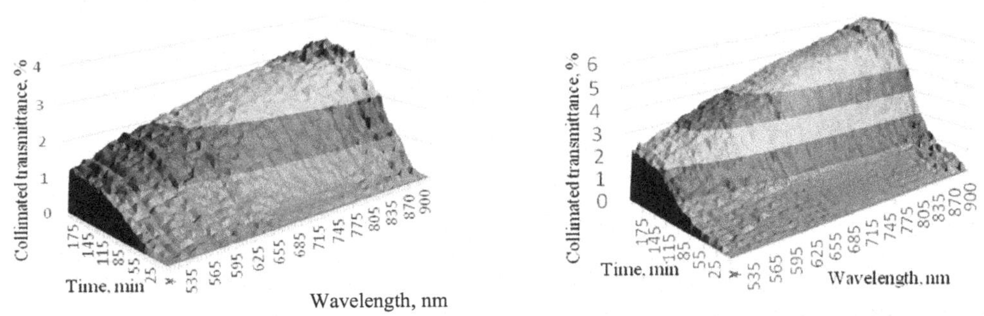

a) b)

FIGURE 1.7 Typical temporal spectra of collimated transmittance of rat skin samples from control (a) and alloxan T1DM (b) groups during the immersion in 70% glycerol solution. (Reprinted with permission from [70].)

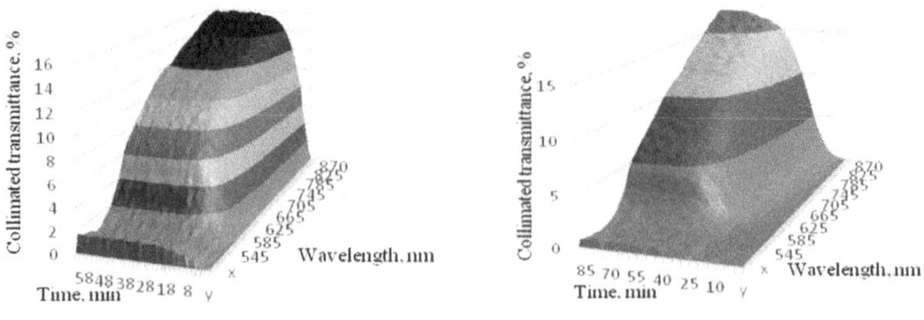

a) b)

FIGURE 1.8 Typical temporal spectra of collimated transmittance of the rat myocardium samples from control (a) and alloxan T1DM (b) groups during the optical clearing using a 70% glycerol solution. (Reprinted with permission from [70].)

phase up to transmittance saturation for the diabetic skin, as compared to that of the control group, in which only about 15 minutes was needed to achieve saturation of optical clearing kinetics.

Figure 1.8 demonstrates typical collimated transmittance kinetics of the myocardium samples from the control rats (Figure 1.8a) and rats with 2 weeks of alloxan T1DM (Figure 1.8b) during their immersion in 70% glycerol solution. The thickness of the control and diabetic samples was 0.58±0.02 mm and 0.76±0.03 mm, respectively. For the control group, collimated transmittance started to grow immediately during the first minute of immersion (unlike skin samples, no "slow" phase), whereas in the diabetic group, the "slow" grow phase took a few minutes. Also, for the diabetic myocardium sample, more time was needed to complete the "fast" grow phase (about 30 minutes), i.e., to achieve saturation of optical clearing, in comparison with the sample from the control group (about 10 minutes only).

Transmittance of skin and myocardium samples increased because of optical clearing, which is due to tissue dehydration and permeation of glycerol molecules into tissues that result in refractive index matching of tissue components and thus decrease in tissue scattering.

Table 1.3 summarizes the mean values of weight and thickness of the tissue samples before and after optical clearing, glycerol diffusion coefficients, permeability of the tissue samples for glycerol, optical clearing efficiency of rat skin and myocardium samples for control and alloxan diabetic groups (2 and 4 weeks duration of diabetes), and skin samples of rats from the streptozotocin diabetic T1DM group (diabetes of 2 weeks duration).

The reduction in the weight of skin and myocardium samples and thickness of myocardium samples was caused by glycerol-induced sample dehydration, contributing to the optical clearing. It is interesting that the weight loss was more expressed in samples from the control group of rats. The loss amounted nearly 29% of the initial weight for samples of both tissues from the control group, while in the diabetic group, the weight loss was about 24% for skin and 26% for myocardium.

Since dehydration is one of the main mechanisms of tissue optical clearing, it is important to quantify tissue water content and amount of water loss. In Ref. [86], dehydration of porcine skin (reduction in total water) was calculated as 33.1% after 30 minutes and as 51.3% after 60 minutes of 70% glycerol solution action on average for 40–200 µm depths measured by confocal Raman microscopy from the epidermal side. Correspondingly, weakly bound water was reduced by 33.2% and 51.1%, strongly bound by 33.0% and 50.4%, tightly bound by 42.3% and 63.6%, and unbound by 28.7% and 55.4% after 30 min and 60 min of glycerol action, respectively.

There was a slight increase in the thickness of skin samples at the level of measurement error; however, for myocardial samples, the decrease in thickness was noticeable. Such a difference in the change in the thickness of the skin and myocardial samples indicates a different structural organization of tissues, different contents of free and bound water in the layers of the tissue, and, accordingly, a different response to the osmotic effect of glycerol. During the prolonged tissue immersion

TABLE 1.3
Mean Values of Weight and Thickness of Tissues Samples before (W_0 and l_0) and after (W and l) Optical Clearing, Glycerol Diffusion Coefficient in Tissues (D), Permeability Coefficient of Tissues (P) for Glycerol, and Optical Clearing Efficiency (OC_{eff}) at Different Wavelengths for the Skin and Myocardium Samples of Rats from Control (c) and Two (d_{a2}) [31] and Four (d_{a4}) [70] weeks of Alloxan Diabetic Groups, and for Skin Samples of Rats from 2 weeks of Streptozotocin Diabetic (d_{s2}) Group [74]

Tissue	c/d	W_0/W, mg	l_0/l, mm	D, cm²/s	P, cm/s	OC_{eff} (600 nm)	OC_{eff} (700 nm)	OC_{eff} (800 nm)
Skin	C	271±68/183±21	0.53±0.11/0.55±0.13	$(0.83\pm0.26)\times10^{-6}$	$(1.68\pm0.88)\times10^{-5}$	0.58±0.09	0.56±0.04	0.57±0.05
	d_{a2}	270±32/203±73	0.56±0.04/0.57±0.07	$(0.68\pm0.21)\times10^{-6}$	$(1.20\pm0.33)\times10^{-5}$	0.60±0.07	0.64±0.04	0.65±0.06
	d_{s2}	198±17/161±1	0.47±0.05/0.57±0.06	$(0.70\pm0.44)\times10^{-6}$	$(1.36\pm0.82)\times10^{-5}$	0.50±0.08	0.57±0.06	0.53±0.08
	d_{a4}	416±105/326±74	0.78±0.16/0.85±0.22	–	$(1.00\pm0.57)\times10^{-5}$	–	–	–
Myocardium	C	210±37/146±35	0.68±0.11/0.51±0.09	$(0.79\pm0.36)\times10^{-6}$	$(11.8\pm6.1)\times10^{-5}$	0.51±0.07	0.59±0.06	0.61±0.07
	d_{a2}	214±41/153±36	0.58±0.06/0.47±0.07	$(0.51\pm0.21)\times10^{-6}$	$(8.60\pm3.21)\times10^{-5}$	0.53±0.02	0.64±0.05	0.66±0.06
	d_{a4}	377±139/315±138	0.96±0.30/0.80±0.26	–	$(1.04\pm0.65)\times10^{-5}$	–	–	–

in the osmotic agent, tissue compression may be replaced by swelling [87,88], and also the osmotic effect in the skin manifests itself in a stronger longitudinal compression, which leads to a slight (within the measurement error) increase in the sample thickness [30]. It is important to note that the method used for measuring collimated transmission at relatively small diameters of the light beam makes it possible to neglect the change in the sample area (longitudinal compression) in the algorithm for calculating the diffusion coefficient.

Both the diffusion coefficient and the permeability coefficient are significantly reduced in the diabetic groups compared with the controls, which means a slower diffusion of glycerol and water in the skin and myocardium of rats with alloxan and streptozotocin T1DM. The decrease in the diffusion rate of glycerol and water in glycated tissues indicates their structural modification, namely the degree of fibril packing, protein crosslinking, and changes in the content of free and bound water [26,28,89–93]. AGE crosslinks and adducts of monovalent lysine sugars induced by glycation of extracellular matrix collagens are the sources of tissue mechanical stiffness [9,10,21]. Consequently, the rate of diffusion of exogenous agents in glycated tissues slows down, so this slowdown can be used as a biomarker for DM. On the other hand, the measurement of diffusion coefficients of molecules in tissues makes the prediction of the hindered diffusion of metabolic endogenous substances during glycation possible, since they are responsible for metabolic processes, and their inhibition contributes to the development of complications.

The increase in the intensity of scattered light and therefore the increase in diffuse reflectance in the glycated tissue were demonstrated in Ref. [91]. The increase in the tissue water content, which is related to tissue modifications at diabetes development, was reported in Refs. [76,94]. Differences in light reflectance of intact and diabetic skin, due to glycation of tissue proteins, were discussed in Ref. [95]. Loss of axial packing of collagen I fibrils due to the distortion and twisting of the matrix by glycation adducts was discussed in Refs. [90–92,96]. The increased water content in the skin of the diabetic patients [94] and intermuscular and intercellular edema [76] may also cause the increase in light scattering in tissues. However, skin dehydration in diabetic foot was detected using a terahertz system in Refs. [38,97]. The difference could be due to different optical and terahertz probing depths, which is the smallest for THz waves, and inhomogeneous distribution of water in tissues depending on location and depth [86,97,98].

Changes in the hydration state of skin, degradation of type I collagen, and greater glycation in diabetic and aging volunteers were observed using Raman spectroscopy [99]. The authors concluded that proteins become hydrophobic, which leads to cutaneous dehydration in people with diabetes, and they reported that the skin of young people has a more compact collagen structure due to thinner diameters of collagen fiber bundles.

Two-photon imaging has shown that diabetes can lead to a significant decrease in the number of collagen fibers and an increase in the degree of disorientation of mouse skin fibrils [93]. Histological analysis also showed that diabetes leads to a change in the filamentous structure of the skin.

Thus, deterioration of tissue permeability for moving molecules due to the formation of interfibrillar links causes metabolic disorders and as a result leads to the development of larger-scale complications of DM.

The effect of alloxan-induced T1DM on the internal organs of rats is discussed in Section 1.2 [57]. In the pancreas, liver, and kidneys of rats on the 15th day after the administration of alloxan, different degrees of morphological changes were achieved compared with the control group of rats. In particular, necrosis of individual cells in the kidneys and pancreas was observed, there was a decrease in the accumulation of glycogen in the liver, fibrous perivascular indurations were found, and a decrease in the number and size of pancreatic islands was observed.

Structural changes in myocardial tissues of rats under conditions of model diabetes were studied in Ref. [76]. Morphological manifestations of myocardial restructuring under alloxan-induced T1DM were detected within 3 weeks after alloxan injection to rats. The authors observed destruction, dissociation, fragmentation, hypertrophy of cardiomyocytes, as well as twisting, thinning, and destructive changes in myocardial fibrils. Also, protein accumulation in muscle cells and an increase in the number of collagen fibers by 2.8 times compared with the control group were noticed, which leads to the increased viscosity of sarcoplasm, strength, and hardness of myofilaments, subjected to the increased mechanical load.

A decrease in the permeability of the skin and myocardium samples for exogenous molecules was observed with an increase in the duration of diabetes, which may be the consequence of more pronounced structural changes in tissues. The data presented in Refs. [57,76,90–92,96] confirm the conclusion that the process of diffusion of molecules, including glycerol, slows down in glycated tissues.

The glycerol diffusion coefficient in rat myocardium for the control group [$(0.79 \pm 0.36) \times 10^{-6}$ cm^2/s] presented in Table 1.3 is close to the glycerol diffusion coefficient in porcine myocardium [$(0.77 \pm 0.46) \times 10^{-6}$ cm^2/s] [100].

In the study of optical clearing of rat muscle using 60% ethylene glycol solution, the diffusion coefficient of water, which flow was induced by the hyperosmolarity of ethylene glycol, was found as $D_{water} = 3.12 \times 10^{-6}$ cm^2/s [71]. While using the aqueous 54% glucose solution as an immersion agent, the diffusion coefficient of the induced water flow in the muscle tissue had a value close to that of 60% ethylene glycol solution, $D_{water} = 3.22 \times 10^{-6}$ cm^2/s [71]. Thus, dehydration seems to be faster than the glycerol diffusion in skin and myocardium. Water has a much higher diffusion rate than glycerol [80].

Table 1.3 presents the optical clearing efficiency of the tissue samples at different wavelengths with 70% glycerol solution. The optical clearing efficiency was higher in tissues from the alloxan group compared with the control ones, which is due to a lower initial optical transmittance of diabetic tissues conditioned by the increased light scattering [30]. The stronger light scattering of diabetic tissue compared with the normal one may be explained primarily by the following three reasons: first, the increased number and irregular packing of collagen fibers in myocardium [76], i.e., the increased number and scattering cross section of scatterers; second, the increased refractive index of collagen and other proteins caused by glycation [25,101]; and third, the increased level of free glucose in tissue, which holds the water in the interfibrillar space. Another reason may be that the hydrated shell of proteins in the presence of diabetes is formed not only via the interaction of water with glycosaminoglycans of the interstitial matrix [102,103], but also via the interaction of water with glucose, each of which is known to bind up to ten water molecules [104].

All the above factors lead to the additional hydration of tissues and correspondingly to the reduction in the refractive index of the interstitial fluid, which increases scattering of light. The optical clearing efficiency of diabetic tissues is higher because the values of the minimal scattering coefficient achieved in skin of rats from both diabetic and control groups are close, whereas the initial scattering coefficient is higher for skin of rats from the control group (Eq. (1.3)). This means that the optical clearing process is slower because of the hindered diffusion of water and agents in diabetic samples. The process of optical clearing of diabetic tissue gets completed later, when the balance between all flows is established, but with nearly the same resulting tissue transparency as in the control group.

The presence of glucose molecules in the diabetic rats also explains the less weight loss achieved in diabetic samples (Table 1.3), since the glucose in these tissues possesses higher hygroscopicity compared with glycerol, each molecule of which binds only six molecules of water [105]. Thus, the glucose contained in diabetic tissues inhibits the osmotic dehydration of tissues caused by glycerol.

Study of glucose diffusion in *ex vivo* skin of mice with alloxan diabetes of duration 11 days and mice from the control (non-diabetic) group was also conducted [30]. Alloxan mixed with saline was subcutaneously injected to mice at the rate of 212 mg/kg body weight. The free glucose level in blood before alloxan injection, on the fourth day after injection, and on the day of the experiment (11th day) was measured using the "Accu-Chek Active" glucometer as 127 ± 18, 276 ± 105, and 322 ± 120 mg/dL (7 ± 1, 15 ± 6, 18 ± 7 mmol/l), respectively. Glucose diffusion coefficients were obtained as described above by measuring collimated transmittance of skin samples immersed in aqueous 30% , 43%, and 56%glucose solutions and further data processing. The rate of glucose

TABLE 1.4
Glucose Diffusion Coefficients Measured in Mouse Skin from Control and 11 days of Alloxan T1DM Groups: D is Estimated from Experimental Data using a Comprehensive Algorithm, and D_1 Using a Simple Algorithm; Corresponding Permeability Coefficients $P = D/l_{aver}$ and $P_1 = D_1/l_{aver}$; τ is Diffusion Time; l_0 and l are Thicknesses of Samples before and after Immersion in Glucose Solutions of Different Concentrations C_{gluc}; OC_{eff} is the Optical Clearing Efficiency of Skin Samples [30]

C_{gluc}, %	D (cm²/sec) / P (cm/sec)	D_1 (cm²/sec) / P_1 (cm/sec)	τ (min)	l_0/l (mm)	OC_{eff}
Control Group of Mice					
30	$(2.87\pm1.53)\times10^{-6}$ / $(1.15\pm0.61)\times10^{-4}$	$(2.60\pm1.15)\times10^{-6}$ / $(1.04\pm0.46)\times10^{-4}$	1.99±1.36	0.27±0.05 / 0.23±0.05	0.34±0.07
43	$(2.70\pm2.22)\times10^{-6}$ / $(1.29\pm1.06)\times10^{-4}$	$(2.36\pm1.85)\times10^{-6}$ / $(1.12\pm0.88)\times10^{-4}$	1.88±0.96	0.23±0.05 / 0.19±0.05	0.52±0.09
56	$(1.40\pm0.96)\times10^{-6}$ / $(0.82\pm0.56)\times10^{-4}$	$(1.26\pm0.86)\times10^{-6}$ / $(0.74\pm0.51)\times10^{-4}$	2.15±0.88	0.19±0.05 / 0.15±0.05	0.62±0.04
Alloxan T1DM Group of Mice					
30	$(1.06\pm0.55)\times10^{-6}$ / $(0.54\pm0.28)\times10^{-4}$	$(1.03\pm0.44)\times10^{-6}$ / $(0.53\pm0.23)\times10^{-4}$	2.87±0.68	0.21±0.05 / 0.18±0.05	0.42±0.06
43	$(1.15\pm0.63)\times10^{-6}$ / $(0.59\pm0.32)\times10^{-4}$	$(0.91\pm0.51)\times10^{-6}$ / $(0.47\pm0.26)\times10^{-4}$	2.98±1.88	0.20±0.05 / 0.19±0.05	0.51±0.05
56	$(1.02\pm0.44)\times10^{-6}$ / $(0.66\pm0.28)\times10^{-4}$	$(0.88\pm0.44)\times10^{-6}$ / $(0.57\pm0.28)\times10^{-4}$	2.80±1.54	0.17±0.05 / 0.14±0.05	0.53±0.12

± standard deviation.

diffusion in skin of mice with alloxan diabetes was up to 2.5 times slower as compared with the control group. Results obtained for both diabetic and nondiabetic skin samples, including glucose diffusion coefficients estimated by comprehensive and simple algorithm, are summarized in Table 1.4.

Glucose diffusion coefficient in diabetic skin is lower than that in skin of mice from the control group for each of the tested glucose concentrations. It should be noted that a comprehensive algorithm and a simpler one gave similar results, both proving the hypothesis that glucose diffuses in skin much more slowly (up to 2.5-fold) in diabetes tissue than in native tissue. Since the diffusion coefficients estimated by these two methods showed a good correlation, a simple processing method can be used for some practical cases where a faster estimation of diffusion or permeability coefficients is needed.

The diffusion coefficient in non-diabetic skin decreases with an increase in the glucose concentration in solution. This can be linked to the specific free/bound/bulk water content in tissue. An approximately twice lower diffusivity at 56% glucose concentration compared with 30% and 43% could indicate that an equilibrium of water fluxes [106,107] out from the skin and from the solution inside the skin is established at this concentration. The amount of free and bulk water in the skin can be estimated as 38%–44% [108–111], which is comparable with the water content of 56% glucose solution. Suppose that the healthy skin contains 54% of water [108] and that there is 10% of protein-bound water in skin; then, an equilibrium in water fluxes can be achieved exactly by the application of 56% glucose solution since it contains an approximately equal free water content as in the skin tissue.

A weak dependence of glucose diffusion coefficient on concentration for diabetic skin could be due to the higher amount of free/bulk water in pathological skin and its better capability to transform from bound/bulk to free in response to osmotic pressure of applied glucose solutions [112,113]. The dependence of the glucose diffusion coefficient in skin on solution concentration indicates the involvement of water flux in the process, which differs for different solutions. The maximal diffusion time τ in skin of non-diabetic mice may be related to the mean total water content in skin (~54%–60%) [108–111]. In this particular case of water balance in skin and applied solution, water does not displace much, and the diffusion is slow and may be related mainly to glucose molecules. Inclusion of water molecules in diffusion should be minimal due to their faster diffusion than glucose.

There is a lack of information about water status in many tissues in the literature, but these data are essential for solving many important problems. Different approaches are welcome to solve the problem experimentally. Measurement of the tissue refractive index using the OCT technique [111] by the application of hyperosmotic agents, mechanical compression, and heating of tissue (tissue drying) may help to quantify free, bulk, and protein-bound water. Multiphoton and Coherent anti-Stokes Raman spectroscopy (CARS) imaging may also be used to evaluate the protein-bound water content in the course of optical clearing agent (OCA) or mechanical compression action and its transforming into bulk and free water [114–116].

Since skin contains a lot of water, it is interesting to compare the obtained glucose diffusion coefficient in skin with the glucose diffusion coefficient in water. The glucose diffusion coefficient in water $D_{glucose/water}=6.7 \times 10^{-6}\,cm^2/sec$ at 25°C for low concentrations [117]. It is few times faster than the glucose diffusion coefficient estimated in skin (Table 1.4). The deceleration of glucose diffusion in skin, which is stronger for diabetic skin, is caused by the interaction of glucose with tissue components.

The slower glucose diffusion in a kidney sample of a diabetic mouse as compared to a non-diabetic one was obtained in Ref. [37]. It was supposed that diabetic kidney has a denser structure due to tissue glycation. The optical clearing efficiency of the kidney, cornea, and skin samples in the THz wavelength range was higher for nondiabetic samples than for diabetic ones with the application of glycerol solution of different concentrations [37], which can be associated with a lower water flux in glycated tissues, as tissue dehydration is the major mechanism of optical clearing for THz waves.

A lower initial collimated transmittance was obtained for diabetic skin compared with normal skin with a similar thickness at the application of 43% and 56% glucose solutions. This corresponds to results of Ref. [91], where the increase in scattering and diffuse reflectance of glycated human tendon was obtained. The results presented in this chapter also prove that diabetic tissue is more scattering compared to non-diabetic ones. The difference in light reflectance for normal and diabetic skin induced by protein glycation was also discussed in Ref. [95].

The increase in glucose concentration in the solution results in an increase in the optical clearing efficiency and reduction in the glucose diffusion coefficient in samples. Thus, the maximum optical clearing of skin samples takes a longer time due to the slower diffusion processes caused by the additional impact of glucose on tissue.

The nonlinear behavior of diffusion time with glucose concentration may be explained by specificity of skin response to the application of glucose solutions of particular concentrations in relation to water content in skin. Strong nonlinear dependence of diffusion time in muscle on glucose concentration was reported in Ref. [107]. pH of glucose solutions and "along" and "transverse" shrinkage of tissue may contribute partially to the nonlinearity of parameters. Different shrinkage degrees were obtained for applied solutions [28].

1.4 OCT *IN VIVO* MEASUREMENTS

The skin of rats with 1 month of alloxan T1DM was also investigated *in vivo* by OCT [70]. The control group of rats was also studied the same way. In *in vivo* experiments, 70% glycerol solution was applied to the rat skin topically, and B-scans of skin were recorded using the Spectral Radar OCT System OCP930SR 022 (Thorlabs Inc., USA) at 930 nm before and during glycerol application. OCT scans of skin were recorded before the application of the agent, then the glycerol solution was applied, and the scans were recorded every 2–5 minutes during the solution application. The recording procedure was repeated for 15 minutes. The obtained data were used for the estimation of the degree and characteristic time of skin optical clearing with the application of 70% glycerol solution in control and alloxan diabetic groups of rats.

First, the recorded scans were used to calculate the kinetics of the light attenuation coefficient in skin by approximating the dependence of the intensity of the reflected light $I\ (z,\ t)$ on the depth of the investigated region z of the A-scan recorded at the time t by the following equation [118]:

$$I(z,t) = A \cdot \exp\left[-\left(\mu_t(t) \cdot z\right)\right] + y_0, \tag{1.4}$$

where A is the maximum intensity of reflected light, $\mu(t)$ is the time-dependent attenuation coefficient, and y_0 is the residual value of reflected light intensity that can be achieved.

For the estimation of degree and characteristic time of skin optical clearing *in vivo* with 70% glycerol solution application, the obtained time dependence of the light attenuation coefficient in rat skin was approximated by the equation obtained from the analysis of agent diffusion in tissues performed in Refs. [66,71]:

$$\mu_{\text{norm}}(t) = \frac{\mu(t)}{\mu(t=0)} = A_D \cdot \exp\left(-\frac{t}{\tau}\right) + y_0, \tag{1.5}$$

where $\mu(t=0)$ and $\mu(t)$ are the attenuation coefficient at time $t=0$ and current time t, respectively; A_D is the maximum degree of sample optical clearing; τ is the characteristic time of sample optical clearing; and y_0 is the residual value of $\mu_{\text{norm}}(t)$ that can be achieved.

OCT scans of *in vivo* rat skin from control and alloxan diabetic groups before and after 5–6 minutes of glycerol solution application are presented in Figure 1.9. It can be clearly observed that as a result of exposure to glycerol on the skin, light began to scatter less in it, and the structure of the tissue became more visible. The areas selected for scanning show that diabetic skin is initially more diffuse and "clears" more slowly than non-diabetic skin.

Figure 1.10 presents the time dependences of the normalized light attenuation coefficient in *in vivo* skin of the control and alloxan diabetic groups with 70% glycerol solution application. According to the obtained time dependences, the degree and characteristic time of skin optical clearing with 70% glycerol solution application were estimated (Table 1.5).

Figure 1.10 shows the decrease in the light attenuation coefficient in skin in both experimental groups after application of glycerol solution on the skin. Thus, under the action of 70% glycerol solution, an increase in light penetration into tissues was observed not only for *ex vivo* spectral measurements, but also for *in vivo* OCT studies.

Table 1.5 shows a tendency of a more prolonged characteristic time of skin optical clearing in diabetes and a slower glycerol diffusion in diabetic skin, which corresponds to results obtained

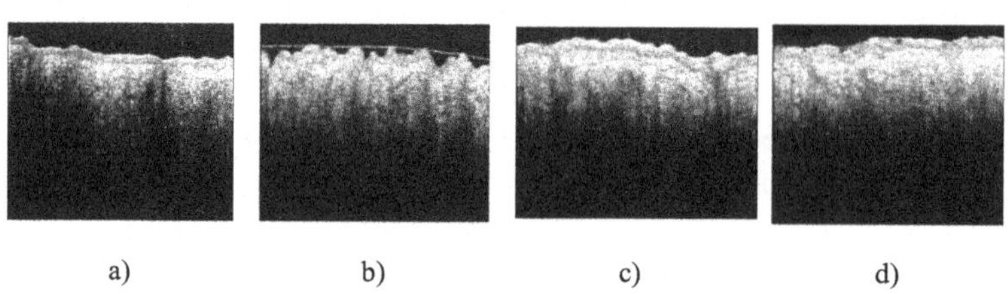

FIGURE 1.9 OCT scans of the rat skin from control (a, b) and alloxan diabetic (c, d) groups before (a, c), after 5 min (b), after 6 min (d) of 70% glycerol solution application.

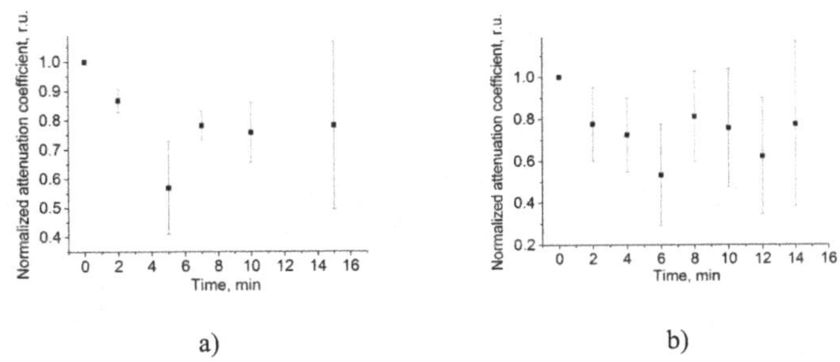

FIGURE 1.10 Time dependence of the normalized light attenuation coefficient in *in vivo* skin of control (a) and alloxan diabetic (b) groups during application of 70% glycerol solution. (Reprinted with permission from [70].)

TABLE 1.5
Mean Values of Degree and Characteristic Time of Optical Clearing of Rat Skin *in vivo* at Application of 70% Glycerol Solution [70] and Estimated Values of Glycerol Diffusion Coefficient in Skin of Control and Alloxan Diabetic Groups

Parameter	Control Group	Diabetic Group
Optical clearing degree (A_D, %)	46±19	37±16
Characteristic time of optical clearing (τ, min)	2.7±0.4	8.9±7.7
Diffusion coefficient (D, cm^2/s)	$(1.53\pm0.74)\times10^{-6}$	$(0.45\pm0.25)\times10^{-6}$

in *ex vivo* studies. Thus, a slowdown of skin optical clearing in the diabetic group of rats was observed in two independent sets of measurements *ex vivo* (spectral collimated transmittance) and *in vivo* (OCT).

1.5 BLOOD PERFUSION PROPERTIES

It has long been known that DM promotes and accelerates the development of cardiovascular disease and macrovascular complications such as atherosclerosis and peripheral arterial disease [15,24,119–121]. It is believed that the negative prognostic value of any traditional risk factor for cardiovascular diseases in combination with DM increases by three times, compared with individuals without DM [122,123]. For example, according to the data of epidemiological studies given in Ref. [122], with a combination of DM and arterial hypertension, the probability of developing fatal coronary heart disease increases by 3–5 times, stroke by 3–4 times, complete loss of vision by 10–20 times, and gangrene of the lower extremities by 20 times [122,124]. Severe complications of DM are functionally associated with an increased concentration of free glucose in blood plasma, clinically defined as hyperglycemia. Prolonged hyperglycemia causes vascular damage in various organs and tissues, which leads to the occurrence and progression of various complications in patients with DM, including foot ulcers [5], retinopathy [125], neuropathy [126], diabetic dermopathy [127], and delayed healing wounds [128]. In addition, DM, as previously mentioned, can not only promote and accelerate the development of cardiovascular diseases [11,129] but also increase the likelihood of many neurovascular diseases, such as stroke, atherosclerosis [130], vascular dementia, and Alzheimer's disease [35]. The most commonly used methods for measuring microcirculation are LDF [1,131], DOCT [1,132], and LSCI [133–138]. These methods make the measurement of microvessel perfusion possible, are non-invasive, and provide continuous measurement [24,137,139].

1.5.1 Laser Speckle Contrast Imaging

LSCI is a fast, full-field, cheap, and relatively simple imaging method that can give 2-D perfusion maps of large tissue surfaces [133]. It is based on the principle that the backscattered light from a tissue that is illuminated with coherent laser light forms a random interference pattern at the detector, the so-called speckle pattern. Movement of particles inside the tissue causes fluctuations in this speckle pattern, resulting in blurring of speckle images when obtained with an exposure time equal to or longer than the speckle fluctuation time scale. This blurring can be related to blood flow if the fluctuations are caused by the movement of red blood cells (RBCs). Initially applied as a slow analogue research tool, LSCI systems can now image blood flow in (near) real time as a result of the rapid increase in cheap computing power. This empowers the translation of LSCI into clinical practice where it is well suited for the assessment of perfusion in a wide range of tissues. LSCI has been applied to image burn wounds, retinal perfusion, cerebral blood flow (CBF), skin microvasculature, liver, esophagus, and large intestine [133]. As a rule, the experimental setup for the speckle contrast imaging technique is simple: diverging laser light illuminates an object located under the CCD camera (or CMOS) detector. The data are transferred to a personal computer for processing using specialized software. The operator has adjustable parameters, such as exposure time, number of pixels, time during which local contrast is calculated, and selection of colors for contrast coding. The choice of the number of pixels for calculating the speckle contrast is very important: using too few pixels will break the statistics, whereas using too many will decrease the spatial resolution. In practice, it has been found that a 7×7 or 5×5 pixel square is generally the best fit. In many works [133,138,139], it is this configuration of the experimental setup for speckle imaging that was used. The methodology of speckle imaging has a number of modifications, including spatial laser speckle imaging (SLSI) and temporal laser speckle imaging (TLSI). In SLSI, the main parameter speckle image contrast is calculated as the ratio of the standard deviation of the pixel intensity of the selected area (N×N pixel area) to the average pixel intensity in the area. TLSI is based on recording time statistics and calculating the contrast of speckle images using a sequence of raw speckle images obtained along several time points instead of a spatial window, i.e., the contrast of a speckle image is calculated as the ratio of the standard deviation of the intensity of one pixel at different times to the average intensity of the pixel [138,139]. But the most useful modification turned out to be the combination of these two techniques, called spatiotemporal speckle imaging. Here, the speckle image contrast is calculated as the ratio of the standard deviation of the intensity over an area of pixels at different times to the average intensity over that area. The calculated contrast of the speckle images can be related to the speed and concentration of RBCs, but the distribution of the speed needs to be determined. The Lorentzian velocity distribution is valid for the Brownian movement or unordered flow, which is not conclusive in a clinical setting. The Gaussian velocity distribution is a better fit for an ordered flow. Notwithstanding, it has become clear that the true velocity distribution is more complex than purely Lorentzian or Gaussian. These issues are discussed in Refs. [138,139]. In Ref. [140], Smausz et al. reconstructed some additional parameters to better fit the theoretical contrast curve using multiple exposures ranging from 0.2 to 500 ms.

As mentioned above, the amount of blurring is dependent on the movement of particles within the object and the exposure time. The first applications of LSCI were single-exposure methods, meaning that the exposure time is kept constant with every measurement. Single exposure has the disadvantage that quantifying the measured perfusion is difficult since sensitivity and quantitative accuracy are highly dependent on the exposure time [141]. In an attempt to make the method more robust and to increase the reproducibility of results, Parthasarathy et al. [142] developed the first multiexposure laser speckle imaging (MESI) setup. This new setup can vary the exposure time while maintaining a constant intensity. In addition to the multiexposure setup, they also derived a new, more complete, mathematical model for speckle imaging by considering the presence of static scattered light. The multiexposure setup shows linearity with relative changes in speed over a broader range of velocities, whereas the single-exposure setup is becoming less accurate for larger variations.

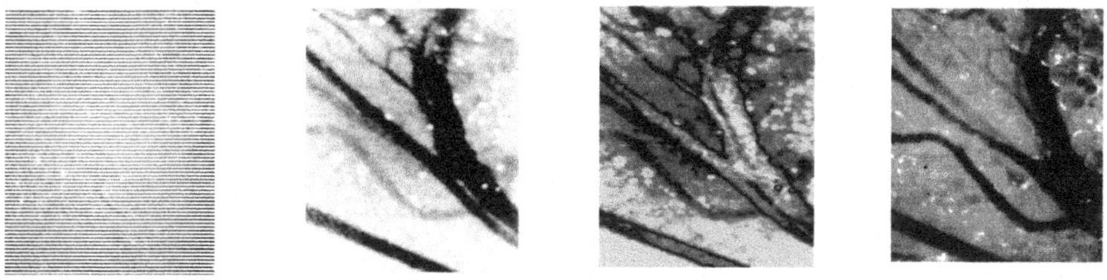

FIGURE 1.11 Images of blood vessels in fat tissue. (Adapted from [143].)

In [143], in order to improve and obtain images of velocity distribution maps over vessels, the authors suggest an alternative data processing algorithm based on the usage of the Gaussian sliding filter for sequential determination of both spatial and temporal parts of the speckle contrast (Figure 1.11). The suggested replacement of the conventional box filter leads to the monotonic damping of high-frequency spectral components, which results in a better elimination of ringing and aliasing effects in the spatiotemporal speckle contrast outputs. Also, in order to improve the method and to obtain an apparatus for assessing the absolute blood flow velocity, the speckle contrast imaging system was calibrated in the work using a phantom simulating blood microcirculation for further analysis of blood flow velocity in the pancreatic vessels of a laboratory rat under conditions of the development of model T1DM [39].

The speckle imaging method is actively used to analyze the functionality of blood vessels in the conditions of the development of diabetes under the influence of various stimuli [144,145]. Ref. [24] presents and describes in detail the existing methods of influence such as iontophoresis and local heating of the skin, which is a stimulus that causes local thermal hyperemia due to vasodilation. Thermal hyperemia is one of the most commonly used patient-friendly reactivity tests [24,146].

In Ref. [147], it was verified that LSCI coupled with physiological postocclusive reactive hyperemia and pharmacological iontophoresis of acetylcholine as local vasodilator stimuli can differentiate between cutaneous microvascular responses in T1DM patients and control patients. In the work of Zhu et al. [36], speckle contrast imaging was combined with hyperspectral monitoring to analyze the cerebral and skin blood flow during the development of DM while studying the vascular response to the effects of sodium nitroprusside and acetylcholine. The norepinephrine response of vascular blood flow with the development of DM was analyzed in a similar way in Ref. [148]. In these works and in Ref. [35], the optical clearing technique was used to improve the visualization of blood vessels during noninvasive monitoring.

In this case, optical clearing can be used not only to improve the visualization of biological tissues by optical methods, but also as a test for reactivity. There are data in the literature on the effect of various optical clearing agents on the vascular system in the development of T1DM models [137,149,150]. In these studies, blood flow parameters were analyzed using speckle imaging. It was found that the reactions of micro-hemodynamics in pancreatic vessels to the effect of clearing agents (70% aqueous solution of glycerol and PEG 300 solution, Omnipaque®-300) in laboratory animals with alloxan/streptozotocin T1DM models differ from the conditionally healthy group of animals. The data obtained in the study are presented in Figure 1.12.

The use of a 70% aqueous solution of glycerol (Figure 1.12a) demonstrates a 50% decrease in blood flow velocity by the 5th minute of exposure to the agent in the group of animals with alloxan T1DM; by the 10th minute, the blood flow velocity was completely restored. In the group of animals

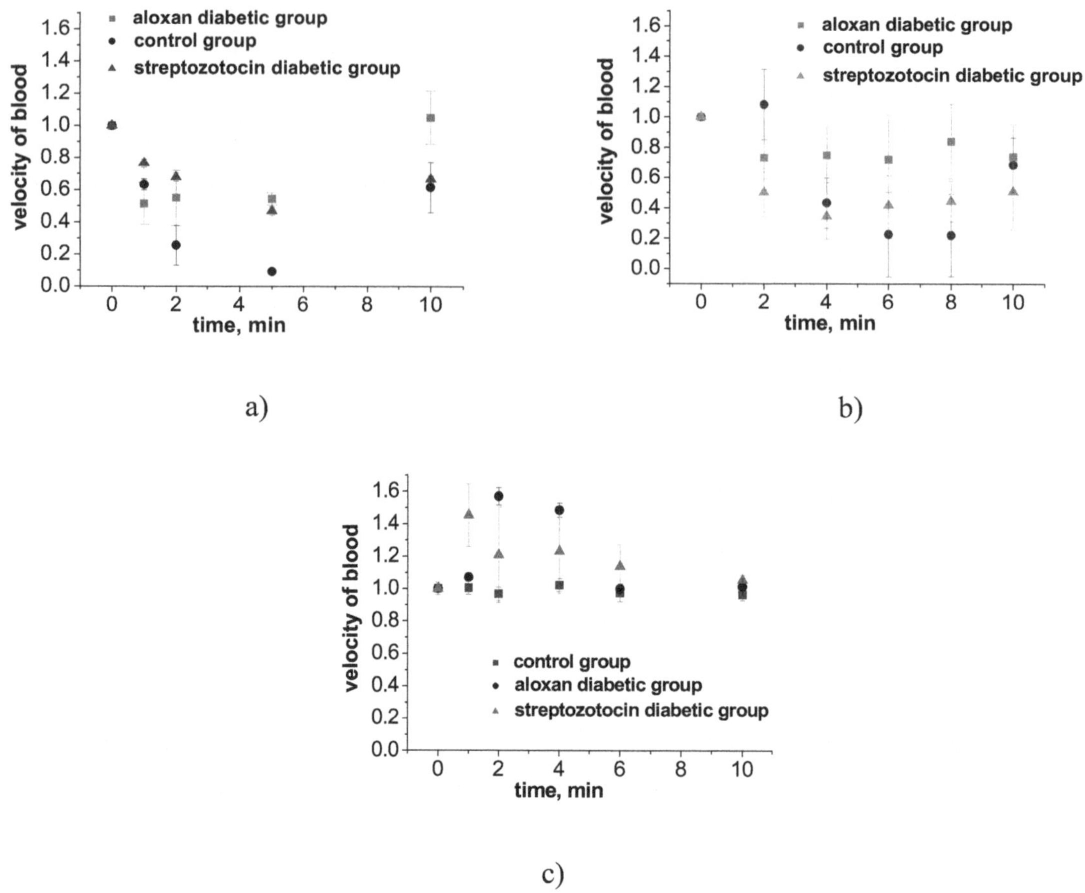

FIGURE 1.12 Temporal behavior of blood flow velocity in the pancreas of rats in control and diabetic (alloxan or streptozotocin T1DM) groups exposed to 70% aqueous glycerol (a), PEG-300 (b), or Omnipaque®-300 (c) solutions. (Adapted from [149].)

with streptozotocin T1DM, the blood velocity in the vessels of the pancreas was reduced by 54% by the 5th minute of exposure to the agent; by the 10th minute, the rate was not restored to the initial values. The blood flow in the control group almost stopped by 5 minutes of exposure to the agent and did not recover until 10 min. The use of a PEG-300 solution (Figure 1.12b) demonstrated a 25% decrease in blood flow by 6 minutes in the group of animals with alloxan T1DM and a 60% decrease in blood velocity in the pancreatic vessels by 6 minutes in the group of animals with streptozotocin T1DM. Analysis of blood flow in the control group shows a 70% decrease in blood velocity in the vessels of the pancreas by 6 minutes. Ten minutes after applying the PEG-300 solution, the velocity was partially restored. Application of Omnipaque®-300 demonstrated a 65% increase in blood flow in the group of alloxan T1DM animals and a 45% increase in blood flow in streptozotocin T1DM animals in the first 2 minutes (Figure 1.12c). In both cases, by 10th minute, blood flow velocity was completely restored. Blood flow in the control group did not show any noticeable changes. Based on these results, a different reaction of blood flow to the effect of OCAs in diabetic and control groups was expressed, which may be associated with a violation of the mechanisms of endothelium-dependent vasodilation in diabetes.

In Ref. [151], the vascular status of diabetic foot ulcers using LSCI was assessed. Thirty-three volunteers with diabetic foot ulcers were included in this prospective, single-center, observational cohort study conducted in the Netherlands. They were classified as non-ischemic, ischemic, or critical ischemic based on the criteria formulated in the international guidelines [151]. The blood flow in the feet of the volunteers with diabetic foot ulcers was recorded using LSCI at various timelines, baseline, during the postocclusive reactive hyperemia test, and during the Buerger's test (leg is raised to reduce blood flow in the foot for 30–60 seconds). Later, in Ref. [152], by the same authors, a fast and reliable microcirculation tracking algorithm for semiautomatic data analysis was presented and tested. The authors concluded that the tracking algorithm developed for the analysis of LSCI data of diabetic foot ulcers has a good to excellent inter-rater reliability in comparison with the current standard of manual assessment. The algorithm shows a tenfold workload reduction compared with the manual approach and may improve clinical applicability of LSCI for the assessment of diabetic foot disease [49,50].

1.5.2 Doppler Optical Coherence Tomography and OCT Angiography

DOCT is one of the types of functional extension of OCT, which combines the Doppler principle with OCT and provides functional *in vivo* imaging of moving samples, flows, and moving components in biological tissues [24,137]. DOCT provides effective suppression of multiple scattered light, provides close to the maximum possible probing depth, and can measure velocity components. It is based on the registration of the interference component of an optical signal formed by mixing a partially coherent probing radiation scattered by an object with a reference partially coherent beam [137]. In DOCT, the value of the random component of the error in measuring the flow velocity and, accordingly, the minimum value of the flow velocity are determined by fluctuations in the frequency of the probing radiation source, dynamic speckle noise due to the scattering of probing radiation in nonstationary randomly inhomogeneous media, and the random component of the movement of scattering centers in the flow (Brownian movement), as well as spurious interference in the optical scheme of the OCT system [131,137]. In this method, the Doppler phase is restored in a complex form from the interference signal using a Fourier transform. The Doppler phase encodes the spatial distribution of moving cells in the vessel and their speed. Thus, the presence of a Doppler shift in the signal is an indicator of the presence of a vessel, and the magnitude of this shift characterizes the distribution of cell movement velocity in the vessel.

In the work of Srinivas et al. [153], the authors evaluated retinal blood flow measurements in normal eyes and eyes with varying levels of DR using Doppler Fourier-domain OCT. The results of the authors' research showed that retinal blood flow was significantly lower in eyes with severe nonproliferative DR and proliferative DR compared with normal eyes. The authors concluded that retinal blood flow determined by DOCT may be a useful parameter for evaluating patients with DR. Similar studies were carried out in Ref. [154].

In Ref. [155], the authors used DOCT to study segmental retinal blood flow after the use of panretinal photocoagulation for the treatment of T2DM with severe DR. The data for five patients with proliferative DR (mean age 51.9 ± 10.5 years) were analyzed. Vessel diameter, mean velocity, and retinal venous blood flow were measured by DOCT before and 4 weeks after panretinal photocoagulation. The authors found that after panretinal photocoagulation, there was a statistically significant decrease in the studied parameters and that DOCT can become a clinically useful tool for noninvasive assessment of changes in blood circulation in the retina during panretinal photocoagulation in patients with DR.

Along with DOCT, OCT angiography (OCTA) can visualize blood flow in capillary-sized retinal vessels by detecting the contrast of moving blood cells, providing a faster, safer, higher-resolution procedure for assessing the retinal microvasculature. Microvascular changes in DR can be reliably detected on OCTA angiograms [156,157]. A recent review [156] has demonstrated the ability of OCTA to identify microvascular features of DR, such as microaneurysms, neovascularization, and capillary nonperfusion.

1.5.3 Laser Doppler Flowmetry

Also, blood perfusion measurements under conditions of the development of diabetes were taken using laser dopplerography. This method, based on probing tissue with laser radiation and analyzing scattered radiation from erythrocytes moving in tissues, allows *in vivo* assessment of perfusion in the microcirculatory network of the bloodstream in patients with DM. This technique has proven itself well in clinical practice and medical research related to the study of human blood perfusion in DM [24,40]. In Ref. [158], the possibility of an integrated approach to study changes in the blood microcirculation system and metabolic processes in the tissue of the lower extremities using optical noninvasive methods, such as LDF, fluorescence spectroscopy, and diffuse reflectance spectroscopy in combination with various modes of thermal stimulus, was evaluated. The study involved 76 patients with T2DM, of which 14 had visible trophic disorders on the soles of the feet and 48 were apparently healthy volunteers. The authors analyzed the parameters of LDF signals, the spectra of fluorescence intensity, and diffuse reflection of the skin of the feet. Statistically significant differences in the registered parameters between the examined groups were obtained. It was concluded that the combined use of noninvasive spectroscopy methods can be used to diagnose complications both in the manifestation of primary signs of diabetes, when pathological changes are still reversible, and in the presence of existing disorders to prevent aggravation of the course of the disease and select an adequate treatment correction [158]. In Ref. [159], with the help of LDF, microvascular perfusion was recorded in the area of the big toe in 63 control subjects and 47 patients with T2DM. The authors found that compared with the control group, patients with T2DM showed a lower perfusion of microvessels at baseline and a greater vasodilatory reserve during local heating. In Ref. [160], the authors introduce a new method of signal processing and data analysis for digital LDF. The methodology was verified using two age groups of healthy volunteers and a group of patients with T2DM.

1.6 SUMMARY

The use of optical methods in combination with the application of external stimuli in the form of OCAs, nanoparticles, or local thermal effects is a promising direction for noninvasive or minimally invasive monitoring of complications in DM. This chapter shows that many optical methods such as spectrophotometry, OCT, DOCT, speckle contrast, and Doppler flowmetry, which are already widely used in clinical trials, can be successfully used to monitor complications in the development of DM. Although we primarily used models of T1DM in this work, the results obtained can be useful for predicting the severity of complications in T2DM as well, since the complications themselves are largely determined by the process of glycation with an increase in blood sugar levels.

The simulation of alloxan/streptozotocin T1DM model in animals showed significant systemic metabolic changes and accumulation of lipoperoxidation products and proinflammatory cytokines in serum of diabetic animals. The most pronounced morphological changes were observed in the kidneys, which were characterized by carbohydrate dystrophy of the tubule epithelium, glomerular fullness, and development of plasma impregnation in the walls of small arterioles.

It was found that IV administration of polyethylene-glycol-coated GNRs does not lead to significant changes in the intensity of lipoperoxidation processes in blood serum and morphological changes in internal organs in control animals, which makes recommendation of their use as diagnostic and therapeutic agents in diabetology possible.

In animals with alloxan/streptozotocin T1DM, a significant decrease in *ex vivo* skin permeability for glycerol and glucose was found. A more prolonged duration of T1DM induced a stronger decrease in *ex vivo* skin and myocardium permeability for glycerol. *In vivo* studies also showed a slower optical clearing and glycerol diffusion in skin in T1DM.

The presented results offer a possibility of using such biocompatible chemicals as glycerol and glucose as biomarkers for early diagnostics of DM complications—in particular, for assessing the degree of myocardial lesion evidence by the skin glycation stage in the process of DM development and treatment.

ACKNOWLEDGEMENTS

The authors express their thanks to Prof. Nikolai G. Khlebtsov and Dr. Boris N. Khlebtsov for their assistance in providing of gold nanoparticles.

This work was supported in part by grants of the Government of the Russian Federation under the Decree No. 220 of 09 April 2010 (Agreement No. 075-15-2021-615 of 04 June 2021) and No. 13.2251.21.0009 of the Ministry of Science and Higher Education of the Russian Federation (Agreement No. 075-15-2021-942).

REFERENCES

1. N. H. Cho, J. E. Shaw, S. Karuranga, Y. Huang, J. D. da Rocha Fernandes, A. W. Ohlrogge, and B. Malanda, "IDF Diabetes Atlas: Global estimates of diabetes prevalence for 2017 and projections for 2045," *Diabetes Res. Clin. Pract.* **138**, 271–281 (2018). DOI: 10.1016/j.diabres.2018.02.023.
2. Y. Zheng, S. H. Ley, and F. B. Hu, "Global aetiology and epidemiology of type 2 diabetes mellitus and its complications," *Nat. Rev. Endocrinol.* **14**(2), 88–98 (2018). DOI: 10.1038/nrendo.2017.151.
3. J. B. Cole and J. C. Florez, "Genetics of diabetes mellitus and diabetes complications," *Nat. Rev. Nephrol.* **16**, 377–390 (2020). DOI: 10.1038/s41581-020-0278-5.
4. H. D. McIntyre, P. Catalano, C. Zhang, G. Desoye, E. R. Mathiesen, and P. Damm, "Gestational diabetes mellitus," *Nat. Rev. Dis. Primers* **5**, 47-1–47-19 (2019). DOI: 10.1038/s41572-019-0098-8.
5. S. M. Grundy, I. J. Benjamin, G. L. Burke, et al., "Diabetes and cardiovascular disease: A statement for healthcare professionals from the American Heart Association," *Circulation* **100**, 1134–1146 (1999). DOI: 10.1161/01.cir.100.10.1134.
6. A. P. Vasiliev and N. N. Streltsova, "Skin microcirculation in patients with arterial hypertension and in patients with a combination of arterial hypertension and type II diabetes mellitus," *Reg. Hemodyn. Microcircul.* **19**(4), 44–52 (2020). DOI: 10.24884/1682-6655-2020-19-4-44-52.
7. G. B. Stefano, S. Challenger, and R. M. Kream, "Hyperglycemia-associated alterations in cellular signaling and dysregulated mitochondrial bioenergetics in human metabolic disorders," *Eur. J. Nutr.* **55**(8), 2339–2345 (2016). DOI: 10.1007/s00394-016-1212–2.
8. R. A. DeFronzo, E. Ferrannini, L. Groop, et al., "Type 2 diabetes mellitus," *Nat. Rev. Dis. Primers* **1**, 15019 (2015). DOI: 10.1038/nrdp.2015.19.
9. J. M. Ashraf, S. Ahmad, I. Choi, et al., "Critical review: Recent advances in detection of AGEs: Immunochemical, bioanalytical and biochemical approaches," *IUBMB Life* **67**(12), 897–913 (2015). DOI: 10.1002/iub.1450.
10. S. Bansode, U. Bashtanova, R. Li, et al., "Glycation changes molecular organization and charge distribution in type I collagen fibrils," *Sci. Rep.* **10**, 3397 (2020). DOI: 10.1038/s41598-020-60250-9.
11. J. B. Tryggestad and S. M. Willi, "Complications and comorbidities of T2DM in adolescents: Findings from the TODAY clinical trial," *J. Diab. Complic.* **29**, 307–312 (2015). DOI: 10.1016/j.jdiacomp.2014.10.009.
12. N. Amin and J. Doupis, "Diabetic foot disease: From the evaluation of the "foot at risk" to the novel diabetic ulcer treatment modalities," *World. J. Diabetes* **7**(7), 153–164 (2016). http://www.wjgnet.com/1948-9358/full/v7/i7/153.htm.
13. J. J. van Netten, D. Clark, P. A. Lazzarini, M. Janda, and L. F. Reed, "The validity and reliability of remote diabetic foot ulcer assessment using mobile phone images," *Sci. Rep.* **7**, 9480 (2017). DOI: 10.1038/s41598-017-09828-4.
14. M. Roustit, J. Loader, D. Baltzis, W. Zhao, and A. Veves, "Microvascular changes in the diabetic foot," in *The Diabetic Foot. Contemporary Diabetes*, A. Veves, J. Giurini, R. Guzman (Eds.), Humana, Cham, 2018. DOI: 10.1007/978-3-319-89869-8_10.
15. V. V. Dremin, E. A. Zherebtsov, V. V. Sidorov, et al., "Multimodal optical measurement for study of lower limb tissue viability in patients with diabetes mellitus," *J. Biomed. Opt.* **22**(8), 085003 (2017), DOI: 10.1117/1.JBO.22.8.085003.
16. J.-L. Zhu, Y.-Q. Cai, S.-L. Long, Z. Chen, and Z.-C. Mo, "The role of advanced glycation end products in human infertility," *Life Sci.* **255**, 117830 (2020). DOI: 10.1016/j.lfs.2020.117830.

17. J. P. Perraudin, "The biochemical aspects of diabetes in oral health," *British J. Diabet.* **19**(21), 93–98 (2019).
18. M. J. L. Verhulst, B. G. Loos, V. E. A. Gerdes, and W. J. Teeuw, "Evaluating all potential oral complications of diabetes mellitus," *Front. Endocrinol.* **10**, 56-1–56-49 (2019). DOI: 10.3389/fendo.2019.00056.
19. K. Byun, Y. C. Yoo, M. Son, et al., "Advanced glycation end-products produced systemically and by macrophages: A common contributor to inflammation and degenerative diseases," *Pharm. Thera.* **177**, 44–55 (2017). DOI: 10.1016/j.pharmthera.2017.02.030.
20. M. Fournet, F. Bonté, and A. Desmoulière, "Glycation damage: A possible hub for major pathophysiological disorders and aging," *Aging and Disease* **9**(5), 880–900 (2018). DOI: 10.14336/AD.2017.1121.
21. K. J. Welsh, M. S. Kirkman, and D. B. Sacks, "Role of glycated proteins in the diagnosis and management of diabetes: Research gaps and future directions," *Diabetes Care* **39**(8), 1299–1306 (2016). DOI: 10.2337/dc15-2727.
22. P.-T. Dong, H. Lin, K.-C. Huang, and J.-X. Cheng, "Label-free quantitation of glycated hemoglobin in single red blood cells by transient absorption microscopy and phasor analysis," *Sci. Adv.* **5**, eaav0561 (2019). DOI: 10.1126/sciadv.aav0561.
23. N. M. Hanssen, J. W. Beulens, S. van Dieren, et al., "Plasma advanced glycation end products are associated with incident cardiovascular events in individuals with type 2 diabetes: A case-cohort study with a median follow-up of 10 years (EPIC-NL)," *Diabetes* **64**, 257–265 (2015).
24. V. V. Tuchin (ed.), *Handbook of Optical Sensing of Glucose in Biological Fluids and Tissues*, CRC Press, Taylor & Francis Group, London (2009).
25. O. S. Zhernovaya, V. V. Tuchin, and I. V. Meglinski, "Monitoring of blood proteins glycation by refractive index and spectral measurements," *Laser Phys. Lett.* **5**(6), 460–464 (2008).
26. J. Blackwell, K. M. Katika, L. Pilon, K. M. Dipple, S. R. Levin, and A. Nouvong, "*In vivo* time-resolved autofluorescence measurements to test for glycation of human skin," *J. Biomed. Opt.*, **13**(1), 014004 (2008).
27. J. Kinnunen, H. T. Kokkonen, V. Kovanen, et al., "Nondestructive fluorescence-based quantification of threose-induced collagen cross-linking in bovine articular cartilage," *J. Biomed. Opt.* **17**(9), 097003 (2012).
28. J.-Y. Tseng, A. A. Ghazaryan, W. Lo, et al., "Multiphoton spectral microscopy for imaging and quantification of tissue glycation," *Biomed. Opt. Express* **2**(2), 218–230 (2011).
29. Y.-J. Hwang, J. Granelli, and J. Lyubovitsky, "Multiphoton optical image guided spectroscopy method for characterization of collagen-based materials modified by glycation," *Analyt. Chem.* **83**(1), 200–206 (2011).
30. D. K. Tuchina, R. Shi, A. N. Bashkatov, E. A. Genina, D. Zhu, Q. Luo, and V. V. Tuchin, "Ex vivo optical measurements of glucose diffusion kinetics in native and diabetic mouse skin," *J. Biophoton.* **8**(4), 332–346 (2015). DOI: 10.1002/jbio.201400138.
31. D. K. Tuchina, A. N. Bashkatov, A. B. Bucharskaya, E. A. Genina, and V. V. Tuchin, "Study of glycerol diffusion in skin and myocardium ex vivo under the conditions of developing alloxan-induced diabetes," *J. Biomed. Photon. Eng.* **3**(2), 020302 (2017).
32. D. K. Tuchina and V. V. Tuchin, "Optical and structural properties of biological tissues under diabetes mellitus," *J. Biomed. Photon. Eng.* **4**(2), 1–22 (2018). DOI: 10.18287/JBPE18.04.020201.
33. E. Zharkikh, V. Dremin, E. Zherebtsov, A. Dunaev, and I. Meglinski, "Biophotonics methods for functional monitoring of complications of diabetes mellitus," *J. Biophoton.* **13**, e202000203 (2020). DOI: 10.1002/jbio.202000203.
34. A. Maslianitsyna, P. Ermolinskiy, A. Lugovtsov, A. Pigurenko, M. Sasonko, Y. Gurfinkel, and A. Priezzhev, "Multimodal diagnostics of microrheologic alterations in blood of coronary heart disease and diabetic patients," *Diagnostics* **11**, 76 (2021). DOI: 10.3390/ diagnostics11010076.
35. J. Zhu, D. Li, T. Yu, and D. Zhu, "Optical angiography for diabetes-induced pathological changes in microvascular structure and function: An overview," *J. Innov. Opt. Health Sci.* **15**(1), 2230002 (2022). DOI: 10.1142/S1793545822300026.
36. W. Feng, S. Liu, C. Zhang, Q. Xia, T. Yu, and D. Zhu, "Comparison of cerebral and cutaneous microvascular dysfunction with the development of type 1 diabetes," *Theranostics* **9**(20), 5854–5868 (2019). DOI: 10.7150/thno.33738.
37. O. A. Smolyanskaya, I. J. Schelkanova, M. S. Kulya, et al., "Glycerol dehydration of native and diabetic animal tissues studied by THz-TDS and NMR methods," *Biomed. Opt. Express* **9**(3) 1198–1215 (2018). DOI: 10.1364/BOE.9.001198.
38. G. G. Hernandez-Cardoso, S. C. Rojas-Landeros, M. Alfaro-Gomez, et al., "Terahertz imaging for early screening of diabetic foot syndrome: A proof of concept," *Sci. Rep.* **7**, 42124 (2017). DOI: 10.1038/srep42124.
39. O. A. Smolyanskaya, E. N. Lazareva, S. S. Nalegaev, et al., "Multimodal optical diagnostics of glycated biological tissues," *Biochemistry (Moscow)* **84**(Suppl. 1), S124–S143 (2019). DOI: 10.1134/ S0006297919140086.

40. D. K. Tuchina and V. V. Tuchin, "*Diabetes mellitus*-induced alterations of tissue optical properties, optical clearing efficiency, and molecular diffusivity," in *Handbook of Tissue Optical Clearing: New Prospects in Optical Imaging*, V. V. Tuchin, D. Zhu, E. A. Genina, Eds., pp. 517–538, Taylor & Francis Group LLC, CRC Press, Boca Raton, FL (2022). https://www.routledge.com/Handbook-of-Tissue-Optical-Clearing-New-Prospects-in-Optical-Imaging/Tuchin-Zhu-Genina/p/book/9780367895099.
41. D. Li, W. Feng, R. Shi, V. V. Tuchin, and D. Zhu, "Tissue optical clearing for in vivo detection and imaging diabetes induced changes in cells, vascular structure, and function," in *Handbook of Tissue Optical Clearing: New Prospects in Optical Imaging*, V. V. Tuchin, D. Zhu, and E. A. Genina, Eds., pp. 539–556, Taylor & Francis Group LLC, CRC Press, Boca Raton, FL (2022). https://www.routledge.com/Handbook-of-Tissue-Optical-Clearing-New-Prospects-in-Optical-Imaging/Tuchin-Zhu-Genina/p/book/9780367895099.
42. P. A. Timoshina, A. B. Bucharskaya, D. A. Alexandrov, and V. V. Tuchin, "Study of blood microcirculation of pancreas in rats with alloxan diabetes," *J. Biomed. Photon. Eng.* **3**(2), 020301 (2017).
43. D. Fuchs, P. P. Dupon, L. A. Schaap, and R. Draijer, "The association between diabetes and dermal microvascular dysfunction non-invasively assessed by laser Doppler with local thermal hyperemia: A systematic review with meta-analysis," *Cardiovasc. Diabetol.* **16**, 11 (2017).
44. G. Walther, P. Obert, F. Dutheil, R. Chapier, B. Lesourd, G. Naughton, et al., "Metabolic syndrome individuals with and without type 2 diabetes mellitus present generalized vascular dysfunction: Cross-sectional study," *Arterioscler. Thromb. Vasc. Biol.* **35**, 1022–1029 (2015).
45. J. W. Patterson, "Diabetic microangiopathy," in *Weedon's Skin Pathology*, J. W. Patterson, G. A. Hosler, and K. L. Prenshaw, Eds., 5th ed., Elsevier, Philadelphia, PA, eBook (2021). http://www.clinicalkey.com/dura/browse/bookChapter/3-s2.0-C20170008388.
46. D. Cabuzu, A. Cirja, R. Puiu, and A. M. Grumezescu, "Biomedical applications of gold nanoparticles," *Curr. Top. Med. Chem.* **15**(16), 1605–1613 (2015). DOI: 10.2174/1568026615666150414144750.
47. M. Venkatachalam, K. Govindaraju, A. M. Sadiq, S. Tamilselvan, V. G. Kumar, and G. Singaravelu, "Functionalization of gold nanoparticles as antidiabetic nanomaterial," *Spectrochim. Acta A: Mol. Biomol. Spectrosc.* **116**, 331–338 (2013). DOI: 10.1016/j.saa.2013.07.038.
48. M. Kleinert, C. Clemmensen, S. M. Hofmann, et al., "Animal models of obesity and diabetes mellitus," *Nat. Rev. Endocrinol.* **14**(3) 140–162 (2018). DOI: 10.1038/nrendo.2017.161.
49. S. Lenzen, "The mechanisms of alloxan- and streptozotocin-induced diabetes," *Diabetologia* **51**(2), 216–226 (2008).
50. A. J. King, "The use of animal models in diabetes research," *Brit. J. Pharmacol.* **166**, 877–894 (2012).
51. T. Szkudelski, "The mechanism of alloxan and streptozotocin action in B cells of the rat pancreas," *Physiol. Res.* **50**, 537–546 (2001).
52. F. Quondamatteo, "Skin and diabetes mellitus: What do we know?" *Cell Tissue Res.* **355**(1), 1–21 (2014).
53. J. Wu, L.-J. Yan, "Streptozotocin-induced type 1 diabetes in rodents as a model for studying mitochondrial mechanisms of diabetic β cell glucotoxicity," *Diabetes Metab. Syndr. Obes.* **8**, 181–188 (2015).
54. A. Rohilla and S. Ali, "Alloxan induced diabetes: Mechanisms and effects," *Int. J. Res. Pharma. Biomed. Sci.* **3**, 819–823 (2012).
55. S. Bukhari, M. Abbasi, and M. Ahmed, "Dose optimization of alloxan for diabetes in albino mice," *Biology* **61**(2), 301–305 (2015).
56. D. C. Damasceno, A. O. Netto, I. L. Iessi, F. Q. Gallego, S. B. Corvino, B. Dallaqua, Y. K. Sinzato, A. Bueno, I. M. Calderon, and M. V. Rudge, "Streptozotocin-induced diabetes models: Pathophysiological mechanisms and fetal outcomes," *BioMed Res. Int.* **2014**, 1–11 (2014). DOI: 10.1155/2014/819065.
57. N. I. Dikht, A. B. Bucharskaya, G. S. Terentyuk, et al., "Morphological study of the internal organs in rats with alloxan diabetes and transplanted liver tumor after intravenous injection of gold nanorods," *Russian Open Med. J.* **3**(3), 0301 (2014). DOI: 10.15275/rusomj.2014.0301.
58. A. V. Alekseeva, V. A. Bogatyrev, B. N. Khlebtsov, A. G. Mel'nikov, L. A. Dykman, and N. G. Khlebtsov, "Gold nanorods: Synthesis and optical properties," *Colloid. J.* **68**, 661–678 (2006). DOI: 10.1134/S1061933X06060019.
59. A. B. Bucharskaya, N. I. Dikht, G. A. Afanasyeva, et al., "The assessment of molecular markers of cell interaction and lipid peroxidation in rats with alloxan diabetes and transplanted liver cancer after intravenous injection of gold nanorods," *Saratov J. Med. Sci. Res.* **11**(2), 107–112 (2015).
60. J. F. Hainfeld, D. N. Slatkin, H. M. Smilowitz, "The use of gold nanoparticles to enhance radiotherapy in mice," *Phys. Med. Biol.* **49**, 309–315 (2004). DOI: 10.1088/0031-9155/49/18/n03.
61. S. BarathManiKanth, K. Kalishwaralal, M. Sriram, S. R. K. Pandian, H.-S. Youn, S. Eom, and S. Gurunathan, "Antioxidant effect of gold nanoparticles restrains hyperglycemic conditions in diabetic mice," *J. Nanobiotechnol.* **8**, 16 (2010). DOI: 10.1186/1477-3155-8-16.

62. H. A. Khan, M. K. Abdelhalim, A. S. Alhomida, and M. S. Al Ayed, "Transient increase in IL-1, IL-6 and TNF-alpha gene expression in rat liver exposed to gold nanoparticles," *Gen. Mol. Res.* **12**(4), 5851–5857 (2013).
63. U. Kanska and J. Boratynski, "Thermal glycation of proteins by D-glucose and D-fructose," *Arch. Immunol. Ther. Exp.* **50**, 61–66 (2002).
64. Y. Dekel, Y. Glucksam, I. Elron-Gross, and R. Margalit, "Insights into modeling streptozotocin-induced diabetes in ICR mice," *Lab. Anim.* **38**, 55–60 (2009).
65. D. K. Tuchina, A. N. Bashkatov, E. A. Genina, and V. V. Tuchin, Patent RF No. 2633494, "Biosensor for noninvasive optical monitoring of the pathology of biological tissues," MPK A61B 5/05, G01N 21/01, Patent holder: N.G. Chernyshevsky Saratov State University, Application No. 2016102046, 22.01.2016, Bul. No. 29 (2017).
66. V. V. Tuchin, *Optical Clearing of Tissues and Blood, PM 154*, SPIE Press, Bellingham, WA (2006).
67. M. G. Ghosn, N. Sudheendran, M. Wendt, A. Glasser, V. V. Tuchin, and K. V. Larin, "Monitoring of glucose permeability in monkey skin *in vivo* using Optical Coherence Tomography," *J. Biophoton.* **3**(1–2), 25–33 (2010).
68. J. Wang, N. Ma, R. Shi, Y. Zhang, T. Yu, and D. Zhu, "Sugar-induced skin optical clearing: From molecular dynamics simulation to experimental demonstration," *IEEE J. Sel. Top. Quant. Electron.* **20**(2), 1–7 (2014).
69. E. A. Genina, A. N. Bashkatov, and V. V. Tuchin, "Tissue optical immersion clearing," *Expert Rev. Med. Devices* **7**(6), 825–842 (2010).
70. D. K. Tuchina, A. N. Bashkatov, A. B. Bucharskaya, and V. V. Tuchin, "Exogenous agent diffusivity in tissues as a biomarker of diabetes mellitus pathology," *Proc. SPIE* **10877**, 108770X (2019).
71. L. M. Oliveira, M. I. Carvalho, E. M. Nogueira, and V. V. Tuchin, "Diffusion characteristics of ethylene glycol in skeletal muscle," *J. Biomed. Opt.* **20**(5), 051019 (2015).
72. J. M. Andanson, K. L. A. Chan, and S. G. Kazarian, "High-throughput spectroscopic imaging applied to permeation through the skin," *Appl. Spectrosc.* **63**(5), 512–517 (2009).
73. M. J. Choi and H. I. Maibach, "Elastic vesicles as topical/transdermal drug delivery systems," *Int. J. Cosmet. Sci.* **27**(4), 211–221 (2005).
74. D. K. Tuchina, A. B. Bucharskaya, and V. V. Tuchin, "Pilot study of glycerol diffusion in ex vivo skin: A comparison of alloxan and streptozotocin diabetes models," *Proc. SPIE* **11457**, 114570J (2020).
75. N. Akhtar, "Vesicles: A recently developed novel carrier for enhanced topical drug delivery," *Current Drug Deliv.* **11**(1), 87–97 (2014).
76. L. M. Shevtsova, O. S. Bykova, N. P. Fedorova, M. V. Grigor'eva, and N. N. Maksimiuk, "Myocardium morphologic reorganization in experimental diabetes," *Vestnik Novgorod State Univ.* **2**(85), 146–150 (2015).
77. A. N. Bashkatov, K. V. Berezin, K. N. Dvoretskiy, et al., "Measurement of tissue optical properties in the context of tissue optical clearing," *J. Biomed. Opt.* **23**(9), 091416 (2018).
78. Glycerine Producers Association, Physical properties of glycerine and its solutions (1963).
79. D. Zhu, K. V. Larin, Q. Luo, and V. V. Tuchin, "Recent progress in tissue optical clearing," *Laser Photon. Rev* **7**(5), 732–757 (2013).
80. V. D. Genin, D. K. Tuchina, A. J. Sadeq, E. A. Genina, V. V. Tuchin, and A. N. Bashkatov, "*Ex vivo* investigation of glycerol diffusion in skin tissue," *J. Biomed. Photon. Eng.* **2**(1), 010303 (2016).
81. T. Son and B. Jung, "Cross-evaluation of optimal glycerol concentration to enhance optical tissue clearing efficacy," *Skin Res. Technol.* **21**(3), 327–232 (2015).
82. A. N. Bashkatov, E. A. Genina, and V. V. Tuchin, "Measurement of glucose diffusion coefficients in human tissues," Chap. 19 in *Handbook of Optical Sensing of Glucose in Biological Fluids and Tissues*, V. V. Tuchin, Ed., 587–621, Taylor & Francis Group LLC, CRC Press, Boca Raton, FL (2009).
83. L. Oliveira and V. V. Tuchin, *The Optical Clearing Method: A New Tool for Clinical Practice and Biomedical Engineering*, Springer Nature, Basel, Switzerland AG, 177 p. (2019).
84. A. Kotyk and K. Janacek, *Membrane Transport: An Interdisciplinary Approach*, Plenum Press, New York (1977).
85. A. N. Bashkatov, E. A. Genina, and V. V. Tuchin, "Optical properties of skin, subcutaneous, and muscle tissues: A review," *J. Innov. Opt. Health Sci.* **4**(1), 9–38 (2011).
86. A. Y. Sdobnov, M. E. Darvin, J. Schleusener, J. Lademann, and V. V. Tuchin, "Hydrogen bound water profiles in the skin influenced by optical clearing molecular agents–quantitative analysis using confocal Raman microscopy," *J. Biophoton.* **12**, e201800283 (2019).
87. V. D. Genin, E. A. Genina, V. V. Tuchin, and A. N. Bashkatov, "Glycerol effects on optical, weight and geometrical properties of skin tissue," *J. Innov. Opt. Health Sci.* **14**(05), 2142006 (2021).

88. D. K. Tuchina, A. N. Bashkatov, E. A. Genina, and V. V. Tuchin, "The effect of immersion agents on the weight and geometric parameters of myocardial tissue *in vitro*," *Biophysics*, **63**(5), 791–797 (2018).
89. E. Selvin, M. W. Steffes, H. Zhu, et al., "Glycated hemoglobin, diabetes, and cardiovascular risk in nondiabetic adults," *N. Engl. J. Med.* **362**(9), 800–811 (2010).
90. S. Tanaka, G. Avigad, B. Brodsky, and E. F. Eikenberry, "Glycation induces expansion of the molecular packing of collagen," *J. Mol. Biol.* **203**(2), 495–505 (1988).
91. B.-M. Kim, J. Eichler, K. M. Reiser, A. M. Rubenchik, and L. B. Da Silva, "Collagen structure and nonlinear susceptibility: Effects of heat, glycation, and enzymatic cleavage on second harmonic signal intensity," *Lasers Surg. Med.* **27**, 329–335 (2000).
92. E. L. Hull, M. N. Ediger, A. N. T. Unione, E. K. Deemer, M. L. Stroman, and J. W. Baynes, "Noninvasive, optical detection of diabetes: Model studies with porcine skin," *Opt. Express* **12**(19), 4496–4510 (2004).
93. W. Feng, C. Zhang, T. Yu, and D. Zhu, "Quantitative evaluation of skin disorders in type 1 diabetic mice by in vivo optical imaging," *Biomed. Opt. Express*, **10**(6), 2996–3008 (2019).
94. H. N. Mayrovitz, A. McClymont, and N. Pandya, "Skin tissue water assessed via tissue dielectric constant measurements in persons with and without *diabetes mellitus*," *Diabetes Technol. Ther.* **15**(1), 60–65 (2013).
95. O. S. Khalil, "Non-invasive glucose measurement technologies: An update from 1999 to the Dawn of the New Millennium," *Diabetes Technol. Ther.* **6**(5), 660–697 (2004).
96. W. Hanna, D. Friesen, C. Bombardier, D. Gladman, and A. Hanna, "Pathologic features of diabetic thick skin," *J. Am. Acad. Dermatol.* **16**, 546–553 (1987).
97. Q. Sun, Y. He, K. Liu, S. Fan, E. P. Parrott, and E. Pickwell-MacPherson, "Recent advances in terahertz technology for biomedical applications," *Quant. Imaging Med. Surg.* **7**(3), 345 (2017).
98. E. Pickwell, B. E. Cole, A. J. Fitzgerald, M. Pepper, and V. P. Wallace, "*In vivo* study of human skin using pulsed terahertz radiation," *Phys. Med. Biol.* **49**(9), 1595 (2004).
99. F. R. Paolillo, V. S. Mattos, A. O. de Oliveira, F. E. Guimarães, V. S. Bagnato, and J. C. de Castro Neto, "Noninvasive assessments of skin glycated proteins by fluorescence and Raman techniques in diabetics and nondiabetics,". *J. Biophot.* **12**(1), e201800162 (2019).
100. D. K. Tuchina, A. N. Bashkatov, E. A. Genina, and V. V. Tuchin, "Quantification of glucose and glycerol diffusion in myocardium," *J. Innov. Opt. Health Sci.* **8**(3), 1541006 (2014).
101. G. Mazarevica, T. Freivalds, and A. Jurka, "Properties of erythrocyte light refraction in diabetic patients," *J. Biomed. Opt.*, **7**(2), 244–247 (2002).
102. E. M. Culav, C. H. Clark, and M. J. Merrilees, "Connective tissue: Matrix composition and its relevance to physical therapy," *Phys. Therapy.* **79**, 308–319 (1999).
103. R. K. Murray, D. K. Granner, P. A. Mayes, and V. W. Rodwell, *Harper's Biochemistry*, Appleton & Lange, Norwalk, CA (1988).
104. C. Molteni and M. Parrinello, "Glucose in aqueous solution by first principles molecular dynamics," *J. Am. Chem. Soc.* **120**, 2168–2171 (1998).
105. J. W. Wiechers, J. C. Dederen, and A. V. Rawlings, "Moisturization mechanisms: internal occlusion by orthorhombic lipid phase stabilizers—a novel mechanism of action of skin moisturization," in *Skin Moisturization*, A. V. Rawlings and J. J. Leyden, Eds., 309–321, Taylor & Francis Inc., Abingdon (2009).
106. L. Oliveira, M. I. Carvalho, E. Nogueira, and V. V. Tuchin, "Optical measurements of rat muscle samples under treatment with ethylene glycol and glucose," *J. Innov. Opt. Health Sci.* **6**(2), 1350012 (2013).
107. L. M. Oliveira, M. I. Carvalho, E. Nogueira, and V. V. Tuchin, "The characteristic time of glucose diffusion measured for muscle tissue at optical clearing," *Laser Phys.* **23**(7), 075606 (2013).
108. P. Ballard, D. E. Leahy, and M. Rowland, "Prediction of in vivo tissue distribution from in vitro data 1. Experiments with markers of aqueous spaces," *Pharm. Res.* **17**(6), 660–663 (2000).
109. P. J. Caspers, G. W. Lucassen, and G. J. Puppels, "Combined *in vivo* confocal Raman spectroscopy and confocal microscopy of human skin," *Biophys J.* **85**(1), 572–580 (2003).
110. N. Nakagawa, M. Matsumoto, and S. Sakai, "*In vivo* measurement of the water content in the dermis by confocal Raman spectroscopy," *Skin Res. Technol.* **16**, 137–141 (2010).
111. A. A. Gurjarpadhye, W. C. Vogt, Y. Liu, and C. G. Rylander, "Effect of localized mechanical indentation on skin water content evaluated using OCT," *Int. J. Biomed. Imag.* **2011**, 1–8 (2011).
112. B. Schulz, D. Chan, J. Backstrom, and M. Rubhausen, "Spectroscopic ellipsometry on biological materials–investigation of hydration dynamics and structural properties," *Thin Solid Films* **455–456**, 731–734 (2004).
113. C. Li, J. Jiang, and K. Xu, "The variations of water in human tissue under certain compression: Studied with diffuse reflectance spectroscopy," *J. Innov. Opt. Health Sci.* **6**(1), 1–9 (2013).
114. V. Hovhannisyan, P.-S. Hu, S.-J. Chen, C.-S. Kim, and C.-Y. Dong, "Elucidation of the mechanisms of optical clearing in collagen tissue with multiphoton imaging," *J. Biomed. Opt.* **18**(4), 1–8 (2013).

115. R. Grüner, I. Latka, J. Lademann, B. Dietzek, and J. Popp, "A fiber coupled and stabilized microscope for analytical CARS micro-spectroscopy," *Laser Phys. Lett.* **10**(6), 065605 (2013).
116. N. Vogler, S. Heuke, D. Akimov, I. Latka, F. Kluschke, H.-J. Rowert-Huber, J. Lademann, B. Dietzek, and J. Popp, "Discrimination of skin diseases using the multimodal imaging approach," *Proc. SPIE* **8427**, 842710 (2012). DOI: 10.1117/12.921748.
117. D. R. Lide, Ed., *Handbook of Chemistry and Physics*, 84th Edition, CRC Press, Boca Raton, FL (2003–2004). https://abu.ut.ac.ir/documents/83901321/98999218/crc.pdf.
118. K. V. Larin, M. G. Ghosn, A. N. Bashkatov, E. A. Genina, N. A. Trunina, and V. V. Tuchin, "Optical clearing for OCT image enhancement and in-depth monitoring of molecular diffusion," *IEEE J. Select. Tops Quant. Electron.* **18**(3), 1244–1259 (2012).
119. D. K. McGuire, W. J. Shih, F. Cosentino, et al., "Association of SGLT2 inhibitors with cardiovascular and kidney outcomes in patients with type 2 diabetes: A meta-analysis," *JAMA Cardiol.* **6**(2), 148–158 (2021) DOI: 10.1001/jamacardio.2020.4511.
120. L. Yazdanpanah, M. Nasiri, and S. Adarvishi, "Literature review on the management of diabetic foot ulcer," *World J. Diabetes.* **6**, 37–53 (2015).
121. N. Sayin, N. Kara, and G. Pekel, "Ocular complications of diabetes mellitus," *World J Diabetes* **6**, 92–108 (2015).
122. A. P. Vasiliev and N. N. Streltsova, "Skin microcirculation in patients with arterial hypertension and in patients with a combination of arterial hypertension and type II diabetes mellitus," *Reg. Hemodynam. Microcirc.* **19**(4), 44–52 (2020) doi: 10.24884/1682-6655-2020-19-4-44-52.
123. A. S. Ametov, I. O. Kurochkin, and A. A. Zubkov, "Diabetes and cardiovascular disease," *RMJ.* **13**, 954 (2014) (In Russ.)
124. C.-Y. Chiang, J.-J. Lee, S.-T. Chien, D. A. Enarson, Y.-C. Chang, Y.-T. Chen, et al. "Glycemic control and radiographic manifestations of tuberculosis in diabetic patients," *PLoS One* **9**(4): e93397. (2014) DOI: 10.1371/journal.pone.0093397.
125. S. Shekar, N. Satpute, and A. Gupta, "Review on diabetic retinopathy with deep learning methods," *J. Med. Imaging*, **8**(6), 060901 (2021).
126. A. I. Vinik, M. L. Nevoret, C. Castellini, and H. Parson, "Diabetic neuropathy," *Endocrinol Metab. Clin. North. Am.* **42**, 747–875 (2013).
127. A. J. Morgan and R. A. Schwartz, "Diabetic dermopathy: A subtle sign with grave implications," *J. Am. Acad Dermatol.* **58**, 447–51 (2008).
128. A. K. Arya, R. Tripathi, S. Kumar, and K. Tripathi, "Recent advances on the association of apoptosis in chronic non healing diabetic wound" *World J. Diabetes* **5**, 756–62 (2014).
129. F. Paneni, S. Costantino, and F. Cosentino, "Insulin resistance, diabetes, and cardiovascular risk," *Curr. Atheroscler Rep.* **16**, 419 (2014).
130. F. Paneni, J. A. Beckman, M. A. Creager, and F. Cosentino, "Diabetes and vascular disease: Pathophysiology, clinical consequences, and medical therapy: Part I," *Eur. Heart J.* **34**, 2436–43 (2013).
131. H. Jonasson, I. Fredriksson, M. Larsson, and T. Strömberg, "Validation of speed-resolved laser Doppler perfusion in a multimodal optical system using a blood-flow phantom," *J. Biomed. Opt.* **24**(9), 095002 (2019). DOI: 10.1117/1.JBO.24.9.095002.
132. V. Doblhoff-Dier, L. Schmetterer, W. Vilser et al., "Measurement of the total retinal blood flow using dual beam Fourier-domain Doppler optical coherence tomography with orthogonal detection planes," *Biomed. Opt. Express* **5**(2), 630–642 (2014).
133. W. Heeman, W. Steenbergen, G. M. van Dam, and E. C. Boerma, "Clinical applications of laser speckle contrast imaging: A review," *J. Biomed. Opt.*, **24**(8), 080901 (2019). DOI: 10.1117/1.JBO.24.8.080901.
134. D. J. Briers, D. A. Zimnyakov, O. V. Ushakova, and V. V. Tuchin, "Speckle technologies for monitoring and imaging tissues and tissue-like phantoms," in *Handbook of Optical Biomedical Diagnostics*, 2nd ed., **vol. 2**: Methods, pp. 429–496, SPIE Press (2016). DOI: 10.1117/3.2219608.ch8.
135. I. Sigal, R. Gad, A. M. Caravaca-Aguirre et al., "Laser speckle contrast imaging with extended depth of field for in-vivo tissue imaging," *Biomed. Opt. Express* **5**(1), 123–135 (2014).
136. P. A. Dyachenko (Timoshina), A. N. Bashkatov, D. A. Alexandrov, V. I. Kochubey, and V. V. Tuchin, "Laser speckle contrast imaging for monitoring of acute pancreatitis at ischemia–reperfusion injury of the pancreas in rats," *J. Innov. Opt. Health Sci.* **15**(1), 2242002 (2021). https://doi.org/10.1142/S1793545822420020.
137. P. A. Dyachenko (Timoshina), A. S. Abdurashitov, O. V. Semyachkina-Glushkovskaya, and V. V. Tuchin, "Blood and lymph flow imaging at optical clearing," in *Handbook of Tissue Optical Clearing: New Prospects in Optical Imaging*, V. V. Tuchin, D. Zhu, and E. A. Genina Eds., pp. 393–408, Taylor & Francis Group LLC, CRC Press, Boca Raton, FL (2022).

138. S. J. Kirkpatrick, "Laser speckle contrast imaging is sensitive to advective flux," *J. Biomed. Opt.* **21**(7), 076001 (2016). DOI: 10.1117/1.JBO.21.7.076001.
139. A. Nadort et al., "Quantitative blood flow velocity imaging using laser speckle flowmetry," *Sci. Rep.* **6**, 25258 (2016). DOI: 10.1038/srep25258.
140. T. Smausz, D. Zölei, and B. Hopp, "Real correlation time measurement in laser speckle contrast analysis using wide exposure time range images," *Appl. Opt.* **48** (8), 1425–1429 (2009).
141. L. M. Richards et al., "Intraoperative multi-exposure speckle imaging of cerebral blood flow," *J. Cereb. Blood Flow Metab.* **37**(9), 3097–3109 (2017). DOI: 10.1177/0271678X16686987.
142. A. B. Parthasarathy et al., "Robust flow measurement with multi-exposure speckle imaging," *Opt. Express* **16** (3), 1975 –1989 (2008).
143. E. B. Postnikov, M. O. Tsoy, P. A. Timoshina, and D. E. Postnov, "Gaussian sliding window for robust processing laser speckle contrast images," *Int. J. Numer. Methods Biomed. Eng.* **35** (4), e3186 (2019).
144. G. Mahe, A. Humeau-Heurtier, S. Durand, G. Leftheriotis, and P. Abraham, "Assessment of skin microvascular function and dysfunction with laser speckle contrast imaging," *Circ Cardiovasc Imaging* **5**, 155–6 (2012).
145. M. Roustit, C. Millet, S. Blaise, B. Dufournet, and J. L. Cracowski, "Excellent reproducibility of laser speckle contrast imaging to assess skin microvascular reactivity," *Microvasc Res.* **80**, 505–112010 (2011)
146. M. Roustit and J. L. Cracowski, "Non-invasive assessment of skin microvascular function in humans: An insight into methods," *Microcirculation* **19**, 47–64 (2012).
147. A. S. de M Matheus, E. L. Silva Clemente, M. de Lourdes Guimarães Rodrigues, D. C. Torres Valença, and M. B. Gomes, "Assessment of microvascular endothelial function in type 1 diabetes using laser speckle contrast imaging," *J. Diabetes Complicat.*, **31**, **4**, 753–757 (2017).
148. W. Feng, R. Shi, C. Zhang, S. Liu; T. Yu, and D. Zhu, "Visualization of skin microvascular dysfunction of type 1 diabetic mice using in vivo skin optical clearing method," *J. Biomed. Opt.* **24**(03), 1 (2018).
149. P. A. Dyachenko (Timoshina), D. A. Alexandrov, A. B. Bucharskaya, and V. V. Tuchin, "Speckle-contrast imaging of pathological tissue microhemodynamics in the development of various diabetes models," *Proc. SPIE* **11457**, 114570K (2020). DOI: 10.1117/12.2564100.
150. P. A. Timoshina, A. B. Bucharskaya, N. A. Navolokin, and V. V. Tuchin, "Speckle-contrast imaging of pathological tissue microhemodynamics at optical clearing," *Proc. SPIE* **10877**, 108770Z (2019). DOI: 10.1117/12.2508794.
151. O. A. Mennes, J. J. Van Netten, J. G. Van Baal, and W. Steenbergen, "Assessment of microcirculation in the diabetic foot with laser speckle contrast imaging," *Physiol. Meas.* **40**(6), 065002 (2019).
152. O. A. Mennes, et al., "Semi-automatic tracking of laser speckle contrast images of microcirculation in diabetic foot ulcers," *Diagnostics (Basel, Switzerland)*, **10**(12), 1054 (2020). DOI: 10.3390/diagnostics10121054.
153. S. Srinivas, O. Tan, M. G. Nittala, et al., "Assessment of retinal blood flow in diabetic retinopathy using Doppler Fourier-domain optical coherence tomography," *Retina* **37**(11), 2001–2007 (2017). DOI: 10.1097/IAE.0000000000001479.
154. B. Lee, E. A. Novais, N. K. Waheed, et al., "En face Doppler optical coherence tomography measurement of total retinal blood flow in diabetic retinopathy and diabetic macular edema," *JAMA Ophthalmol.* **135**(3), 244–251(2017). DOI: 10.1001/jamaophthalmol.2016.5774.
155. Y. Song, T. Tani, T. Omae, A. Ishibazawa, T. Yoshioka, K. Takahashi, et al., "Retinal blood flow reduction after panretinal photocoagulation in type 2 diabetes mellitus: Doppler optical coherence tomography flowmeter pilot study," *PLoS One* **13**(11), e0207288 (2018). DOI: 10.1371/journal.pone.0207288.
156. D. Gildea, "The diagnostic value of optical coherence tomography angiography in diabetic retinopathy: A systematic review," *Int. Ophthalmol.* **39**(10), 2413–2433 (2019). DOI: 10.1007/s10792-018-1034-8.
157. H. Akil, S. Karst, M. Heisler, M. Etminan, E. Navajas, and D. Maberley, "Application of optical coherence tomography angiography in diabetic retinopathy: A comprehensive review," *Can. J. Ophthalmol.* **54**(5), 519–528 (2019). DOI: 10.1016/j.jcjo.2019.02.010.
158. E. V. Potapova, V. V. Dremin, E. A. Zherebtsov, I. N. Makovik, E. V. Zharkikh, A. V. Dunaev, O. V. Pilipenko, V. V. Sidorov, and A. I. Krupatkin, "A complex approach to noninvasive estimation of microcirculatory tissue impairments in feet of patients with diabetes mellitus using spectroscopy," *Opt. Spectrosc.* **123**(6), 955–964 (2017).
159. M. Sorelli, P. Francia, L. Bocchi, A. De Bellis, and R. Anichini, "Assessment of cutaneous microcirculation by laser Doppler flowmetry in type 1 diabetes," *Microvasc. Res.* **124**, 91–96 (2019). DOI: 10.1016/j.mvr.2019.04.002. PMID: 30959000.
160. I. Kozlov, E. Zherebtsov, G. Masalygina, K. Podmasteryev, and A. Dunaev, "Laser Doppler spectrum analysis based on calculation of cumulative sums detects changes in skin capillary blood flow in type 2 diabetes mellitus," *Diagnostics* **11**, 267 (2021). DOI: 10.3390/diagnostics11020267.

2 Optical Methods for Diabetic Foot Ulcer Screening

Robert Bartlett, Gennadi Saiko, and Alexandre Yu. Douplik

CONTENTS

DOI: 10.1201/9781003112099-2

2.1 INTRODUCTION

Shifts in diet, lifestyle, and exercise are driving the growth of diabetes in all countries. In 2018, 34.2 million Americans, or 10.5% of the population, had diabetes. In senior citizens older than 65 years, the prevalence is 27%.

Diabetic foot ulceration (DFU) is a major complication of diabetes. Its annual prevalence is estimated to be 4%–10%, and the lifetime risk of developing these ulcers in people with diabetes is estimated to be anywhere from 15% to 25%. Foot ulceration increases the risk of lower limb amputation, one of the most debilitating complications of diabetes. If untreated, DFU may become infected and require total or partial amputation of the affected limb. Around 50% of nontraumatic lower limb amputations are attributable to diabetic ulcers [1].

Even when ulcers have healed, clinicians should view DFU as a lifelong relapsing condition that requires monitoring to minimize the annual recurrence frequency. Among patients with a history of ulceration, 60% will develop another ulcer within 1 year of wound healing [2]. Diabetes increases the incidence of foot ulcer admissions by 11-fold, accounting for more than 80% of all amputations and increasing hospital costs more than tenfold over the 5 years [3].

In this chapter, we briefly overview the pathophysiology of diabetic foot, current diagnostic methods, and novel optical modalities, which have been translated into the clinic but have not yet received widespread clinical adoption.

We will focus on diagnostic methods used (the current methods) and those with the potential to be used (novel optical methods) by primary care practitioners for screening purposes. Efficient screening modalities allow earlier interventions in a patient population that already presents clinically with late-stage complications, significant morbidity, and mortality risk.

Thus, we have included optical modalities based on their relevance to primary diabetic foot care screening. Consequently, we have excluded several modalities such as optical coherence tomography, photoacoustic imaging, Raman spectroscopy, microscopy, and video capillaroscopy as they have a limited clinical impact on diabetic wound care and primary care as screening tools. These modalities are useful research tools but of limited use in clinical care due to their expense, complexity, and time requirements for screening processes. Our focus being on the extraction of physiological parameters, we also excluded planimetric and stereometric applications, which can be used for wound margin delineation and assessment of wound area, its length, width, and depth. In addition, we excluded pure image processing techniques, which classify wound (epithelial vs. granulation vs. eschar vs. slough) or periwound (e.g., erythema [4]) tissue types.

2.2 PATHOPHYSIOLOGY OF DIABETIC FOOT

Wounds that fail to progress through the normal stages of healing and are still open at 1 month are considered "chronic wounds" [5]. Failure to heal is typically multifactorial, with several different disease processes (comorbidities) interacting. Arterial disease, diabetes, obesity, immune deficiencies, and malnutrition are common comorbidities that delay healing [6].

DFU is one of the most common types of chronic wounds. The diabetic foot pathophysiology that leads to ulceration, impaired healing, and limb loss is a triad of polyneuropathy, vasculopathy, and immune defects.

2.2.1 Polyneuropathy

Polyneuropathy is a common complication of diabetes and encompasses several neuropathic syndromes. Neuropathy can be symmetric or asymmetric, and the symptoms can be attributed to impaired sensation or motor function, or a combination of both. Diabetic neuropathy may affect various combinations of sensory, motor, and autonomic neuropathy. The clinical presentation is varied (Table 2.1).

The distal symmetric neuropathy with sensory impairment is the most common form affecting 30% of all people with diabetes [7]. Because the longest nerves are the first to be affected, the earliest symptoms appear in the feet. It is important to understand that many patients with distal sensory neuropathy may not be aware of early sensory impairment and motor loss. For these individuals, their first symptom maybe a relatively painless, nonhealing ulcer of the foot.

The astute clinician should take care to avoid "anchoring bias." Diabetic neuropathy is a diagnosis of exclusion. Nondiabetic, treatable neuropathies may be present in patients with diabetes (Table 2.2).

TABLE 2.1
Diabetic Neuropathic Presentations

- Chronic sensory neuropathy
- Symmetric distal motor neuropathy
- Proximal motor neuropathy
- Focal vascular neuropathy
- Truncal radiculopathy
- Acute painful neuropathy of poor glycemic control
- Acute painful neuropathy of rapid glycemic control
- Cranial mononeuropathy (CN III)

TABLE 2.2
Differential Diagnosis of Distal Symmetric Neuropathy

Drugs
- Alcohol
- Chemotherapy
- Nitrofurantoin
- Isoniazid
- Colchicine
- Amiodarone
- Dapsone

Genetic
- Hereditary sensory neuropathies
- Charcot-Marie-Tooth disease

Metabolic
- Diabetes
- Uremia
- Myxedema
- Porphyria
- Paraproteinemia
- Vitamin deficiency
- Mercury
- Arsenic
- Neoplastic syndromes
- Bronchial carcinoma
- Gastric carcinoma
- Lymphoma

Inflammatory
- Guillain Barre Syndrome
- Polyarteritis Nodosa Inflammatory Demyelinating Polyneuropathy

Infectious
- Leprosy
- Lyme borreliosis
- HIV

Diabetic nerve damage is multifactorial. The damage is attributed to endothelial dysfunction affecting the vasa nervorum, which provides blood to larger nerves, oxidative stress from hyperglycemia, impaired antioxidant production, and advanced glycation end products. Epidemiology studies indicate neuropathy is more likely with increased age, duration of diabetes, poor glycemic control, and vascular disease.

A thorough foot examination is important to detect disease early. Screening for peripheral neuropathy and peripheral arterial disease (PAD) can help identify patients at risk of foot ulcers. A history of ulcers or amputations and poor glycemic control increase the risk of future ulceration.

2.2.1.1 Neuropathy Consequences

Impaired sensory functions are associated with painful symptoms and loss of protective sensation. Impaired motor function leads to atrophy and loss of muscle tone. Because the classical form of diabetic neuropathy is distal, the foot's intrinsic muscles are affected more than proximal muscles. When this occurs, tendons from the large proximal muscles are no longer balanced by normal muscle tone from the intrinsic muscles of the feet. This leads to numerous foot deformities such as "hammertoes" and buckling of the metatarsal heads, "clawed foot." The deformed foot is less efficient and subject to mechanical stress and subsequent "painless" breakdown of the soft tissue due to loss of protective sensation.

Impaired autonomic function leads to altered vascular tone and malperfusion of the foot. Additionally, the autonomic system governs oil and sweat glands. Neuropathy is associated with the loss of these glands, leading to dry, cracked skin prone to infection.

2.2.2 Vasculopathy

Diabetes-associated vasculopathy occurs at macrovascular and microvascular levels. The macrovascular disease affects large blood vessels, whereas the microvascular disease affects the very small vessels, less than 300 μm. Some of these vessels contain smooth muscle, which regulates blood flow to the capillary beds. The terminal vessels of the microcirculation are the capillaries that lack smooth muscles.

2.2.2.1 Macrovascular Disease

The macrovascular disease can be atherosclerotic occlusive disease or medial calcinosis, which is nonocclusive. The proatherogenic changes associated with diabetes include increases in vascular inflammation and derangements in the cellular components of the vasculature and alterations in blood cells and hemostatic factors. Patients with diabetes have a fourfold increase in the prevalence of atherosclerosis and are more likely to have an accelerated course. They are also more sensitive to moderate changes in perfusion compared with patients without diabetes. Patients with diabetes have a propensity for diffuse disease below the knee with a minimal disease in the pedal arteries – specifically the anterior tibia artery, the posterior tibia artery, and peroneal artery. In contrast, nondiabetics have a more discreet plaque-like disease of the proximal arteries. These anatomical observations are important when planning surgical interventions.

The most common symptom of PAD is claudication, defined as pain, cramping, or aching in the calves, thighs, or buttocks, which appears with walking exercise and is relieved by rest. When significant neuropathy is present, these symptoms will be masked despite significant arterial occlusion. Patients with extremely sedentary life styles may not walk far enough to experience classical claudication ischemic pain.

As PAD progresses over time, pain will occur at rest and may awaken patients during the night. Oftentimes, these patients find relief by dangling the legs over the side of the bed with gravity assisting the arterial flow. In extreme cases, the tissue dies and turns black (dry gangrene) or may become infected (wet gangrene).

2.2.2.2 Microvascular Disease

Diabetes causes structural and functional changes in microcirculation. The most prominent structural change is thickening of the basement membrane, which supports the endothelial cells. These microvascular changes are more pronounced in the lower extremities. Changes in the basement membrane affect multiple functions such as vascular permeability, cellular adhesion, and proliferation, which delay wound healing.

Membrane thickening impairs oxygen diffusion and migration of carbon dioxide, nutrients, and leukocytes from the lumen to the interstitial space outside the capillaries. This limits the ability to control infection. Blood vessels become less elastic with thicker basement membranes, limiting vasodilation in response to injury and stress – decreasing the delivery of oxygen, nutrients, and the immune response. It is important to note thickening is not an occlusive process as the luminal diameter is unchanged.

The impaired vasodilation of the microcirculation in response to injury is described as functional ischemia. This is to say there is a mismatch of supply to local demand at the capillary level. The loss of appropriate dilation of the met-arterioles results in a functional arteriovenous shunting of blood past the tissue beds in need [8,9].

DFU is a chronic complication of diabetes associated with neuropathy and/or arteriosclerotic peripheral vascular diseases in the lower limb, alongside microvascular abnormalities [10,11]. There are two physiological theories pertaining to diabetic microvascular disease. The capillary steal premise is based on the sympathetic autonomic neuropathy in the lower limb. The loss of appropriate vasoconstrictor control and alteration in blood flow increase arteriovenous shunting and reduce nutritive blood flow.

The hemodynamic premise is supported by the evidence of hyperglycemic oxidative stress and cell damage, which limit nitric oxide (NO) production [12,13]. Nitric oxide is an important signaling molecule for vasodilation and other cell processes. Downstream consequences of the hemodynamic model are microvascular remodeling, thickening of capillary basement membranes, microvascular sclerosis, and loss of autoregulation and nutritive flow [14–17].

2.3 CURRENT DIAGNOSTIC METHODS

2.3.1 Neuropathy Examination

The American Diabetes Association 2020 guidelines recommend assessing diabetic peripheral neuropathy starting at the diagnosis of type 2 diabetes and 5 years after diagnosing type 1 diabetes and at least annually thereafter. Assessment for distal symmetric polyneuropathy should include a careful history and assessment of either temperature or pinprick sensation (small fiber function) and vibration sensation using a 128-Hz tuning fork (large fiber function). All patients should undergo an annual 10-g monofilament testing to identify feet at risk of ulceration and amputation. Up to 50% of diabetic peripheral neuropathies may be asymptomatic. These patients are at risk of injuries to their insensate feet if the sensory impairment is not recognized early with preventive foot care.

Symptoms vary according to the type of sensory fibers affected. Typically, small fibers are affected early, producing pain and dysesthesia (burning sensation). Large fiber involvement is associated with tingling, numbness, progressive loss of protective sensation, and poor balance [18].

It is the loss of the "gift of pain" that places patients at risk of recurring injury and foot ulceration over pressure points on foot. Because of the loss of painful feedback, it is common for patients to walk on these open ulcers for weeks before they are discovered [19].

2.3.1.1 Small Fiber Tests

- Pinprick – Gently touch the skin with a pin or the back end of the pin and ask the patient whether it feels sharp or dull. Begin at the forefoot and move proximally. Record the level where sharp/dull discrimination begins.

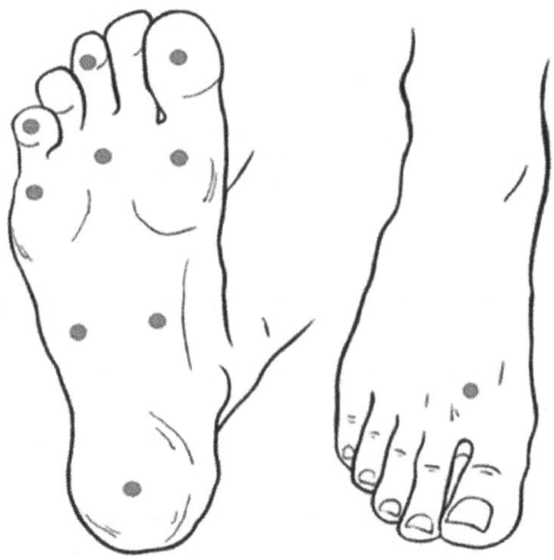

FIGURE 2.1 Ten-point Semmes-Weinstein examination. Nine points are located on the plantar surface, and one point is on the dorsal surface. The examination is positive for neuropathy if the patient fails to perceive 3 or more locations. (Daly and Leahy [51]. Copyright Wiley-VCH GmbH. Reproduced with permission.)

- Temperature – Use two test tubes, one filled with cold water and one with warm water. Touch the skin for 2 seconds and ask the patient "hot or cold?". Begin at the forefoot and move proximally. Record the level where hot/cold discrimination begins.

2.3.1.2 Large Fiber tests

- Vibration – Use a 128-Hz tuning fork. The test is considered positive when the patient loses vibratory sensation over the metatarsal heads but the examiner still perceives it when applied to their distal radius at the wrist. Begin using the first dorsal metatarsal-phalangeal head. If that location is positive for neuropathy, assess over the medial and lateral malleoli at the ankle.
- Light touch – This test is also known as the Semmes-Weinstein monofilament (SWM) examination. Use a 10-gram monofilament to perform a 10-point assessment, as illustrated in Figure 2.1. Hold the monofilament in place for about 2 seconds. Press the monofilament to the skin, so it buckles at one of two times as you say, "Time one" or "Time two." Have the patient identify at which time he or she was touched. Randomize the sequence of applying the filament throughout the examination. If there is an ulcer, callus, or scar on the foot, the monofilament is applied to an adjacent area.

The SWM examination provides a convenient screening for the loss of protective sensation. The inability to perceive 10 g of force or a 5.07 monofilament applied is associated with clinically significant large fiber neuropathy and a high risk of foot complications. In three studies, the SWM examination identified foot ulceration risk with a sensitivity of 66%–99%, a specificity of 34%–86%, a positive predictive value of 18%–39%, and a negative predictive value of 94%–95% [20–23].

2.3.2 Macrovascular Assessment

Early diagnosis of PAD is important because it signals the need to evaluate more central disease of the coronary, renal, and cerebral arteries. The absence of peripheral pulses on the physical examination would support the diagnosis of PAD. However, patients can have significant disease before the pulses disappear. The physical examination as a "test" lacks sensitivity and has a poor agreement between examiners. Better studies are available for PAD screening. The ankle brachial index (ABI) and the handheld Doppler are the commonly used procedures. They have good sensitivity, are inexpensive, and can be performed well at the bedside [24].

2.3.2.1 ABI

The ABI is evaluated by measuring the systolic blood pressures in the ankles (dorsalis pedis and posterior tibia arteries) and arms (brachial artery) using a handheld Doppler and then calculating a ratio. It is measured by placing the patient in a supine position for 5 minutes. Systolic blood pressure is measured in both arms, and the higher value is used as the denominator of the ABI. Systolic blood pressure is then measured in the dorsalis pedis and posterior tibia arteries by placing the cuff just above the ankle. The higher value is used as the numerator of the ABI in each limb (Table 2.3).

The ABI is a simple, noninvasive, quantitative measurement of the patency of the lower extremity arterial system. Compared with pulse palpation, the ABI is much more accurate. It has been validated against contrast arteriography and found to be 95% sensitive and almost 100% specific [25].

2.3.2.2 Toe Brachial Index

The toe brachial index (TBI) is evaluated when the ABI is abnormally high due to calcification of the arteries (calcinosis) in the leg which makes them difficult to compress (an abnormally high ABI is > 1.3). When this is the case, the TBI is used as an alternative because the digital arteries for toes do not calcify.

The TBI is calculated similarly to the ABI. The blood pressure of the great toe is divided by the systolic brachial blood pressure. Clinicians can measure the toe pressure by placing a small toe cuff around the great toe and attaching a photoplethysmography probe at the pulp of the great toe tip. The TBI is considered a better test; however, it is not routinely used as a screening tool because of the need for specialized toe cuffs and photoplethysmography probes, limiting widespread TBI use as a "simple" screen.

2.3.2.3 Doppler Probe

Simple auscultation with a Doppler ultrasound probe also serves as a more reliable screening tool than palpation on physical examination. However, it does require moderate experience. On auscultation, a normal artery should have a triphasic sound. The first phase is the systolic rush of blood, followed by the second phase: the elastic recoil of the arterial wall and a final small phase produced by the shock wave of the aortic valve closure. As the arterial disease progresses, the sound changes to a subdued biphasic pattern. It becomes a quite monophasic sound with a severer disease, and with a severe disease, no sound is detectable.

TABLE 2.3
Diagnostic Criteria for PAD using ABI Measurements

ABI Value	Interpretation
0.91–1.30*	Normal
0.70–0.90	Mild obstruction
0.40–0.69	Moderate obstruction
<0.40	Severe obstruction

* An ABI value >1.3 is unreliable and suggests poorly compressible arteries at the ankle level due to the presence of medial arterial calcification.

The purpose of arterial screening examinations is to determine whether the macrovascular disease is present. If the screening tests are positive, more sophisticated examinations are needed to localize the disease and determine whether the patient can be helped by vascular bypass or endovascular interventions such as balloon stents. These "anatomic" examinations are performed using contrast angiography, magnetic resonance angiography, or computerized tomographic angiography, which have various merits that go beyond the scope of this text.

2.3.3 Microvascular Assessment

There is a large array of options for evaluating macrocirculation. Studying macrocirculation is easier due to the size of the vessels and the small number of "named" vessels. In contrast, evaluation of the microcirculation is more difficult because it comprises vessels measured in microns and the vessels number in the millions (capillaries). Technology drivers for macrocirculation studies are the numerous therapeutic options that exist for endovascular procedures to improve blood flow. However, there are few options for local control of microcirculation.

From a clinical perspective, it is important to understand that macrocirculation is merely a conduit for delivering oxygen and nutrients. The real "biology" of exchange occurs at the microcirculatory level. Because diabetes impairs microcirculation, it has been observed that technical correction of macroflow does not reliably lead to clinical healing of DFU. An increase in macroflow does not necessarily translate to an increase in microflow (nutritive) at the capillary level because of significant arteriovenous shunting and the diffusion barrier of a thick basement membrane. Therefore, the objective marker for success should be a measurable change in tissue oxygen.

From a clinical perspective, there are only three available methods for evaluating microcirculation. They are skin perfusion pressure (SPP) measurement, indocyanine green (ICG) angiography, and transcutaneous oxygen pressure ($TcPO_2$) measurement.

2.3.3.1 Transcutaneous Oxygen Pressure ($TcPO_2$)

$TcPO_2$ measurement is a noninvasive test that assesses the partial pressure of oxygen diffusing through the skin. A "Clark electrode" is used to measure oxygen using a platinum cathode and a silver anode covered with a thin membrane permeable to oxygen. The electrode is submerged in a small, ringed, normal saline water bath on intact skin. Voltage is applied to the electrodes, and the oxygen molecules are reduced to water:

$$O_2 + 4e^- + 4H+ \rightarrow 2H_2O.$$

The flow of electrons (current) is directly proportional to the number of oxygen molecules expressed as gas pressure in mmHg [26].

The water bath serves to capture the transdermal diffusion of oxygen. It also hydrates the skin to improve dermal permeability. The Clarke electrode is heated to 44.5°C to induce local hyperthermia and soften the stratum corneum. Hyperthermia creates maximum capillary vasodilatation for a uniform baseline, removing any "local" variability from vasoconstrictive signals. Because the electrodes must be placed on intact skin, the measurements are always periwound. $TcPO_2$ interpretation assumes that the values inside the wound are the same or worse than the values in the periwound tissue. An ideal device would provide measurement in the wound itself; however, this is not possible.

A complete transcutaneous examination of both lower extremities typically requires 1 hour. The study is done with the patient in the supine position, and care must be taken to avoid placing the electrodes over boney prominences, veins, scar tissue, cellulitis, or areas of edema as these areas are associated with lower values. Performing accurate $TcPO_2$ tests requires skill on the part of the operator [27]. Clinical decisions are based on the room air measurements. A normal healthy value in the foot being >50 mmHg, and a value <40 mmHg is sufficient hypoxia to impair wound healing.

Naturally, the lower the value, the less likely the healing. $TcPO_2$ values <30 mmHg indicate that critical limb ischemia (CLI) is present, although other conditions, such as anemia, pulmonary disease, edema, and severe heart failure, should be considered as contributing factors for low values.

Provocative maneuvers are frequently used to improve clinical predictions. One maneuver is to breathe 100% oxygen for 15 minutes and evaluate the change in oxygen pressure. Patients with severe disease will have a "blunted" response to oxygen with little change. Because transcutaneous testing is frequently used to screen patients for hyperbaric oxygen (HBO) therapy, measurements may be performed in the hyperbaric chamber at treatment pressure. Diabetic patients whose $TcPO_2$ values during a hyperbaric session are >200 mmHg have a significant likelihood of benefitting from HBO therapy, whereas patients whose "in-chamber" $TcPO_2$ values are <50mmHg are not likely to benefit [28–31].

Measurements during hyperbaric therapy appear to be the best predictor for selecting patients who will or will not respond to HBO [32].

$TcPO_2$ measurement can be used for:

- Assessment of periwound oxygenation
- Screening for PAD
- Diagnosis of CLI
- Prediction of the likelihood for healing based on oxygen-dependent mechanisms
- Evaluation and monitoring of compromised flaps
- Prediction of amputation
- Selection of optimal amputation level
- Identification of patients who would benefit from HBO
- Assessment of response to HBO

2.3.3.2 Skin Perfusion Pressure

SPP is a noninvasive test that quantitatively analyzes the cutaneous blood flow of the microcirculation by measuring the "opening pressure" of blood vessels using a red laser light source. Blood flow to the selected region is stopped using a blood pressure cuff. The cuff is inflated to a pressure above systolic in the microcirculation (typically 90 mm Hg); the pressure is held for 10 seconds or longer until the laser flow indicator reads less than 0.1%.

As the pressure in the blood pressure cuff is gradually reduced, a laser Doppler is used to detect the first movement of red blood cells (RBCs) in the circulation. RBCs reflect the laser beam, and the very moment flow resumes, the reflected signal changes (Doppler shift). The pressure associated with the first movement of RBCs is the opening pressure for the microcirculation.

Unlike photoplethysmography, which detects the intensity of light scattered from RBCs, the laser Doppler sensor determines the presence of a Doppler frequency shift generated by the motion of RBCs. A visual recording shows the cuff pressure (x-axis) vs. the detected RBC motion or perfusion (y-axis).

SPP has several advantages:

- Less time-consuming than $TcPO_2$
- No calibration required
- Anatomical evaluation of vasculosomes
- Not affected by arterial calcinosis
- Not affected by moderate edema
- Not affected by thick skin

SPP has been studied extensively in both diagnostic and prognostic capacities. It provides quantitative measurement to guide the need for revascularization and the probability of healing lower extremity ischemic ulcers, skin incisions, or amputation sites. The following observations are made in clinical practice [33–36]:

- <30 mmHg –critical limb ischemia
- 30–40 mmHg –impaired wound healing based on perfusion
- 40–50 mmHg –abnormal, impaired healing
- >50 mmHg –normal skin perfusion.

SPP has been correlated with other noninvasive wound diagnostic methods to determine whether it is more sensitive for some conditions or locations. It is found to be a satisfactory substitute for toe pressure, with high correlation coefficients regardless of whether patients have diabetes (Pearson values 0.85 and 0.93, respectively) [37]. A large clinical study found a high correlation between SPP and ankle or toe blood pressure or $TcPO_2$ (respective coefficients of 0.75, 0.85, and 0.62) [38]. When healing of 94 limbs with gangrene or ulcers was predicted using a receiver operating curve and a cutoff value of 40 mmHg, the sensitivity and specificity obtained using SPP were 72% and 88%, respectively.

2.3.3.3 Fluorescence Imaging

Exogenous fluorescent imaging provides a convenient method to visualize the microcirculation using ICG. Developed for near-infrared (NIR) photography by the Kodak Research Laboratories in 1955, ICG was approved for clinical use in 1957 and has a well-established safety profile.

The methodology for imaging is a straightforward fluorescent process in which the fluorophore (ICG) is excited by a wavelength of 750–820 nm. ICG absorbs a portion of the light within the blood vessels of the target tissue. The absorbed light undergoes quantum transformation and is released as fluorescent emissions at longer wavelengths around 820–900 nm. The change from an excited state back to a ground state occurs within a nanosecond. Emission filters at the camera sensor are used to prevent mixing the unabsorbed excitation light (strong), which is reflected, and the fluorescing light (weak) [39].

Although the fluorescent emission light is only a fraction of the intensity of the excitation light, a surprisingly good signal-to-noise ratio is attained. The image is a brightly fluorescing tissue area comprised of blood vessels containing ICG, which can be seen against an almost black background because the filters remove light waves outside the fluorescent range. Without the filters, it would be impossible to see the weak fluorescence image because of the strong reflection of the excitation light "bouncing back" to the sensor.

The ICG dye is administered intravenously and has a good safety profile. The liver quickly clears it with a half-life of about 3–4 minutes. Liver clearance is important as patients with advanced diabetes are prone to renal impairment, which is frequently a contraindication for those agents dependent on renal clearance. One in 42,000 cases of ICG use may experience minor side effects such as hot flushes, hypotension, tachycardia, dyspnea, and urticaria. The frequencies of mild, moderate, and severe side effects were only 0.15%, 0.2%, and 0.05%. For the competitor substance fluorescein, the proportion of people with side effects is 4.8% [40].

ICG has several useful properties for studying microcirculation:

- It has been thoroughly verified during its prolonged clinical use
- It binds efficiently to blood lipoproteins and stays in the vascular space; it does not leak from circulation like fluorescein, which makes it ideal for angiogram
- It has an excellent safety profile and is nontoxic and nonionizing, with 60 years of clinical experience
- It has a short half-life, which permits repeated examinations if needed
- It has an excellent signal-to-noise ratio because there is little tissue autofluorescence from the excitation light
- It can do deep imaging, to a depth of 2 cm.

ICG angiography is frequently used to monitor flaps in compromised diabetic patients who require amputations of the foot or leg. Such flaps are used to close amputation sites. In recent years, ICG shows promise as a means of monitoring healing. Microangiogenesis is cited as a beneficial mechanism of HBO therapy. A prospective study found ICG angiography to be a useful biomarker for the early identification of responders after their first two HBO sessions. The authors found 100% of the wounds that demonstrated improved perfusion from session 1 to session 2 went on to heal within 30 days of HBO therapy completion, compared with none in the subgroup that did not demonstrate improved perfusion ($p<0.01$). This study demonstrates a beneficial impact of HBO therapy on perfusion in chronic wounds by ameliorating hypoxia, improving angiogenesis, and the potential role of ICG angiography in identifying those who would benefit from HBO therapy [41].

2.3.4 Microbiological Studies

All chronic wounds, diabetic wounds particularly, are prone to bacterial infections. Wound healing always occurs in the presence of bacteria, whose role depends on their concentration, species composition, and host response.

There are several distinct levels of bacteria present in the wound: contamination (the presence of nonreplicating organisms), colonization (replicating microorganisms adherent to the wound in the absence of injury to the host), and infection. It is believed that contamination and colonization by low concentrations of microbes are normal and do not inhibit wound healing. However, critical colonization and infection are considered to be associated with a significant delay in wound healing.

The switching point from colonization to local infection is a matter of ongoing debate. However, most studies suggest that this transition happens in the range of 10^4–10^5 CFU/g [42], where CFU stands for colony-forming units. For example, Breidenbach et al. [43] demonstrated that a bacterial load must exceed 10^4 CFU/g to cause infection in complicated lower limb wounds. Thus, quantifying the bacterial presence (at least semiquantitative or categorical) is of great importance for a proper diagnosis and treatment selection.

The microbial species composition in wounds changes over time. Quite often, infections are polymicrobial. Group *B Streptococcus* and *S. aureus* are common organisms found in DFU.

Early infection diagnostics represents a significant clinical challenge. The current diagnostic approach involves a bedside visual assessment to detect the clinical signs and symptoms (CSS) of wound infection, which may often be supported by semiquantitative microbiological analysis [44]. Currently, the gold-standard specimen collection method is to perform a tissue biopsy or needle aspirate of the wound's leading edge after debridement. However, the practical standard for sample collection is a microbiological swab (Levine or Z technique). After specimen collection, the sample is analyzed using culture methods (e.g., pour or spread plate). This method has several significant drawbacks, including (1) a sample can be contaminated by normal skin or mucosa flora, (2) swabs frequently yield too small a specimen for accurate microbiologic examination [45], and (3) the duration of incubation of cultures can be relatively long. While most aerobic bacteria will grow over 2 days, anaerobes take longer and frequently may not grow at all.

2.4 NOVEL OPTICAL METHODS

An adequate macro- and microcirculation are necessary for healing. Insufficient perfusion impairs angiogenesis, collagen deposition, and epithelialization and leads to sustained inflammation. Hypoxia is a reduction in oxygen delivery against cellular demand. In contrast, ischemia is a state in which perfusion is lacking, resulting in hypoxia and a diminished supply of nutrients needed to repair tissues [46–49]. Because oxygen is the "rate-limiting" step for healing, the ideal measurement system would provide a noninvasive, direct measurement of tissue oxygen levels in both the wound and the periwound tissues.

Perfusion is only one determinant of outcome; wound size, wound depth, wound tissue composition, and the presence of infection also govern the prognosis for healing or amputation. To facilitate communication and research, the Society for Vascular Surgery (SVS) created a new classification

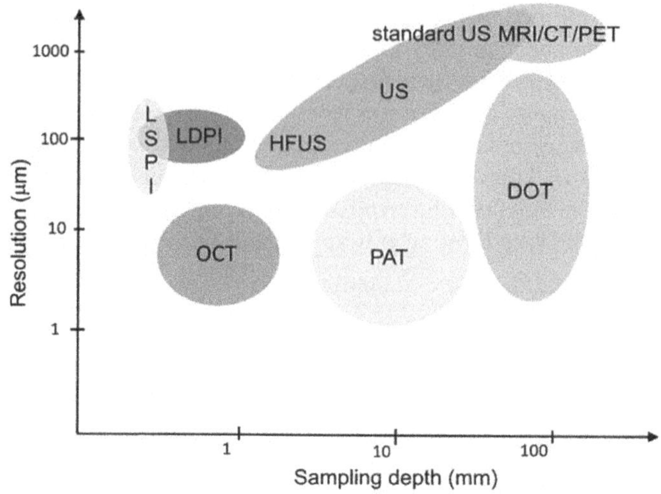

FIGURE 2.2 The relative domains occupied by various imaging modalities in terms of resolution and sampling depth. (Modified from [51], copyright Wiley-VCH GmbH, reproduced with permission. DOT – diffuse optical tomography, US – ultrasound, LDPI – laser Doppler perfusion imaging, PAT – photoacoustic tomography, LSPI – laser speckle perfusion imaging, OCT – optical coherence tomography.)

system of the threatened lower extremity, incorporating these important considerations. This new system is referred to as the SVS Wound, Ischemia, and Foot Infection classification system or "WIFI classification" for short. The WIFI system has been clinically validated as a predictor for limb salvage and wound healing [27,50].

The following sections of this chapter will review novel and practical methods to evaluate wound features, wound perfusion, and wound infection. These methods have been used in clinical research; however, they have not received wide adoption in clinical practice yet. Most of these technologies have the potential to be used in point-of-care settings and thus can be used for DFU screening.

In Figure 2.2, several techniques are depicted in resolution-sampling depth coordinates.

2.4.1 Laser Doppler

Photons undergo a frequency change if mobile scatterers are illuminated with coherent light (the Doppler effect). RBCs are primary mobile scatterers in cutaneous circulation; thus, the Doppler effect can be used to evaluate their velocities.

Laser Doppler flowmetry (LDF) is a general class of noninvasive techniques that utilize the optical Doppler effect to assess microcirculatory blood perfusion. In practice, LDF often refers to a single-point measurement with a fiberoptic (contact) probe (experimental modality), while Laser Doppler perfusion imaging (LDPI) refers to a clinical imaging modality.

Laser Doppler imaging is a relatively mature diagnostic modality, which has been used in many clinical situations. Primarily, it is used to assess burn depth, but it is also applied in surgery, wound healing, and general vascular diagnostics. However, it has very limited applications in diabetic wound care.

In a comparison study [52], LDF and $TcPO_2$ measurements were performed in 25 patients with chronic lower limb ulcers with various etiologies (experimental arm) and 25 healthy individuals (control arm). A statistically significant difference ($p<0.05$) was found between the LDF values of the two groups. No statistically significant differences were found between the two groups via the $TcPO_2$ measurements.

One study [53] used LDPI to measure the circulation in ischemic ulcers. Ten out of the 25 included patients were diabetic foot patients. They found that changes in the circulation measured by LDPI may coincide with changes in the number of visible capillaries within an ischemic ulcer.

Some studies used LDPI to measure the effect of different healing techniques for DFU and showed that LDPI could measure the circulation in DFU and that this technique could be used to assess microcirculation [54].

2.4.2 Laser Speckle Imaging

A speckle pattern is formed by reflecting coherent light from a rough surface or by reflecting or transmitting the light through a medium with refractive index distribution. This phenomenon results from the interference of different reflected portions of the incident beam with random relative optical phases. Determinations of RBC velocity may be obtained by assessing the temporal statistical behavior of speckles.

Laser speckle (perfusion or contrast) imaging (LSPI or LSCI) is a relatively mature diagnostic modality, which has been used in many clinical situations, including the noninvasive assessment of blood flow of cutaneous wounds [55]. However, it has very limited applications in diabetic wound care.

In [56], laser speckle imaging was used to study whether iontophoresis increases skin microcirculation in diabetic patients (iontophoresis is a process of transdermal drug delivery). Typically, a vasodilator such as acetylcholine is used to measure a tissue's "microvascular capacity." The authors found that LSCI could be used to diagnose small changes in microcirculation in specific areas, such as the skin on the foot or ankle.

2.4.3 Spectroscopic Methods

Light in the visible and near-infrared (NIR) ranges of the spectrum delivered to biological tissue undergoes multiple scattering and absorption events. Hemoglobin, water, fat, and melanin are the primary absorbers in the tissue (Figure 2.3). It is well established that this reflected light carries quantitative information about tissue pathology.

Near-infrared spectroscopy (NIRS) typically refers to a single-point contact measurement using a fiber probe.

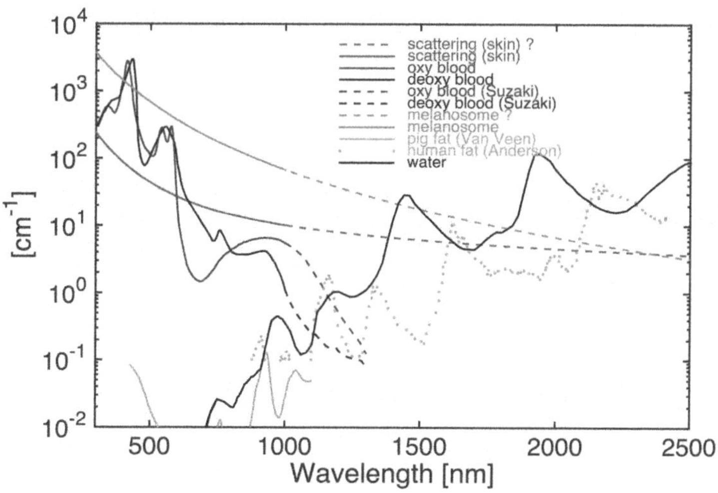

FIGURE 2.3 Absorption and scattering spectra of skin constituents. (Courtesy of Scott Prahl, Oregon Institute of Technology, 2022.)

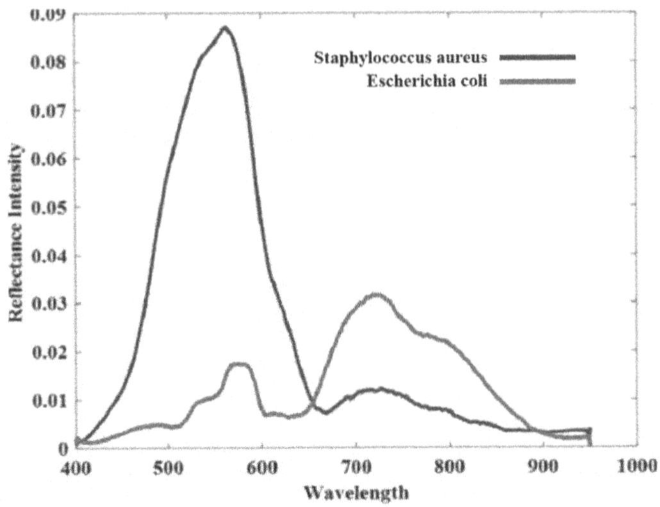

FIGURE 2.4 Reflectance spectral intensity used as reference spectral signatures of *Staphylococcus aureus* and *Escherichia coli*. (Modified from [60].)

Based on spectral parameter measurement in animal experiments with rats, it was found that the diabetic group has a significantly higher average value of absorption coefficient (~100%) at wavelength 780 nm compared with a healthy group during the wound healing process. Compared with healthy rats, diabetic rats have a higher value of reduced scattering coefficient (by ~ 30%) [57]. A clinical study deals with the application of diffuse reflectance intensity ratios based on oxyhemoglobin (HbO_2) bands (R542/R580), ratios of oxy- and deoxyhemoglobin (RHb) bands (R580/R555), total hemoglobin (tHb) concentration, and hemoglobin oxygen saturation (SO_2) between normal and DFU sites. Considerable differences in the values of RHb and HbO_2 were observed for ulcer regions when compared with control sites in clinical studies. The total hemoglobin (tHb) concentration was higher in foot ulcers [58]. Higher values of SO_2 were also observed during the healing process. This follows from the fact that an increase in blood supply and, in due course, SO_2 reduces as the wound heals [59]. A retrospective clinical study was conducted to determine whether NIRS can be used to evaluate wounds and adjacent soft tissues to identify patterns involved in tissue oxygenation and wound healing and to predict which wounds may or may not heal. Most of the wounds that were progressing toward healing showed a wispy and gradual decrease in SO_2 as the distance away from the wound increased. Landsman [60] described this as having an appearance resembling a ray of sunlight, with the central red region moving to a combination of yellow and red as the distance away from the wound increased. This was probably the most predictive sign that the wound was starting to heal.

Biospectroscopy can be used for other clinical indications. Poosapadari et al. [61] used spectroscopy to discriminate between the two most prevalent pathogens in DFU patients (*S. aureus* and *E. coli*), with 100% sensitivity and 75% specificity in detecting the presence of these infections and with a 100% negative predicted value in excluding the infection in such wounds (Figure 2.4).

2.4.4 Hyperspectral and Multispectral Imaging

In recent years, the progress in cameras, image analysis techniques, and computational power makes it possible to extend the advances in biospectroscopy into clinical imaging modalities.

Biomedical hyperspectral imaging (HSI) aims to record the spectrum for each pixel of the image and extract the concentration of tissue chromophores. In this sense, HSI is the natural extension of color (RGB) imaging and biospectroscopy (a single-point measurement). Spectrum at each pixel as a function of a wavelength (λ, nm) can be considered a spectroscopic input, which can be decomposed and spectral signatures can be found.

2.4.4.1 Hyperspectral vs. Multispectral

Oftentimes, HSI and "multispectral" (MSI) imaging are used interchangeably. The distinction between them is based on an arbitrary "number of bands." MSI deals with several images at discrete and somewhat narrow bands, placed at particular spectral points (e.g., isosbestic or absorption maxima). Thus, multispectral images do not produce the "spectrum" of an object but rather sample the spectrum at several points. HSI implies narrow equally spaced spectral bands over a continuous spectral range, which can be considered a spectrum.

In biomedical applications, HSI or MSI is used primarily to extract data about components of the blood, which are chromophores in the visible and NIR spectrum [62]. The primary outcome of HSI/MSI methods is tissue oxygen saturation (SO_2) maps, which are indicative of abnormalities in blood circulation. Occasionally, HbO_2, RHb, and total hemoglobin (tHb) or their proxies are also reported.

HSI/MSI is increasingly being used within different clinically diagnostic areas. In line with many reported clinical applications [62], HSI demonstrated its utility in wound care in general and DFU in particular. For example, HSI has been used to assess tissue viability and health in diabetes patients at risk of foot ulceration [63,64]. The applications of HSI/MSI in wound care were reviewed in Ref. [65]. Here, we briefly present major findings for DFU.

Greenman et al. [66] used HSI to measure blood oxygen saturation in healthy, diabetic, and diabetic neuropathic individuals. They found that in the foot at resting, SO_2 was lower in the neuropathic group (30 ± 12; $p=0.027$) than in the control (38 ± 22) and non-neuropathic groups (37 ± 12).

Serena et al. [67] compared SO_2 assessed by MSI with $TcPO_2$ measurements in 12 locations in 10 patients. They found that $TcPO_2$ provides higher readings in general. However, SO_2 extracted by the MSI system and $TcPO_2$ are strongly correlated ($r=0.74$).

2.4.4.2 Blood Oxygen Saturation and Wound Healing

Khaodhiar et al. [63] measured RHb and HbO_2 in nondiabetic controls ($n=14$) as well as diabetic patients with ($n=10$) and without DFU ($n=13$) four times over 6 months using HSI. HbO_2 and RHb measurements in the periwound of nonhealing ulcers were lower than in healing ulcers ($p<0.01$) and contralateral foot ($p<0.001$).

Nouvong et al. [64] measured HbO_2 and RHb in diabetic patients ($n=44$) over 24 weeks using HSI. They found that in the periwound area, HbO_2 and SO_2 were higher around DFUs that healed (85 ± 21 and $66\pm9\%$, respectively) vs. nonhealed (64 ± 22 and $60\pm10\%$, respectively).

Jeffcoate et al. [68] compared SO_2 measurements using HSI against blood gas analysis in blood samples (*in vitro*) in patients with DFUs ($n=43$). They found a negative association between SO_2 and healing by 12 weeks ($p=0.009$) and a significant positive correlation between oxygenation assessed by HSI and time to healing ($p=0.03$). A strong correlation between SO_2 and blood gas measurements ($r=0.994$) was found.

2.4.4.3 Prediction of DFU Development

Yudovsky et al. [69], using the dataset in Ref. [64], assessed the utility of HSI in predicting DFU development. Diabetic patients ($n=54$) at risk of DFU were monitored and evaluated retrospectively for ulceration. They found that the difference in the HbO_2 (and RHb at a minor degree) value between wound and periwound was a predictor of ulceration. The authors constructed an index, which could predict the tissue at risk of ulceration with a sensitivity and specificity of 95% and 80%, respectively, for images taken, on average, 58 days before skin breakdown.

2.4.4.4 Other Clinical Applications

Yudovsky et al. [70] demonstrated the utility of biospectroscopy and HSI/MSI to determine the thickness of the epidermis layer.

Several studies demonstrated the utility of HSI/MSI to detect tissue water content, which can be helpful for several clinical indications, including deep tissue injury and preclinical edema identification. It also can have relevance to the diabetic population. For example, in Ref. [71], the authors found differences in skin water content between diabetic and nondiabetic populations. In Ref. [72], the authors used the 1400–1500 nm band to find the skin's hydration. In Ref. [73], this approach was extended to the 970-nm range, which allows deeper light penetration into the tissue.

2.4.5 Spatial Frequency Domain Imaging (SFDI)

An original method described by Tromberg et al. [74] is a development of a rapid, noncontact imaging for quantitative, wide-field characterization of optical absorption and scattering properties of turbid media based on spatial frequency domain imaging (SFDI). This group experimentally demonstrated that by projecting sinusoidal patterns of light onto tissue, one can determine the tissue's optical properties by measuring the relative decay of spatial patterns of differing frequencies. An algorithm was proposed for reconstructing three-dimensional images directly from measurements made by illuminating tissue with sinusoidal patterns [75] and experimentally proven as a quantitative optical tomography of subsurface heterogeneities (e.g., blood vessels) using spatially modulated structured light [76] and collecting images separated in terms of absorption and reduced scattering coefficients [77]. SFDI has been demonstrated preclinically to track wound healing in a diabetes model [78]. A clinical study has been conducted applying SFDI to measure perfusion in lower extremities for the prediction of both healing and formation of DFU [79,80].

2.4.6 Orthogonal Polarization Spectral Imaging

Orthogonal polarization spectral (OPS) imaging is based on illuminating skin tissue with linearly polarized light and collecting resultant depolarized photons scattered by the tissue components using a polarizer positioned orthogonal to the plane of illuminating light [81]. It increases sensitivity by discarding single scattered photons, mostly from specular (Fresnel) reflection on surfaces. In OPS imaging, the microvasculature is illuminated with polarized green light. The technique can be used to visualize vessels and RBCs (which absorb light in the green range of spectrum) and can be particularly helpful in dermatology and rheumatology. In wound care, the method has been mainly applied to detect burn depth [82]. However, it was used on wounds with other etiologies as well (e.g., chronic venous insufficiency [83]).

2.4.7 Thermography

Thermography has been tried in wound care for a long time. In recent years, it started getting adopted into clinical practice. There are two primary temperature monitoring modalities in diabetic wound care: noncontact (thermal camera) and contact (temperature monitoring mat and temperature monitoring wearables [84]).

We briefly consider thermal cameras as a noncontact imaging modality. Thermal sensors operate in the middle-wavelength infrared (MWIR, 3–5 μm) or long-wavelength infrared (LWIR, 8–14 μm) spectral ranges. The most common materials are amorphous Si (a-Si), InSb, InGaAs, HgCdTe, and quantum well-integrated photodetectors (QWIP) arranged into focal plane arrays or FPA. While for many applications, room temperature operations are sufficient, for specific applications, the cryogenic cooling of IR detectors is required to achieve high performance. Uncooled detectors are mostly based on pyroelectric and ferroelectric materials or microbolometer technology. Thermal imaging sensors for biomedical research and clinical applications use low-cost, uncooled FPA microbolometers operating in the LWIR range. The resolution of thermographic sensors is much lower than those of normal cameras. Currently, it is in the range of 60×80 or 120×160 for regular applications and 640×480 for high-end applications.

The utility of thermography in wound care is known since the early 1960s when Lawson et al. [85] used infrared (IR) scanning to predict burn depth with an accuracy of 90%, as confirmed by histology.

Thermography has significant potential as an adjuvant technique in diabetic foot assessment. For example, the increased temperature is a reliable marker of inflammation and can thus predict the risk of ulceration, infection, and amputation [86]. Similarly, the decreased temperature may be a sign of insufficient blood supply and indicate ischemia.

2.4.7.1 Inflammation Detection

Several approaches based on plantar temperature distributions have been proposed. Benbow et al. [87] assessed the risk of ulceration and ischemic foot disease based on the mean foot temperature, determined from eight standard sites on the plantar surface. They found that an elevated mean foot temperature was associated with an increased risk of neuropathic foot ulceration. Diabetic patients with normal or low mean foot temperature were at risk of ischemic foot disease.

Another approach is to compare temperature maps of the individual's contralateral (left foot to their right) sites (an asymmetry analysis). This method has the advantage of being specific to the patient. However, it is dependent on geometrical symmetry between feet [88]. Preventive care is recommended when a patient is observed with temperature asymmetry exceeding 2.2°C (4°F) for at least two consecutive days between contralateral sites [89]. Using a remote temperature monitoring mat and the 2.2°C asymmetry approach, Frykberg et al. [90] predicted 97% of all nonacute plantar DFU on average 35 days before clinical presentation with a specificity of 43%.

Nagase et al. [91] identified 20 patterns of thermal distribution to aid in diabetic foot assessment and surgical procedures. Bharara et al. [92] proposed a wound inflammatory index or temperature index for diabetic foot assessment, which is based on the difference in mean foot temperature and the wound bed, the area of the wound bed, and the area of the isotherm (the highest or the lowest temperature area).

Laveryet al. [93], in a multicenter study on 129 patients in remission, found that unilateral once-daily foot temperature monitoring can predict 91% of impending nonacute plantar foot ulcers on average 41 days before clinical presentation.

2.4.7.2 Infection Detection

In addition to detecting inflammation, the authors in a case study [94] found that thermography may be able to predict osteomyelitis, a severe complication of the diabetic wound, before visible signs of infection are shown.

2.4.7.3 Blood Perfusion Assessment

Astavio-Picado et al. in a study on a group of 277 patients with diabetic pathology [95] found that there were lower temperatures under the first metatarsal head, the fifth metatarsal head, the heel, and the pulp of the big toe of both left and right feet of the patients in the neuropathy, vasculopathy, and neurovasculopathy groups relative to the group with neither pathology.

In Ref. [96], the authors compared laser Doppler imaging vs. LDF vs. thermographic imaging for blood perfusion assessment of the skin. They found $r=0.577$ ($p<0.01$, $n=38$) correlation between the thermography imager and the laser Doppler imager in seven normal volunteers. They also found $r=0.358$ ($p<0.01$, $n=60$) correlation between the IR thermography imager and the laser Doppler imager in ten patients with scleroderma.

At least in one study [97] with eight patients with lower extremity arterial occlusive disease, the authors noted a difference in temperature distribution before and after undergoing endovascular intervention or a surgical bypass procedure. The improved flow was confirmed by arterial duplex ultrasound.

Time series of thermographic images can be used for blood flow imaging and assessments. In such methods (see for example Ref. [98]), the temporal variations of the temperature can be used for the assessment of different physiological mechanisms, e.g., vasomotion. Signals in five frequency bands (0.005–0.02 Hz, 0.02–0.05 Hz, 0.05–0.15Hz, 0.15–0.4 Hz, and 0.4–2.0 Hz) have endothelial (metabolic), neurogenic, myogenic, respiratory, and cardiac origins, respectively [99].

2.4.7.4 Thermography with Interventions

While thermography is typically used as a passive methodology, active dynamic thermography was proposed [100,101] to characterize tissue components' thermal diffusivity in wound regions. In this approach, the thermal images of the target area are captured before (baseline) and after intervention (typically heating the tissue by a light source, e.g., a halogen lamp).

In Ref. [102], the temperature observations in 51 patients with PAD were taken before and after a 6-minute walking test. They found that the postexercise temperature dropped in the lower extremities with arterial stenosis but was maintained or increased slightly in the extremities with patent arteries (temperature changes at sole in PAD vs. non-PAD patients: 1.25°C vs. 0.15°C; $p<0.01$).

One known problem with thermography is distortion in the image caused by evaporative water loss in the wound bed. This problem can be solved either by allowing the wound to dry completely (which may delay the timing of the assessment) or by applying a nonpermeable covering to the wound bed, which eliminates the problem of evaporation [103].

2.4.8 Endogenous Bacterial Fluorescence

Bacterial presence can be visualized using fluorescence imaging. Most clinically relevant bacteria produce porphyrins [104] (*S. aureus, S. epidermidis, Candida, S. marcescens, Viridians streptococci, Corynebacterium diphtheriae, S. pyogenes, Enterobacte*r, and *Enterococcus*) or pyoverdine [105] (*P. aeruginosa*), which fluoresce while excited in the 405-nm range (red and bluish-green fluorescence, respectively).

The utility of fluorescence imaging for the visualization of bacterial load in wounds has been demonstrated in numerous clinical studies, including studies with DFUs *in vivo* [106–109] and *in vitro* [110].

Fluorescence imaging may be used to distinguish between particular strains in the wound (e.g., *P. aeruginosa*), to assess (qualitatively or semiquantitatively) the presence of bacteria in the wound (infection detection) or to guide sampling, debridement, or antimicrobial selection.

2.4.8.1 Infection Detection

Fluorescence imaging demonstrated its utility in identifying bacterial loads more than 10^4 CFU/g close to real time. In particular, Farhan et al. [44] identified 11 clinical studies (including 613 wounds with various etiologies), which aimed to assess the diagnostic accuracy of detection of bacterial loads of $\geq 10^4$ CFU/g and calculated weighted averages for sensitivity (74%), specificity (88%), positive predictive value (91%), negative predictive value (53%), and overall diagnostic accuracy (75%). It represents a three- to fourfold increase in the sensitivity compared with CSS alone.

2.4.8.2 Treatment Selection

The ultimate utility of fluorescence imaging is to assist in treatment selection. Studies have shown that fluorescence imaging can prompt changes in the proposed treatment plans, including alterations in antimicrobial prescribing [111], decisions around negative pressure wound therapy [112], and timing of grafting or applications of skin substitutes [113]. It may lead to changes in the treatment plan in as many as 73% of cases.

2.4.8.3 Sampling and Debridement Guidance

Another practical application of fluorescent imaging is to guide debridement and sampling (swabbing or biopsy). A pilot evaluation compared standard Levine swab results with fluorescence-guided curettage samples and found that the Levine technique gave a 36% false-negative laboratory report [114].

2.4.9 Other Techniques

Similar to orthogonal polarization imaging, several techniques (typically contact modalities) exploit the idea of discarding surface reflectance. Dark field microscopy is one of such modalities. The side-stream dark field (SDF) imaging device discards the surface-scattered photons by optically isolating the light guide from the illuminating outer ring. The SDF device consists of a central light guide in contact with the skin, surrounded by concentrically placed light-emitting diodes that emit green light. Goedhart et al. [115] developed and validated the SDF technology against OPS on nailfold capillaries.

The technology subsequently evolved into the incident dark field (IDF) technique. The IDF is based on the method initially developed in Ref. [116]. Gilbert-Kawai et al. [117] compared IDF with SDF in sublingual microcirculatory imaging and found the former superior.

In addition to dark field microscopy, several other contact imaging techniques have been proposed: lensless microscopy [118], which uses multicore fibers and numerical aperture-gated microscopy [119].

However, it should be noted that the above-mentioned techniques are microscopic techniques, which are limited to areas with thin epidermis or mucosa. Their applications to wound care, particularly diabetic foot, can be quite limited.

In Ref. [120], the authors proposed speckle plethysmography (SPG), a wearable technology based on speckle imaging for characterizing microvascular flow and resistance. In Ref. [121], the authors compared SPG with ABI, TBI, and clinical presentation of patients per Rutherford category on 167 limbs (90 patients). Qualitatively, they found that SPG is analogous to Doppler velocity measurements, and waveform phasicity and amplitude degradation were observed with increasing PAD severity.

2.5 SUMMARY

DFU is one of the severest chronic microvascular complications of DM, associated with significant deterioration of a patient's life quality. Early detection and proper management are critical factors in improvements of clinical outcomes. Diagnostics modalities, which can be deployed in primary care, can significantly impact patient outcomes.

In this chapter, we have briefly reviewed the pathophysiology of DFU, current diagnostic methods, and novel optical modalities, which are at different stages of translation into routine medical practice.

Several novel optical modalities can help medical professionals with diagnostics at an earlier time point. HSI/MSI, thermography, and endogenous bacterial fluorescence imaging are among the closest to the wide adoption in diabetic foot care. They can detect the changes to the blood perfusion, presence of inflammation and infection, and guide sampling and debridement.

Optical diagnostic modalities will improve the quality of care. Currently, many aspects of medicine, particularly in primary care, heavily rely on the practitioner's experience. For example, many aspects of patient evaluation rely on visual observations, which are highly subjective. So, to some extent, medicine is more art than science. Hippocrates, the "father of medicine," taught the importance of assessing "rubor" (redness) and "calor" (warmth) in all patients. The use of optical methods will transform these important but subjective findings into reliable and quantifiable measurements, which will lead to more accurate clinical decision making.

The novel optical modalities with their quantifiable outputs may gradually change this paradigm. In the first step, these quantitative results may increase objectivity.

In the next step, the accumulated objective data may be used to train AI models. Thus, one can expect that soon multiple AI tools will be available to assist in diagnostics.

In the further step, the accumulated data will be integrated with treatment data. Thus, AI-based tools will help with diagnostics and assist in selecting the appropriate treatment.

At this point, AI-assisted clinical decision-making will achieve a high level of objectivity and become genuinely evidence based. It will also significantly democratize the medical field and allow any citizen access to world-class healthcare.

In summary, various advanced optical imaging techniques developed recently demonstrate a high potential to move into everyday clinical practice and improve patient outcomes by providing objective diagnostics at an earlier time point. They also can be helpful tools to assess the tissue physiology of wounded and nonwounded diabetic limbs, which would further enhance our understanding of the pathophysiology of the disease and guide our development of innovative therapeutics to improve patient outcomes.

Even though the potential for novel optical modalities to aid in diabetic foot screening is clear, the exact methods by which they are best employed are not. The advantages of optical modalities, such as their safety, ease, and speed of use, have driven their numerous research applications. However, several limiting factors [86], including price, difficulties in patient/camera positioning, visual/thermal image registration, and lack of quantitative tools slow down their integration into routine clinical assessments. Thus, how the diabetic foot is to be analyzed or assessed by optical modalities has yet to be solidified and validated. Still, as noncontact, objective measures, optical modalities can enhance the prevention and clinical management of diabetic ulcers.

REFERENCES

1. A. Raghav, Z.A. Khan, R.K. Labala, et al., "Financial burden of diabetic foot ulcers to world: A progressive topic to discuss always," *Therap. Adv. Endocrinol. Metab.* **9**, 29–31 (2018).
2. B.J. Petersen, S.A. Bus, G.M. Rothenberg, et al., "Recurrence rates suggest delayed identification of plantar ulceration for patients in diabetic foot remission". *BMJ Open Diabetes Res. Care* **8**(1), 1–8 (2020).
3. D.G. Armstrong, A.J.M. Boulton, and S.A. Bus, "Diabetic foot ulcers and their recurrence," *N. Engl J. Med.* **376**, 2367–2375 (2017).
4. M. Ptakh and G. Saiko, "Developing a robust estimator for remote optical erythema detection," *BIOSTEC* **2**, 115–119 (2021).
5. G.S. Lazarus, D.M. Cooper, D.R. Knighton, et al., "Definitions and guidelines for assessment of wounds and evaluation of healing," *Wound Repair Regen.* **2**, 165–170 (1994).
6. R. Waaijman, M. de Haart, M.L. Arts, et al., "Risk factors for plantar foot ulcer recurrence in neuropathic diabetic patients," *Diabetes Care* **37**(6), 1697–1705 (2014).
7. A. Veves, J.M. Giurini, and F.W. LoGerfo, Eds, *The Diabetic Foot*, 2nd ed., Humana Press, Totowa, NJ (2006).
8. M.D. Flynn and J.E. Tooke, "Diabetic neuropathy and the microcirculation". *Diabet Med.* **12**(4), 298–301 (1995).
9. A. Behroozian and J.A. Beckman, "Microvascular disease increases amputation in patients with peripheral artery disease," *Arterioscler. Thromb. Vasc. Biol.* **40**(3), 534–540 (2020).
10. A.J. Boulton, "The pathogenesis of diabetic foot problems: An overview." *Diabet Med.* **13**, S12–S16 (1996).
11. M.D. Flynn and J.E. Tooke, "Aetiology of diabetic foot ulceration: A role for the microcirculation?" *Diabet Med.* **9**, 320–329 (1992).
12. H.H. Parving, G.C. Viberti, H. Keen, et al., "Hemodynamic factors in the genesis of diabetic microangiopathy," *Metabolism* **32**, 943–949 (1983).
13. D.D. Sandeman, A.C. Shore, and J.E. Tooke, "Relation of skin capillary pressure in patients with insulin-dependent diabetes mellitus to complications and metabolic control," *N. Engl. J. Med.* **327**, 760–764 (1992).
14. C.Y.L. Chao and G.L.Y. Cheing, "Microvascular dysfunction in diabetic foot disease and ulceration," *Diabetes Metab. Res. Rev.* **25**, 604–614 (2009).
15. L. Uccioli, L. Mancini, A. Giordano, et al., "Lower limb arterio-venous shunts, autonomic neuropathy and diabetic foot," *Diabetes Res. Clin. Pract.* **16**, 123–130 (1992).
16. J.E. Tooke, "Peripheral microvascular disease in diabetes," *Diabetes Res. Clin. Pract.* **30**, S61–S65 (1996).
17. D. Lowry, M. Saeed, P. Narendran, et al., "The difference between the healing and the nonhealing diabetic foot ulcer: A review of the role of the microcirculation," *J. Diabetes Sci. Technol.* **11**, 914–923 (2017).
18. J.W. Albers and R. Pop-Busui, "Diabetic neuropathy: Mechanisms, emerging treatments, and subtypes," *Curr. Neurol. Neurosci. Rep.* **14**, 473 (2014).

19. R. Pop-Busui, A.J.M. Boulton, E.L. Feldman, et al., "Diabetic neuropathy: A position statement by the American Diabetes Association," *Diabetes Care* **40**(1), 136–154 (2017).
20. E.J. Boyko, J.H. Ahroni, V. Stensel, et al., "A prospective study of risk factors for diabetic foot ulcer. The Seattle diabetic foot study," *Diabetes Care* **22**(7), 1036–1042 (1999).
21. S.J. Rith-Najarian, T. Stolusky, and D.M. Gohdes, "Identifying diabetic patients at high risk for lower-extremity amputation in a primary health care setting. A prospective evaluation of simple screening criteria," *Diabetes Care* **15**(10), 1386–1389 (1992).
22. H. Pham, D.G. Armstrong, C. Harvey, et al., "Screening techniques to identify people at high risk for diabetic foot ulceration: A prospective multicenter trial," *Diabetes Care* **23**(5), 606–611 (2000).
23. R. Yong, T.J. Karas, K.D. Smith, et al., "The durability of the Semmes-Weinstein 5.07 monofilament," *J. Foot Ankle Surg.* **39**(1), 34–38 (2000).
24. A.T. Hirsch, M.H. Criqui, D. Treat-Jacobson, et al., "Peripheral arterial disease detection, awareness, and treatment in primary care," *JAMA* **286**, 1317–1324 (2001).
25. E.F. Bernstein and A. Fronek, "Current status of non-invasive tests in the diagnosis of peripheral arterial disease," *Surg. Clin. North Am.* **62**, 473–487 (1982).
26. J.W. Severinghaus and P.B. Astrup, "History of blood gas analysis. IV. Leland Clark's oxygen electrode," *J Clin Monit.* **2**(2), 125–139 (1986).
27. J.L. Mills, M.S. Conte, D.G. Armstrong, et al., "The society for vascular surgery lower extremity threatened limb classification system: Risk stratification based on wound, ischemia, and foot infection (WIfI)," *J. Vasc. Surg.* **59**(1), 220–234 (2014).
28. C.E. Fife, C. Buyukcakir, G.H. Otto, et al., "The predictive value of transcutaneous oxygen tension measurement in diabetic lower extremity ulcers treated with hyperbaric oxygen therapy; A retrospective analysis of 1144 patients," *Wound Rep. Regen.* **10**, 198–207 (2002).
29. C.E. Fife, D.R. Smart, P.J. Sheffield, et al., "transcutaneous oximetry in clinical practice: Consensus statements from an expert panel based on evidence," *UHM* **36**(1), 43e53 (2009).
30. D.R. Smart, M.H. Bennett, and S.J. Mitchell, "Transcutaneous oximetry, problem wounds and hyperbaric oxygen therapy," *Diving Hyperb. Med.* **36**, 72–86 (2006).
31. K.A. Arsenault, A. Al-Otaibi, P.J. Devereaux, et al., "The use of transcutaneous oximetry to predict healing complications of lower limb amputations: A systematic review and meta-analysis," *Eur. J. Vasc. Endovasc. Surg.* **43**, 329–336 (2012).
32. O. Kawarada, Y. Yokoi, A. Higashimori, et al., "Assessment of macro and microcirculation in contemporary critical limb ischemia," *Catheter. Cardiovasc. Interv.* **78**, 1051–1058 (2011).
33. G. Urabe, K. Yamamoto, A. Onozuka, et al., "Skin perfusion pressure is a useful tool for evaluating outcome of ischemic foot ulcers with conservative therapy," *Ann. Vasc. Dis.* **2**, 21–26 (2009).
34. M.V. Marshall, J.C. Rasmussen, I.C. Tan, et al., "Near-infrared fluorescence imaging in humans with indocyanine green: A review and update," *Open Surg.Oncol. J.* **2**, 12–25 (2010).
35. B.D. Lepow, D. Perry, D.G. Armstrong, "The use of SPY intra-operative vascular angiography as a predictor of wound healing," *Podiatry Manage* **30**, 141–148 (2011).
36. K. Igari, T. Kudo, T. Toyofuku, et al., "Quantitative evaluation of the outcomes of revascularization procedures for peripheral arterial disease using indocyanine green angiography," *Eur. J. Vasc. Endovasc. Surg.* **46**, 460–465 (2013).
37. T. Lo, R. Sample, P. Moore, et al., "Prediction of wound healing outcome using skin perfusion pressure and transcutaneous oximetry: A single center experience in 100 patients," *Wounds* **21**, 310–316 (2009).
38. T. Yamada, T. Ohta, H. Ishibashi, et al., "Clinical reliability and utility of skin perfusion pressure measurement in ischemic limbs – comparison with other noninvasive diagnostic methods," *J. Vasc. Surg.* **47**, 318–323 (2008).
39. B. Yuan, N.G. Chen, and Q. Zhu, "Emission and absorption properties of indocyanine green in intralipid solution," *J. Biomed Opt.*, **9**(3), 497–503 (2004).
40. M. Hope-Ross, L.A. Yannuzzi, E.S. Gragoudas, et al., "Adverse reactions due to indocyanine green," *Ophthalmology* **101**(3), 529–533 (1994).
41. B. Hajhosseini, G.J. Chiou, S.S. Virk, et al., "Hyperbaric oxygen therapy in management of diabetic foot ulcers: Indocyanine green angiography may be used as a biomarker to analyze perfusion and predict response to treatment," *Plast Reconstr. Surg.*, **147**(1), 209–214 (2021).
42. P.G. Bowler, B.I. Duerden, and D.G. Armstrong, "Wound microbiology and associated approaches to wound management," *Clin. Microbiol. Rev.* **14**, 244–269 (2001).
43. W.C. Breidenbach and S. Trager, "Quantitative culture technique and infection in complex wounds of the extremities closed with free flaps," *Plast. Reconstr. Surg.* **95**, 860–865 (1995).

44. N. Farhan and S. Jeffery, "Diagnosing burn wounds infection: The practice gap & advances with moleculight bacterial imaging," *Diagnostics* **11**, 268 (2021).
45. J.A. Washington, "Principles of diagnosis," in *Medical Microbiology*, S. Baron, Ed., 4th ed., University of Texas Medical Branch at Galveston, Galveston, TX (1996) Available from: https://www.ncbi.nlm.nih.gov/books/NBK8014/.
46. J. Kluz, R. Malecki, and R. Adamiec, "Practical importance and modern methods of the evaluation of skin microcirculation during chronic lower limb ischemia in patients with peripheral arterial occlusive disease and/or diabetes," *Int. Angiol.* **32**, 42–51 (2013).
47. O. Bongard and B. Fagrell, "Discrepancies between total and nutritional skin microcirculation in patients with peripheral arterial occlusive disease (PAOD)," *Vasa* **19**, 105–111 (1990).
48. G. Jörneskog, K. Brismar, and B. Fagrell, "Skin capillary circulation is more impaired in the toes of diabetic than non-diabetic patients with peripheral vascular disease," *Diabet. Med.* **12**, 36–41 (1995).
49. G. Jörneskog, "Why critical limb ischemia criteria are not applicable to diabetic foot and what the consequences are," *Scand J. Surg.* **101**, 114–118 (2012).
50. L.X. Zhan, B.C. Branco, D.G. Armstrong, et al., "The society for vascular surgery lower extremity threatened limb classification system based on Wound, Ischemia, and foot Infection (WIfI) correlates with risk of major amputation and time to wound healing," *J. Vasc. Surg.* **61**(4), 939–944 (2015).
51. S.M. Daly and M.J. Leahy, "Go with the flow': A review of methods and advancements in blood flow imaging," *J. Biophotonics* **6**(3), 217–255 (2013).
52. E. Raposio, N. Bertozzi, R. Moretti, et al., "Laser doppler flowmetry and transcutaneous oximetry in chronic skin ulcers: A comparative evaluation," *Wounds* **29**(7), 190–5 (2017).
53. M.E. Gschwandtner, E. Ambrózy, B. Schneider, et al., "Laser doppler imaging and capillary microscopy in ischemic ulcers," *Atherosclerosis* **3**(142), 225–232 (1999).
54. N. Morimoto, N. Kakudo, P.V. Notodihardjo, et al., "Comparison of neovascularization in dermal substitutes seeded with autologous fibroblasts or impregnated with bFGF applied to diabetic foot ulcers using laser Doppler imaging," *J. Artif. Organs* **16**(17), 352–357 (2014).
55. T.M. van Vuuren, C. van Zandvoort, S. Doganci, et al., "Prediction of venous wound healing with laser speckle imaging," *Phlebology* **32**, 658–664 (2017).
56. M. Hellmann, M. Roustit, F. Gaillard-Bigot, et al., "Cutaneous iontophoresis of treprostinil, a prostacyclin analog, increases microvascular blood flux in diabetic malleolus area," *Eur. J. Pharmacol.* **758**, 123–128 (2015).
57. M. Neidrauer, L. Zubkov, M.S. Weingarten, et al., "Near infrared wound monitor helps clinical assessment of diabetic foot ulcers," *J. Diabetes Sci. Technol.* **4**(4), 792–798 (2010).
58. S. Anand, N. Sujatha, V.B. Narayanamurthy, et al., "Diffuse reflectance spectroscopy for monitoring diabetic foot ulcer – A pilot study," *Opt. Lasers Eng.* **53**, 1–5 (2014).
59. S.M. Rajbhandari, N.D. Harris, S. Tesfaye, et al., "Early identification of diabetic foot ulcers that may require intervention using the micro lightguide spectrophotometer," *Diabetes Care* **22**(8), 1292–1295 (1999).
60. A. Landsman, "Visualization of wound healing progression with near infrared spectroscopy: A retrospective study," *Wounds* **32**(10), 265–271 (2020).
61. S.P. Arjunan, A.N. Tint, B. Aliahmad, et al., "High-resolution spectral analysis accurately identifies the bacterial signature in infected chronic foot ulcers in people with diabetes," *Int. J. Low Extrem Wounds* **17**(2), 78–86 (2018).
62. G. Lu and B. Fei, "Medical hyperspectral imaging: A review," *J. Biomed. Opt.* 19, 010901 (2014).
63. L. Khaodhiar, T. Dinh, K.T. Schomacker, et al., "The use of medical hyperspectral technology to evaluate microcirculatory chaves in diabetic foot ulcers and to predict clinical outcomes," *Diabetes Care* **30**(4), 903–910 (2007).
64. A. Nouvong, B. Hoogwerf, E. Mohler, et al., "Evaluation of diabetic foot ulcer healing with hyperspectral imaging of oxyhemoglobin and deoxyhemoglobin," *Diabetes Care* **32**(11), 2056–2061 (2009).
65. G. Saiko, P. Lombardi, Y. Au, et al., "Hyperspectral imaging in wound care: A systematic review," *Int Wound J.* 1–17 (2020).
66. R.L. Greenman, S. Panasyuk, X. Wang, et al., "Early changes in the skin microcirculation and muscle metabolism of the diabetic foot," *Lancet* **366**(9498), 1711–1717 (2005).
67. T.E. Serena, R. Yaakov, L. Serena, et al., "Comparing near infrared spectroscopy and transcutaneous oxygen measurement in hard-to-heal wounds: A pilot study," *J. Wound Care* **29**(Sup6), S4–S9 (2020).
68. W.J. Jeffcoate, D.J. Clark, N. Savic, et al., "Use of HSI to measure oxygen saturation in the lower limb and its correlation with healing of foot ulcers in diabetes," *Diabet Med.* **32**(6), 798–802 (2015).

69. D. Yudovsky, A. Nouvong, K. Schomacker, et al., "Assessing diabetic foot ulcer development risk with hyperspectral tissue oximetry," *J. Biomed Opt.* **16**(2), 026009 (2011).
70. D. Yudovsky and L. Pilon, "Rapid and accurate estimation of blood saturation, melanin content, and epidermis thickness from spectral diffuse reflectance," *Appl Opt.* **49**(10), 1707–1719 (2010).
71. H.N. Mayrovitz, A. Clymont, N. Pandya, "Skin tissue water assessed via tissue dielectric constant measurements in persons with and without diabetes mellitus," *Diab. Tech. Theur.* **15**(1), 1–6 (2013).
72. M. Attas, T. Posthumus, B. Schattka, et al., "Long-wavelength near-infrared spectroscopic imaging for in-vivo skin hydration measurements," *Vib. Spectrosc.* **28**(1), 37–43 (2002).
73. G. Saiko, "On the feasibility of skin water content imaging adjuvant to tissue oximetry," *Adv. Exp. Med. Biol.* **1269**, 191–195 (2021).
74. D.J. Cuccia, F.P. Bevilacqua, A.J. Durkin, et al., "Quantitation and mapping of tissue optical properties using modulated imaging," *J. Biomed. Opt.* **14**(2), 024012 (2009).
75. V. Lukic, V.A. Markel, and J.C. Schotland, "Optical tomography with structured illumination," *Opt. Lett.* **34**(7), 983–985 (2009).
76. S. Konecky, A. Mazhar, D. Cuccia, et al., "Quantitative optical tomography of sub-surface heterogeneities using spatially modulated structured light," *Opt. Express* **17**, 14780–14790 (2009)
77. S. Gioux, A. Mazhar, and D.J. Cuccia, "Spatial frequency domain imaging in 2019: Principles, applications, and perspectives," *J. Biomed. Opt.* **24**(7), 071613 (2019).
78. M. Saidian, J.R.T. Lakey, A. Ponticorvo, et al., "Characterisation of impaired wound healing in a preclinical model of induced diabetes using wide-field imaging and conventional immunohistochemistry assays," *Int. Wound J.* **16**(1), 144–152 (2018).
79. G.A. Murphy, R.P. Singh-Moon, A. Mazhar, et al., "Quantifying dermal microcirculatory changes of neuropathic and neuroischemic diabetic foot ulcers using spatial frequency domain imaging: A shade of things to come?" *BMJ Open Diabetes Res. Care* **8**(2), e001815 (2020).
80. https://clinicaltrials.gov/ct2/show/NCT03341559 as was visited on May 12, 2021.
81. W. Groner, J.W. Winkelman, A.G. Harris, et al., "Orthogonal polarization spectral imaging: A new method for study of the microcirculation," *Nat Med.* **5**(10), 1209–1212 (1999).
82. O. Goertz, A. Ring, A. Kohlinger, et al., "Orthogonal polarization spectral imaging: A tool for assessing burn depths?," *Ann. Plast. Surg.* **64**(2), 217–221 (2010).
83. C.E. Virgini-Magalhaes, C.L. Porto, F.F. Fernandes, et al., "Use of microcirculatory parameters to evaluate chronic venous insufficiency," *J. Vasc. Surg.* **43**(5), 1037–1044 (2006).
84. J. Martín-Vaquero, A. Hernández Encinas, A. Queiruga-Dios, et al., "Review on wearables to monitor foot temperature in diabetic patients," *Sensors (Basel)* **19**(4), 776 (2019).
85. R.N. Lawson, G.D. Wlodek, and D.R. Webster, "Thermographic assessment of burns and frostbite," *Can. Med. Assoc. J.* **84**, 1129–1131 (1961).
86. M. Bharara, J. Schoess, and D.G. Armstrong, "Coming events cast their shadows before: Detecting inflammation in the acute diabetic foot and the foot inremission," *Diabetes Metab. Res. Rev.* **28**(1), 15–20 (2012).
87. S.J. Benbow, A.W. Chan, D.R. Bowsher, et al., "The prediction of diabetic neuropathic plantar foot ulceration by liquid-crystalcontact thermography," *Diabetes Care* **17**(8), 835–839 (1994).
88. N. Kaabouch, W.C. Hu, Y. Chen, et al., "Predicting neuropathic ulceration: Analysis of static temperature distributions in thermal images," *J. Biomed. Opt.* **15**(6), 061715 (2010).
89. L.A. Lavery, K.R. Higgins, D.R. Lanctot, et al., "Home monitoring of foot skin temperatures to prevent ulceration," *Diabetes Care* **27**, 2642–2647 (2004).
90. R.G. Frykberg, I.L. Gordon, A.M. Reyzelman, et al., "Feasibility and efficacy of a smart mat technology to predict development of diabetic plantar ulcers," *Diabetes Care* **40**, 973–980 (2017).
91. T. Nagase, H. Sanada, K. Takehara, et al., "Variations of plantar thermographic patterns innormal controls and non-ulcer diabetic patients: Novel classification using angiosome concept," *J. Plast. Reconstr. Aesthet. Surg. JPRAS* **64**(7), 860–866 (2011).
92. M. Bharara, J. Schoess, A. Nouvong, et al., "Wound inflammatory index: A "proof of concept" study to assess wound healingtrajectory," *J. Diabetes Sci. Technol.* **4**(4), 773–779 (2010).
93. L.A. Lavery, B.J. Petersen, D.R. Linders, et al., "Unilateral remote temperature monitoring to predict future ulceration for the diabetic foot in remission," *BMJ Open Diab. Res. Care* **7**, e000696 (2019).
94. M. Oe, R.R. Yotsu, H. Sanada, et al., "Thermographic findings in a case of type 2 diabetes with foot ulcer and osteomyelitis," *J. Wound Care* **21**(6), 274–278 (2012).
95. Á. Astasio-Picado, E.E. Martínez, and B. Gómez-Martín, "Comparison of thermal foot maps between diabetic patients with neuropathic, vascular, neurovascular, and no complications," *Curr. Diabetes Rev.* **15**(6), 503–509 (2019).

96. A.M. Seifalian, G. Stansby, A. Jackson, et al., "Comparison of laser doppler perfusion imaging, laser doppler flowmetry, and thermographic imaging for assessment of blood flow in human skin," *Eur. J. Vasc. Surg.* 8, 65–69 (1994).
97. P.H. Lin and M. Saines, "Assessment of lower extremity ischemia using smartphone thermographic imaging," *J. Vasc. Surg. Cases Innov. Tech.* **3**(4), 205–208 (2017).
98. A.A. Sagaidachnyi, A.V. Fomin, D.A. Usanov, et al., "Thermography-based blood flow imaging in human skin of the hands and feet: A spectral filtering approach," *Physiol. Meas.* **38**, 272–288 (2017).
99. M.J. Geyer, Y.K. Jan, D.M. Brienza, et al., "Using wavelet analysis to characterize the thermoregulatory mechanisms of sacral skin blood flow," *J. Rehab. Res. Dev.* **41**, 797–805 (2004).
100. A. Renkielska, A. Nowakowski, M. Kaczmarek, et al., "Burn depths evaluation based on active dynamic IR thermal imaging–a preliminary study," *Burns* **32**(7), 867–875 (2006).
101. N.J. Prindeze, P. Fathi, M.J. Mino, et al., "Examination of the early diagnostic applicability of active dynamic thermography for burn wound depth assessment and concept analysis," *J. Burn Care Res.* **36**(6), 626–635 (2014).
102. C.L. Huang, Y.W. Wu, C.L. Hwang, et al., "The application of infrared thermography in evaluation of patients at high risk for lowerextremity peripheral arterial disease," *J. Vasc. Surg.* **54**(4), 1074–1080 (2011).
103. R.P. Cole, P.G. Shakespeare, H.G. Chissell. et al., "Thermographic assessment of burns using a nonpermeable membrane as wound covering," *Burns* **17**(2), 117–22 (1991).
104. B. Kjeldstad, T. Christensen, and A. Johnsson, "Porphyrin photosensitization of bacteria," *Adv. Exp. Med. Biol.* **193**, 155–159 (1985).
105. Y.S. Cody and D.C. Gross, "Characterization of pyoverdin (pss), the fluorescent siderophore produced by *Pseudomonas syringae* pv. Syringae," *Appl. Environ. Microbiol.* **53**, 928–934 (1987).
106. L. Le, M. Baer, P. Briggs, et al., "Diagnostic accuracy of point-of-care fluorescence imaging for the detection of bacterial burden in wounds: Results from the 350-Patient FLAAG trial," *Adv. Wound Care* **10**(3), 123–136 (2020).
107. R. Hill and K.Y. Woo, "A prospective multi-site observational study incorporating bacterial fluorescence information into the UPPER/LOWER wound infection checklists," *Wounds* **32**, 299–308 (2020).
108. T.E. Serena, K. Harrell, L. Serena, et al., "Real-time bacterial fluorescence imaging accurately identifies wounds with moderate-to-heavy bacterial burden," *J. Wound Care* **28**, 346–357 (2019).
109. K. Ottolino-Perry, E. Chamma, K.M. Blackmore, et al., "Improved detection of clinically relevant wound bacteria using autofluorescence image-guided sampling in diabetic foot ulcers," *Int. Wound J.* **14**, 833–841 (2017).
110. L.M. Jones, D. Dunham, M.Y. Rennie, et al., "Invitro detection of porphyrin-producing wound bacteria with real-time fluorescence imaging," *Future Microbiol.* **15**, 319–332 (2020).
111. R. Hill, M.Y. Rennie, and J. Douglas, "Using bacterial fluorescence imaging and antimicrobial stewardship to guide wound management practices: A case series," *Ostomy Wound Manag.* **64**, 18–28 (2018).
112. R. Raizman, "Fluorescence imaging guided dressing change frequency during negative pressure wound therapy: A case series" *J. Wound Care* **28**, S28–S37 (2019).
113. B. Aung, "Can fluorescence imaging predict the success of CTPs for wound closure and save costs?," *Today's Wound Clin.* **13**, 22–25 (2019).
114. R. Raizman, D. Dunham, L. Lindvere-Teene, et al., "Use of a bacterial fluorescence imaging device: Wound measurement, bacterial detection and targeted debridement," *J. Wound Care* **28**, 824–834 (2019).
115. P.T. Goedhart, M. Khalilzada, R. Bezemer, et al., "Sidestream Dark Field (SDF) imaging: A novel stroboscopic LED ring-based imaging modality for clinical assessment of the microcirculation," *Opt. Exp.* **15**(23), 15101 (2007).
116. H. Sherman, S. Klausner, and W.A. Cook, "Incident dark-field illumination: A new method for microcirculatory study," *Angiology* **22**, 295–303 (1971).
117. E. Gilbert-Kawai, J. Coppel, V. Bountziouka, et al., "A comparison of the quality of image acquisition between the incident dark field and sidestream dark field video-microscopes," *BMC Med. Imaging* **16**, 10 (2016).
118. I. Schelkanova, A. Pandya, G. Saiko, et al., "Spatially resolved, diffuse reflectance imaging for subsurface pattern visualization toward development of a lensless imaging platform: Phantom experiments," *J. Biomed. Opt.* **21**(1), 015004 (2016).
119. A. Pandya, I. Schelkanova, and A. Douplik, "Spatio-angular filter (SAF) imaging device for deep interrogation of scattering media," *Biomed. Opt. Exp.* **10**(9), 4656–4663 (2019).
120. M. Ghijsen, T.B. Rice, B. Yang, et al., "Wearable speckle plethysmography (SPG) for characterizing microvascular flow and resistance," *Biomed. Opt. Exp.* **9**, 3937–3952 (2018).
121. M.K. Razavi, D.P.T. Flanigan, S.M. White, et al., "A real-time blood flow measurement device for patients with peripheral artery disease," *J. Vasc. Interv. Rad.* **32** (3), 453–458 (2021).

3 The Use of Capillaroscopy and Aggregometry Methods to Diagnose the Alterations of Microcirculation and Microrheology in Diabetes

Andrei E. Lugovtsov, Yury I. Gurfinkel, Petr B. Ermolinskiy, Anastasia A. Fabrichnova, and Alexander V. Priezzhev

CONTENTS

3.1 INTRODUCTION

Diabetes mellitus (DM) is a group of metabolic diseases based on absolute or relative insulin deficiency, leading to the development of chronic hyperglycemia. The high prevalence of DM, constituting 8.35% in the world for 2013 according to the International Diabetes Association, determines the relevance of the study of this pathology. Commonly, DM leads to macrovascular complications and different comorbidities as an increased risk of cardiovascular disasters, the development of renal failure, blindness, amputations, etc.

The most common and important metabolic diseases are type 1 and type 2 DM. The basis of the pathogenesis of type 1 DM (T1DM) is the autoimmune destruction of the ß-cells of the pancreas, which leads to an absolute deficiency of insulin and, as a result, severe hyperglycemia.

Type 2 DM (T2DM) is a group of diseases, the leading link in the pathogenesis of which is insulin resistance, accompanied by a relative deficiency of insulin. In most cases, the basis for the development of insulin resistance (consequently, T2DM) is such a pathological condition as obesity

DOI: 10.1201/9781003112099-3

[1]. The lack of insulin and hyperglycemia cause disorders of the carbohydrate and other types of metabolisms with the development of acute and chronic (late) specific complications of DM [1].

One of the main pathophysiological mechanisms that process in DM is the glycosylation of proteins leading to changes in the conformation of proteins and their functions [2]. Activation of non-enzymatic glycosylation processes leads to the accumulation of glycosylation end products. It is believed that these end products play a key role in the development of diabetic complications [2,3]. Thus, non-enzymatic glycosylation of the basement membranes and proteins of the extracellular matrix of the outer and middle layers of blood vessels leads to the thickening of membranes, activation of inflammation, the development of endothelial dysfunction, vasoconstriction, and, as a result, the appearance of angiopathy. End products of glycosylation interact with receptors on the surface of endothelial cells and macrophages activating the production of growth factors, adhesion molecules, and interleukins. Their negative impact is realized not only in the walls of blood vessels, but also in neurons and bone tissue [3,4]. The end products of glycation can change the structure of circulating blood proteins, gene transcription, and intracellular glycation products that disrupt cell functions.

Accumulation of the end products of glycosylation in DM is believed to be based on the phenomenon of metabolic memory. It is considered that the early glycemic background is "remembered" in tissues and target organs, whose damage persists despite a further decrease in glucose levels. It is proved that the level of glycosylation products determined by skin biopsy reliably correlates with the prevalence of such complications of DM as retinopathy and nephropathy. The phenomenon of glycemic memory has also been proven in large-scale studies of patients with T1DM and T2DM [3,5].

Prostaglandins are also involved in the development of complications of DM as far as their metabolic disorder causes tissue hypoxia. A decrease in the synthesis of prostaglandins leads to vasoconstriction, platelet aggregation, and an increase in ischemia and hypoxia [6].

Metabolic changes in the blood plasma in DM have a direct impact on the rheological properties of the blood [7]. In DM, there are both microrheological disorders, manifested by an increase in the viscosity of whole blood and plasma, and microrheological disorders, manifested by increasing the red blood cell (RBC) aggregation, reducing their deformability, and increasing the adhesion of RBC to endothelial cells [8,9]. In addition, platelet and white cell disorders are observed in DM. They are usually precede and accompany microrheological changes [10].

3.2 BLOOD MICRORHEOLOGICAL PROPERTIES IN DIABETES

3.2.1 Blood Viscosity

Blood is a two-phase system that includes blood plasma and blood cells. Normally, blood cells occupy 30%–50% of the blood volume (i.e., hematocrit), mainly due to the RBC content. As a result, whole blood exhibits the properties of a non-Newtonian fluid. It means that the viscosity of blood in the laminar flow is not constant and varies depending on the shear rate, the diameter of the vessel, and the hematocrit [11,12].

In general, blood viscosity at low shear rates (such as $1.0\,s^{-1}$) *in vivo* is generally consistent with the diastole cycle, while at high shear rates ($300\,s^{-1}$), it is consistent with the systole cycle. The shear rate is the velocity gradient that, in the vascular bed, has maximum values at the vessel wall and minimum (i.e., around zero) values in the center of the vessel [12].

Low shear rates in the center of large vessels determine the special role of RBC aggregation in large vessels, even at high flow rates. In capillaries, the aggregation and deformation of RBCs are also of significant importance. Changes in RBC aggregation and deformation are expressed by the changes in the quality of oxygen supply [13].

Although one of the most important parameters determining blood viscosity is hematocrit, RBC aggregation and RBC deformability play crucial roles. The viscosity values at low shear rates are primarily determined by RBC aggregation, and the viscosity values at high shear rates are strictly dependent on the ability of RBCs to deform [14]. In addition, viscosity at low shear rates correlates with the plasma concentrations of fibrinogen and globulins [14].

Numerous studies show that the viscosities of whole blood, as well as those of plasma, measured at different shear rates in patients suffering from DM are significantly different from the norm [9,15,16]. Hyperglycemia increases blood osmolarity, which leads to increased capillary permeability, which increases hematocrit and blood viscosity. In addition, hyperglycemia can cause osmotic diuresis, thus reducing plasma volume and increasing hematocrit. High hematocrit levels have been shown to be associated with a decreased blood flow in the retina [17]. According to Ref. [18], hematocrit and blood viscosity decrease with an improved glycemic control in patients suffering from DM.

There is sufficient evidence indicating that increased blood viscosity is a pathogenetic factor in the development of diabetic microangiopathy, microcirculation disorders, and reduced tissue perfusion in DM [19]. Increasing blood viscosity may also be particularly important in the etiology of diabetic retinopathy [15,20,21]. The development of diabetic angiopathy has been shown to be associated with abnormalities in hematocrit, plasma viscosity, and RBC aggregation [22,23] and reduced deformity of RBCs [24,25]. Thus, patients suffering from DM have a higher blood viscosity than people without DM [25].

3.2.2 RBC Aggregation and Disaggregation

RBCs interact with each other forming linear or more complex structures (i.e., RBC aggregates (Figure 3.1)) under low shear rates. These structures can be reversibly dispersed into singe cells under high shear rates. Normally, RBCs are in dynamic equilibrium, constantly aggregating and disaggregating *in vivo* [26]. It is crucial to note that RBC aggregation occurs only in solutions with macromolecules of high molecular weight. In the case of blood plasma, they are fibrinogen, albumin, etc. [26].

Currently, there are several theories explaining the mechanism(s) of RBC aggregation. According to one of them, RBCs aggregate due to the formation of "cross-bridges" between them from fibrinogen and/or other macromolecules, which are connected to the RBC membrane using hydrogen and Van der Waals interactions [27,28]. According to an alternative point of view, aggregation occurs under the influence of osmotic forces arising from the fact that the configuration of polymer molecules cannot directly approach the cell–cell surface, and a "depletion" layer between two adjacent cells is formed [29].

The forces that cause aggregates to disaggregate include flow shear force, electrostatic repulsion between cells, and the elastic forces of the RBC membrane [28,30]. Thus, the aggregation of RBC mainly depends on hemodynamics, the composition of the cell membrane, and the composition

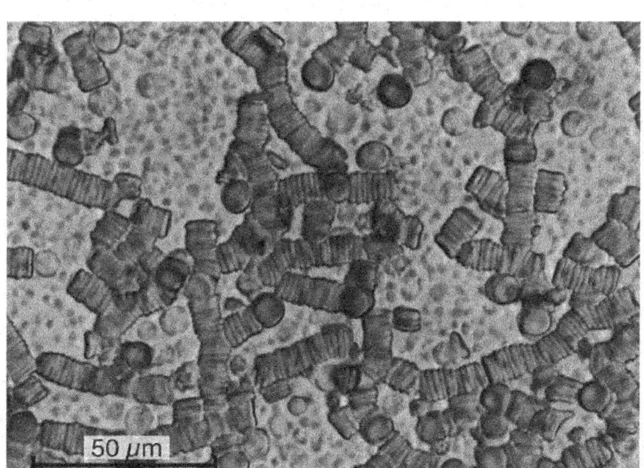

FIGURE 3.1 Microphotograph of RBC aggregates in plasma solution. The small dots are platelets.

of the blood plasma. The largest contribution to the aggregation of RBC is made by such important blood plasma proteins as fibrinogen, albumin, and globulins [26]. Hyperaggregation has been shown to be associated with increased concentrations of fibrinogen and 2-alpha macroglobulin, and conversely, aggregation decreases with increasing albumin concentrations [31]. Also, the dependence of aggregation on ESR (erythrocyte sedimentation rate) has also been revealed [26].

Many pathological conditions that are characterized by microvasculatory disorders are accompanied by the hyperaggregation of RBC [26]. At the same time, large aggregates, which have a mesh and lumpy structure, are formed and have an increased strength of adhesion between cells. Lump aggregates retain their structure even at very shear rates.

Pathological hyperaggregation negatively affects the local blood supply to organs and tissues, significantly increasing the uneven distribution of RBCs in microvessel networks [31]. In severe cases of pathological aggregation, blood flow rate may decrease, and RBC aggregates can be detected even at the level of arterioles. This prevents cells from entering narrow capillaries and promotes their bypass through wider vessels, bypassing capillary networks, increasing the likelihood of developing zones of local hypoxia [32]. Thus, pathological aggregation of RBCs entails the development of microvasculature pathology.

It has been shown that the aggregation of RBCs in DM increases using a variety of techniques [8,33–35]. RBC hyperaggregation is considered to be one of the most important pathophysiological consequences in diabetic patients with poor glycemic control [17]. Hyperaggregation is caused by several mechanisms, including plasma and cellular factors.

Thus, RBC hyperaggregation is associated with several pathological changes that are a consequence of hyperglycemia. With an increase in glucose levels, glycosylation of the RBC membrane, and intracellular proteins and the change in the lipid composition of the cell membrane, there occur a violation of lipid asymmetry in the membrane bilayer and lipid packaging [8]. All these pathological mechanisms lead to an increased RBC aggregation [26]. Also, in DM, the concentration of sialic acids carrying a negative charge on the membrane of the RBC is reduced, which leads to a decrease in the forces of electrostatic repulsion of cells and their increased aggregation [36]. Additionally, changes in the asymmetry of the RBC lipid bilayer increase their adhesion ability to endothelial cells [37].

Experiments proved that during the incubation of blood cells in the plasma of a patient with DM, RBC hyperaggregation occurs more often when compared with the control group [38]. In T2DM, the increase in fibrinogen levels appears to be due to increased insulin content [39].

It was shown that the aggregation of RBCs increases with an increased content of glucose, triglycerides, alipoprotein B, total cholesterol, C-reactive protein in the blood plasma, ESR in the blood. However, the dependence of aggregation on glycated hemoglobin levels, which is often used as a criterion for metabolic control of diabetes, has not been identified in all studies [40,41]. This may be likely because glycated hemoglobin reflects an average metabolic control of 3 months by the time of its measurement, while glucose levels are a more direct indicator of diabetes compensation.

Hemorheological indicators in T1DM and T2DM differ from each other. Thus, according to the papers [8,42], RBC aggregation increases with T1DM and T2DM, but with T2DM, it significantly exceeds both the values in the control group and the values in T1DM.

RBC hyperaggregation in the blood of patients suffering from T1DM and T2DM is reduced by the incubation of blood *in vivo* and *in vitro* with insulin, regardless of other factors [34]. Additionally, RBC hyperaggregation is observed in patients with DM both without and in the presence of macro- and microvascular complications.

In patients without vascular complications, RBC hyperaggregation is the cause of hemorheological disorders rather than ischemia and hypoxia complications [43]. These disorders may further lead to tissue acidosis and increased platelet aggregation, contributing to endothelial dysfunction [9].

Normally, in large vessels (e.g., the coronary or carotid artery), the aggregation forces may exceed the disaggregation forces, except at the bifurcation sites, where the flux recirculates and stops opposite the bifurcation of the vascular wall, due to the counter pressure gradient. It has been shown that in such zones, in patients suffering from DM, there is an increase in the adhesion of RBCs to the endothelium, which increases the risk of atherosclerotic plaque formation.

3.2.3 RBC Deformability

The ability of RBCs to deform is a feature that helps perform their major function, i.e., the delivery of oxygen to the tissues. The average size of RBCs (≈8 μm) is almost twice the size of the smallest capillaries (3–5 μm). It means that RBCs are deformed when passing through the capillaries. Deformation occurs due to the shear stress arising from the shifting layers of blood plasma. The ability of RBCs to deform depends on the cellular shape (surface-to-volume ratio), the mechanical properties of the membrane and cytoskeleton, and the intracellular viscosity, which is associated with the concentration of hemoglobin.

It has been established that a decrease in RBC deformability is observed in most patients with DM. A correlation has been shown between the severity of complications of DM and the degree of impaired deformability of RBCs [9]. There is also a point of view that the violation of tissue perfusion observed in DM is primarily associated with a violation of the deformability of RBCs [43]. In DM, the shape of the RBC is somewhat flattened, increasing the perimeter of the cell [44].

Glycosylation of membrane and intracellular proteins associated with hyperglycemia significantly reduces the deformation properties of membranes and the viscoelastic properties of the whole cell in DM [45,46]. Increased intracellular glucose concentration leads to an increase in glycosylated hemoglobin, which leads to the decrease in RBC deformability [47]. According to some researchers, reduced membrane deformability is associated with an increase in the binding of glycosylated hemoglobin to its inner surface [48]. Oxidative stress inflicts damage to the membrane proteins of RBCs, even with a relatively short exposure time. Increasing the cholesterol/phospholipid ratio of RBC membranes in diabetic patients also decreases their deformability [49].

Historically, the first methods for studying deformability were based on retracting a part of the RBC membrane with a pipette and filtering through pores of diameter 3–5 μm. The method of assessing the deformation during the passage of cells through flowing microchannels was also used. Later, the deformability of RBCs began to be assessed by the optical method proposed by M. Bessis and N. Mohandas [50]. The method is based on the diffraction of the laser beam on a suspension of RBCs in a shear flow, and the registration of the axial ratio of ellipse-shaped curves approximating the isointensity lines of the corresponding diffraction patterns. Different flow shear rates make it possible to achieve experimental conditions that are closer to the physiological ones. The estimated elongation index (IE) of the diffraction pattern is directly related to the ability of RBCs to deform [51,52]. In practice, the most used modifications of the method are based on the registration of forward light scattering, in particular the method of ektacytometry.

Many studies have shown that RBC deformability decreases in DM. Such data were obtained in the study of RBCs by ektacytometry [53] and filtration [54] as well as by analyzing video images of RBCs flowing in microchannels obtained with a high-speed camera [55]. When incubating RBCs in a solution of glucose *in vitro* (50 mmol/l), the deformability of RBCs was shown to decrease [56]. Deformability has also been shown to depend on blood glucose levels *in vivo*: with an increase in glycemia, the ability of RBCs to deform decreases [57]. Thus, using various methods, it was found that RBC deformation is disturbed in DM. Incubating RBCs with insulin, as well as infusing insulin in patients suffering from DM, regardless of glucose levels, on the contrary, improves the deformation properties of RBCs [58].

It was found that RBCs of patients suffering from DM are slower in restoring their shape, compared with the RBCs of people without diabetes [59]. According to nail bed capillaroscopy, there is a decrease in tissue perfusion in DM, due to a violation of the deformability of RBC [60].

Summarizing, in DM, a violation of microrheological characteristics of blood can occur, consisting of an increase in the viscosity of whole blood and plasma, increased aggregation of RBCs, decrease in the deformability of RBCs, and increase in their adhesion to endothelial cells. There may also be a change in the capillary bed, leading to a decrease in tissue perfusion. The severity of such disorders is mainly dependent on the duration of the course of the disease, its compensation, the presence of specific complications of diabetes mellitus, other clinical parameters that reflect the presence of various pathological changes in the body (the level of fibrinogen, globulins, ESR, etc.).

3.3 OPTICAL METHODS FOR ASSESSING AGGREGATION AND DEFORMABILITY OF RBCS AND BLOOD MICROCIRCULATION

All measurements on whole human blood samples and RBC suspensions were performed *in vitro*; *in vivo* measurements were performed on patients (see Section 3.3.4 on vital digital capillaroscopy). Measurements of the aggregation parameters of RBCs were carried out during the first 3 hours after blood sampling from the cubital veins of healthy volunteers or patients on an empty stomach. Ethylenediaminetetraacetic acid was used as the anticoagulant in all blood samples to prevent blood clotting. The *in vitro* measurements were carried out at the temperature of 37°C (22°C in case of laser tweezers measurments), which corresponds to the physiological conditions in the human body.

Recommendations for hemorheological laboratories developed by the International Expert Group for Hemorheological Research Standardization [61] were followed. The study was approved by the ethics committees of M.V. Lomonosov Moscow State University. All volunteers were informed on the purpose of the study and gave written informed consent in accordance with the Declaration of Helsinki.

Optical methods allow fast and and efficiently study the aggregation and deformation parameters of RBCs based on the analysis of the kinetics of diffuse scattering of laser radiation by dense suspensions of RBCs, in particular whole-blood samples or the diffraction patterns obtained from dilute suspensions of RBCs [62,63]. Diffuse light scattering has been implemented in several commercially available systems: laser optical rotary cell analyzer LORCA (Mechatronics, the Netherlands) [62], laser aggregometer and deformometer of erythrocytes LADE (Reomedlab, Russia) [64], RBC aggregation analyzer RheoScan (RheoScan-AnD300, RheoMeditech, Republic of Korea) [65], and automatic RBC aggregometer FAEA (Myrenne, Germany) [66]. Several methods have also been developed to study aggregation and deformation parameters at the level of individual cells, the most common among which are the trapping and retention of cells using micropipettes and atomic force microscopy [67,68]. It should be noted that in these methods, the studied objects are subjected to direct mechanical action from the measuring device.

Laser tweezers (LT) are an alternative approach to characterize RBCs and their interaction with each other at the single-cell level [69,70]. A distinctive feature of LT is the ability to measure the interaction forces of individual cells without direct mechanical contact with the measuring equipment [64]. The ability to manipulate cells using one or multiple optical traps makes it possible to measure the forces in the range of 0.1–100 pN, which represents the interaction forces of RBCs.

We assessed the changes in the aggregation and deformation parameters of RBCs in the blood of diabetic patients in comparison with those of the control group. We used four optical techniques for assessing the microrheological and microcirculatory parameters of blood: laser aggregometry, optical trapping and manipulation of individual red cells, laser diffractometry technique, and vital digital capillaroscopy [71].

Statistical analysis of the obtained experimental results was carried out using the Statistica software. For the analysis, the nonparametric statistical Mann–Whitney U-test was used, which is suitable for assessing the significance of differences between two independent samples – between the control group of healthy donors and the group of patients suffering from DM [72]. Differences between the measured values of aggregation parameters were considered statistically significant at the level of statistical significance $p < 0.05$.

3.3.1 Laser Aggregometry

Measurements of blood aggregation properties on large ensembles of cells were carried out using the method of diffuse scattering of laser radiation (i.e., laser aggregometry). For these purposes, the RheoScan laser aggregometer of RBCs was used [73–75]. Two types of disposable cuvettes made of transparent plastic were used for measurements on the RheoScan.

To measure such aggregation parameters of blood as the characteristic time of cell aggregation, amplitude of aggregation, and aggregation index (AI), cuvettes of the first type were used (Figure 3.2a). The cuvette is small, flat, 0.5 cm in diameter, containing a thin metal stirring rod that can be rotated by an external magnetic field generated inside the thin reservoir of whole blood to mix its contents. Whole blood (8 μl) was placed into the cuvette using a micropipette.

After placing the microcuvette of the first type filled with whole blood into the device, the measurement process was initiated. The device registered the dependence of the intensity of laser radiation (wavelength, 633 nm) scattered forward by the blood sample on time, i.e., aggregation kinetics (Figure 3.3). At the very beginning of the measurement, the magnetic rod rotated at high speed and, thereby, destroyed all the aggregates formed in the blood sample while in stasis. At this point, the RBCs in the sample were subjected to the maximum shear stress. Under these conditions, completely disaggregated RBCs experienced deformation, and the intensity of light scattered forward by the blood was close to the minimum (in the aggregation kinetics, within the time range designated as $t<0$, the signal value was close to zero). At $t=0$, the rotation of the rod inducing stirring was abruptly stopped. A short-lasting additional decrease in the signal was observed, which is associated with a decrease in the intensity of the forward-scattered radiation, corresponding to the changes in the orientation of the RBCs. Then, the process of spontaneous aggregation of RBC at rest (without external shear stress) began in the blood sample. The laser beam propagating through the reservoir with blood was scattered on RBCs and newly formed aggregates. Over time, the intensity of the light transmitted forward increased due to the increasing average size of the radiation scattering centers. The process of spontaneous aggregation took about 2 minutes, after which the intensity reached its maximum value, indicating that almost all RBCs in the sample aggregated (Figure 3.3). The magnitude of the intensity change from the beginning to the end of the RBC aggregation process is the amplitude of aggregation (AMP) characterizing the total aggregation state.

Aggregation kinetics were used to determine the characteristic time of aggregate formation $T_{1/2}$ in seconds (i.e., the time during which the intensity of light scattered forward reaches half the intensity value at maximum aggregation after 120 s), which characterizes the rate of aggregate formation, and AI. AI characterizes the number of RBCs aggregated during the first 10 seconds as a percentage. This value is calculated as a ratio of the area under the curve describing the aggregation

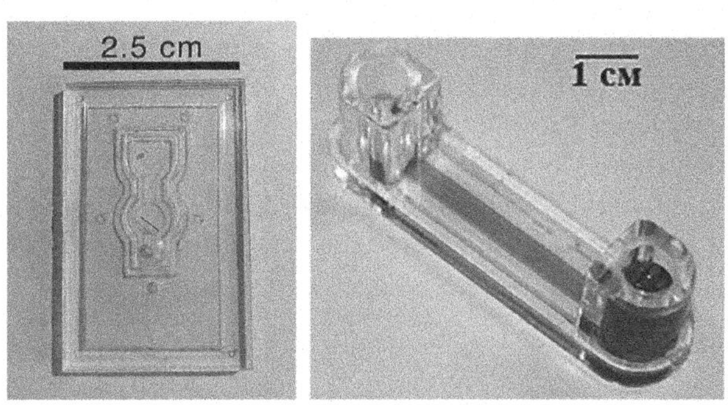

(a) (b)

FIGURE 3.2 (a) Disposable microcuvette with a magnetic rod for measuring RBC aggregation parameters on the RheoScan device and (b) disposable cuvette with two reservoirs and a microchannel for measuring the critical shear stress of RBC on the RheoScan device.

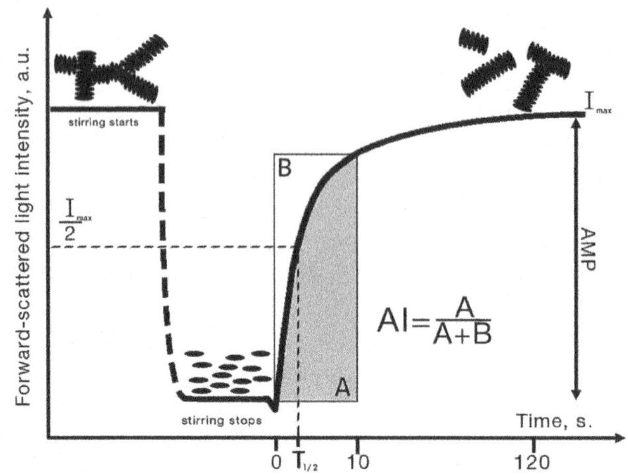

FIGURE 3.3 Kinetics of spontaneous RBC aggregation recorded by a RheoScan. I_{max} is the intensity of light scattered forward from the whole-blood layer at the maximum RBC aggregation; AI is the aggregation index; AMP is a parameter characterizing the total aggregation state; and $T_{1/2}$ is the characteristic time of aggregate formation. (This figure has been adapted from Ref. [88].)

kinetics to the total area above and below the curve for the first 10 s of the spontaneous aggregation process (see Figure 3.3). The higher the AI and AMP and the lower the $T_{1/2}$, the higher the RBC aggregation.

Cuvettes of the second type (Figure 3.2b), which were used to assess the hydrodynamic strength of aggregates (i.e., critical shear stress), contain a microchannel (0.2 mm high × 4.0 mm wide × 40 mm long) and whole-blood reservoirs at both ends (0.5 ml). This type of cuvette is used to measure the critical shear stress (CSS).

Whole blood in the microchannel is subjected to changing shear stress under different values of hydrodynamic pressure. To measure the CSS, which characterizes the hydrodynamic strength of aggregates formed in the blood, 0.5 ml of whole blood was placed into the cuvette. The forces of viscous friction arising in the microchannel led to the destruction of RBC aggregates under the shear stresses from 0 to 20 Pa. The CSS parameter is measured in Pascals (Pa) units. A schematic representation of the measurement process is shown in Figure 3.4. CSS measurements were carried out as follows. During measurements, blood is passed through a narrow microchannel (≈200 μm) with a shear stress that monotonically decreases with time (the initial shear stress is 20 Pa). Since the shear stress at the beginning of the measurements (during the first 10 s) was high, the intensity of the light scattered in the opposite backward direction increased due to the destruction of the aggregates formed earlier. However, when a certain shear stress equal to CSS was reached, the stress arising in the microchannel would no longer be sufficient to destroy the formed aggregates. The intensity of the backscattered light then began to decrease. Thus, the CSS stress value was used to characterize the hydrodynamic strength of RBC aggregates in the flow. In other words, CSS is the minimum shear stress that must be applied to the cell aggregates to initiate the process of their forced disaggregation.

3.3.2 Laser Tweezers

To measure the interaction forces of RBCs, a home-made two-channel LT system was used [76]. The schematic layout of the experimental setup is shown in Figure 3.5.

Nd:YAG laser with a wavelength of 1064 nm and a power of 1 W was used as the radiation source. The position of the laser, the beam-splitting cube, and the lens system along the beam

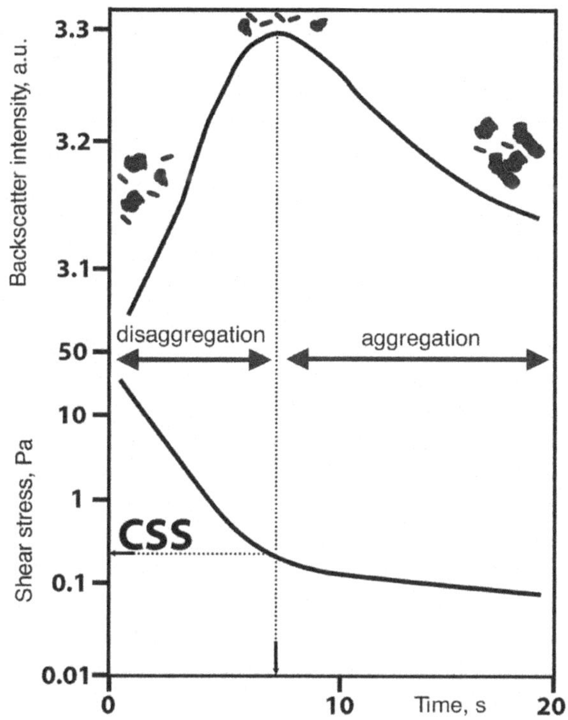

FIGURE 3.4 Schematic representation of the critical shear stress (CSS) measurement process. CSS characterizes the equilibrium point between aggregation and disaggregation.

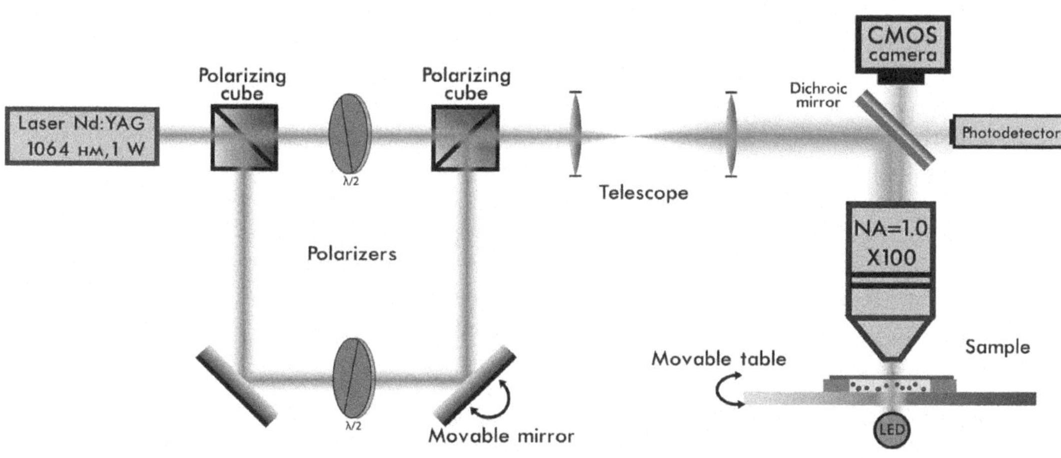

FIGURE 3.5 Schematic layout of the two-channel laser tweezer experimental setup.

propagation path were adjusted with high accuracy to achieve a high intensity gradient in the beam focusing region. A dichroic mirror was installed at the beam exit from the lens system, with the help of which the beam was headed to the rear entrance aperture of the Olympus objective (×100, NA=1, water immersion), and was also partially passed to the photodetector for the power measurement. The beam power was adjusted using half-wave plates installed directly after the lasers. Electric motors capable of precisely rotating them with a small rotation step were connected to the plates. One beam was always stationary; a mirror was installed in the path of the second beam, the rotation of which allowed for moving the beam focusing area, thereby changing the position of the trapping point. Thus, two trapping areas (two traps) were used: one fixed and the other that can move inside the sample cuvette. In the vertical part of the system, a lens and a CMOS camera were installed, to detect the image of the trapped cell illuminated by the white light of a lamp installed behind the objective and the sample cuvette.

The cuvette (Figure 3.5) was placed upon a motorized platform and consisted of a slide and a cover slip, placed one above the other. The distance between them was 100 μm due to the double-sided adhesive tape between the slides at two edges. The sample was placed into the cuvette cavity between the glasses using a micropipette. The cavity was then sealed with Vaseline at the two untapped edges to prevent unwanted flows and sample evaporation.

Before measuring the interaction forces, the calibration procedure was performed, which is necessary to determine the unambiguous relationship between the values of the optical trapping force (F_{trap}) and the laser beam power. Calibration was carried out by comparing F_{trap} with the viscous friction force acting on the trapped cell when the platform with the cuvette was moved at a given speed relative to the stationary trap, for different values of the laser beam power. This procedure is described in detail in our previous works [77,78]. In general, while conducting experiments with optical trapping of live cells, one should account for a possible heating effect of the laser trap. In our experiments, RBC heating by the laser trapping beam was excluded due to the very low absorption by RBCs at the wavelength of the laser used for trapping (hemoglobin absorption coefficient at 1064 nm laser beam $\mu_a = 10\,cm^{-1}$) and the presence of a large volume of heat-removing fluid (plasma) around the cell. Theoretical estimates show that the heating of trapped cells does not exceed 1°C for every 10 mW of laser beam power [79]. As far as the typical maximum laser beam power in the waist does not exceed 40 mW, the difference between cell temperature and surrounding medium does not exceed 4°C. It was shown that the properties of RBC aggregation did not change at the temperatures of 20°C and 37°C when using LT [80]. Therefore, measurements with laser tweezers were carried out at room temperature.

LT force measurements require a highly diluted RBC suspension with relatively small number of cells in the cuvette. The dilution was carried out in autologous plasma with a concentration of cells in the volume of the final sample of about 0.05%. The plasma was obtained using a two-step centrifugation: centrifugation for 10 minutes at 200 g followed by centrifugation for 10 minutes at 3000 g.

The aggregation force (AF) measurement procedure consisted of 4 steps (Figure 3.6).

During the first operation, two RBCs were trapped using LT and oriented in parallel by moving the platform upon which the cuvette with the RBC suspension was located. One of the RBCs was trapped by the fixed trap while the other was trapped by the movable trap. The interaction forces during aggregation and disaggregation depend on the area of initial contact [81]; therefore, to achieve adequate repeatability of the results, visual control of the invariability of the overlap area during the primary contact of cells was performed during all measurements. At the second step, the

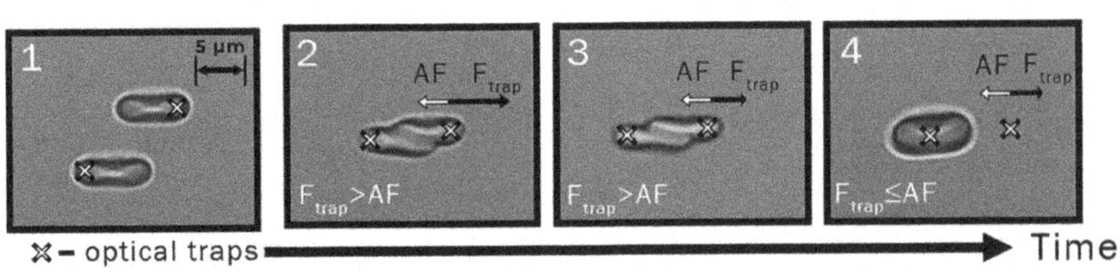

FIGURE 3.6 Schematic representation of the step-by-step protocol for the measurement of the aggregation force (AF). Step (1): Two RBCs are trapped with two laser traps (the trap position is designated with crosses). Step (2): RBCs are brought into contact. Step (3): The trapping beam power is decreased (i.e., a decrease in the trapping force F_{trap}). F_{trap} prevents aggregation ($F_{trap} > AF$). Step (4): F_{trap} is not enough to prevent the aggregation of RBCs ($F_{trap} \leq AF$).

cells were brought into contact with the help of the movable trap in such a way that the interaction area was about 40% of the surface area of each cell. Further, at the third step, the power of the movable laser beam was gradually decreased, thereby reducing the optical trapping force F_{trap}, which keeps the RBCs in the trap. The beam power decreased until the F_{trap} was insufficient to prevent spontaneous aggregation of RBCs, the cell broke out of the trap, and the process of spontaneous aggregation began ($F_{trap} \leq AF$). At this moment, the value of the beam power was fixed and the value of the optical trapping force was calculated, which was equated to the aggregation force AF. The higher the AF, the higher the RBC aggregation.

The procedure for measuring the disaggregation force (DF) also consisted of four steps. Steps 1 and 2 were identical to those of the process of measuring AF. At the third step, the position of the movable laser beam (trap) was changed to separate the formed aggregate at the previous step. In this case, the magnitude of F_{trap} of the movable beam was gradually reduced incrementally to find the minimum force required to separate the aggregate. This value of optical power was fixed and considered equal to the power of DF. The higher the DF, the harder it is to separate RBC aggregates.

Also, the time of RBC aggregation, which is the time interval from the "point contact" of their membranes to the formation of a pair aggregate, was measured by LT. Two RBCs were brought together to the state of a "point contact", i.e., the minimal overlap allowed by the setup. Then the laser beams forming the traps were shut off, which led to the spontaneous aggregation of RBCs.

These parameters measured by LT have been described in more detail in Ref. [76].

For each sample, the RBC aggregation time and aggregation and disaggregation forces were measured for at least 15 pairs of different RBCs. The results were obtained after averaging the measured values to eliminate subjective errors during the experiment.

3.3.3 Laser Diffractometry

Laser diffractometry performed with a RheoScan diffractometer (i.e., ektacytometry) was used to obtain the shear-induced deformation parameters of RBCs by processing the light intensity distribution in the diffraction pattern [82,83]. The dimensions of the channel were 0.2 mm high × 4.0 mm wide × 40 mm long and identical to the chamber in Figure 3.2b. The method consisted in obtaining and analyzing the diffraction patterns resulting from irradiating a suspension of RBCs in the flow chamber under different shear stresses (from 20 to 0.5 Pa) (Figure 3.7). When the shear stresses

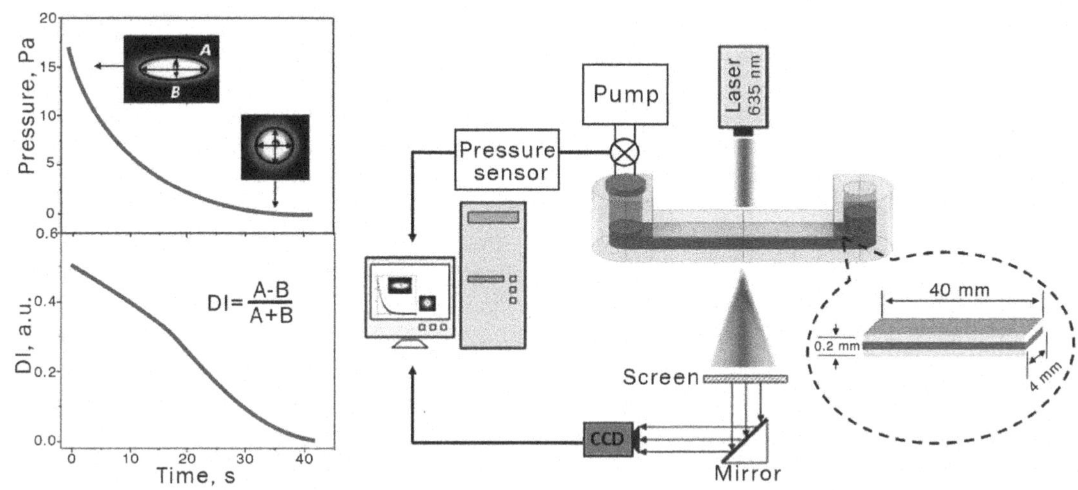

FIGURE 3.7 Schematic layout of the laser ektacytometry of RBCs using a RheoScan AnD 300 device. (Redrawn from Ref. [84].)

in the channel were small, the RBCs were not deformed, but were oriented so that the shape of their cross-section resemble a circle. The isointensity lines of the diffraction pattern obtained by illuminating a suspension of such cells with a laser, respectively, have the shape of a circle. When the shear stress increases, RBCs are pulled along the shear flow, taking on an elongated ellipsoidal shape. The isointensity lines of the corresponding diffraction pattern on the screen also take on an elliptical appearance as this pattern is a superposition of diffraction patterns originating from many individual cells.

The degree of ellipticity of the diffraction pattern was analyzed by approximating the diffraction pattern with an ellipse along an isointensity line (points of the pattern at which the intensity is approximately the same). Integral deformability can be characterized by the RBC deformability index (DI), which is determined by the formula [83]:

$$DI = \frac{A - B}{A + B},$$

where A is the large axis of the ellipse and B is the small axis.

The higher the DI, the higher is the ability of RBCs to deform in flow.

3.3.4 Vital Digital Capillaroscopy

Vital digital nailfold capillarocopy (VDC) is a non-invasive and safe method for the *in vivo* analysis of microcirculation of blood in capillaries [85].

Following a minimum 15 minute of long-seated rest, the *in vivo* microvascular measurements were conducted between 9 am and 11 am in a quiet temperature-controlled room (the temperature was maintained between 22°C and 23.5°C), with the subject in the seated position and the left hand kept at the heart level. All participants were required to refrain from smoking and caffeinated drinks for 1 day before the examination. Capillary blood velocity (CBV) was measured in the eponychium of the fourth or third finger of the left hand.

Skin temperature was measured at the dorsal middle phalangeal area of the tested finger of the left hand by medical precision thermometry; the mean skin temperature was 33.2°C ± 1.7°C with no significant differences in the studied group.

Nail fold capillaries were visualized using a digital capillaroscope Kapillaroskan-1 (AET, Russia) equipped with a high-speed CCD camera (1/3″ monochrome progressive-scan IT CCD sensor, resolution 640×480 px, frame rate 200 fps full frame), TM-6740GE (JAI, Japan). The nail bed illumination was achieved with a LED-based illuminating system. Two ranges of the total magnification (125×) and (400×) were used to visualize the nail bed capillaries. Panoramic images of the capillaries were obtained at the 125× magnification, while more detailed imaging of single capillaries was performed at the 400× total magnification and included the measurement of static parameters (capillary length and diameters in different parts) and CBV in different parts of the capillary. We also recorded the presence or absence of blood aggregates in the capillaries (Figure 3.8) using the digital analysis of the obtained microscopy images and videos.

For determining the CBV, after recording a video fragment, the program stabilizes the dynamic images of the capillaries and then processes the images in the specified region of interest in the offline regime. The tracks of specific spots (RBCs) differ in the level of light intensity. The program marks them, recognizes them in the next frame, and determines the average velocity along the axis of the capillary over 5-second intervals (500 frames). The CBV is estimated at least in 6 capillaries, and the results are averaged. We estimated CBV only in the capillaries of the first line, where the capillaries are located within one layer. Thus, the obtained values of CBV are not affected by the movement of blood in the vessels located above and below the investigated capillary. No patients were subjected to caffeine-containing substances and drinks on the eve of the study, and finger temperature during the measurements was in the range of 32°C–34°C. A detailed procedure of carrying out CBV measurements is described in Ref. [86]. Usually, at rest in healthy people not taking caffeine-containing substances and drinks on the eve of the study, the average CBV varies in the range of 800–1500 µm/s.

In this paper, we distinguish the two states – the presence (aggregates = 1) or absence (aggregates = 0) – of RBC aggregates in the capillaries based on digital processing of the nail bed images and videos obtained with the VDC technique *in vivo*. The criteria for distinguishing the presence of RBC aggregates in the capillary blood flow can be formulated as follows:

1. Blood aggregates are clusters of blood cells that form autonomous conglomerates separated by plasma gaps.
2. Conglomerates do not merge while moving along the capillary bed.
3. The appearance of aggregates in most cases is accompanied by a reduction in CBV.

Therefore, the determination of the presence of RBC aggregates in the capillary blood flow was performed by a digital assessment of the images and videos of nail bed capillaries according to the following scale: the absence of aggregates in the capillaries — 0, and the presence of aggregates in the capillaries — 1. In the case of digital detection of at least several distinguished clusters of RBCs in the flow, we considered the state as Aggregates = 1, otherwise the Aggregates state was indicated by 0 (Figure 3.8).

In this work, the registration the dependence of CBV on time was used to reveal the presence of a stasis, as a sign of a pronounced deterioration in the rheological properties of blood. The minimum duration of the blood flow halting in the capillary, which we classified in our study as stasis, was 0.25 seconds. Generally, the duration of stasis can reach several seconds.

Figure 3.9 demonstrates a typical dependence of CBV on time during the 30 seconds of measurements in a patient suffering from arterial hypertension (AH) and T2DM. In the center of the graph, one can see a drop in CBV to zero. The duration of this stasis was 1.16 seconds. In our study, the presence of a stasis in the capillaries was assessed as 1 if CBV remained about zero for longer than 0.25 seconds in at least one visualized capillary; otherwise, the absence of a stasis was evaluated as 0. Accordingly, the higher the CBV, the lower the probability of the stasis.

3.4 EXPERIMENTAL STUDY

The results of measurements on large ensembles of cells using diffuse light scattering are presented in Figure 3.10. They were averaged for ten patients suffering from T2DM and ten healthy donors. The characteristic time of formation of RBC aggregates significantly decreases from 8.8 ± 2.1 s in the control group to 5.3 ± 1.2 s in patients with T2DM (Figure 3.10a). This indicates an accelerated

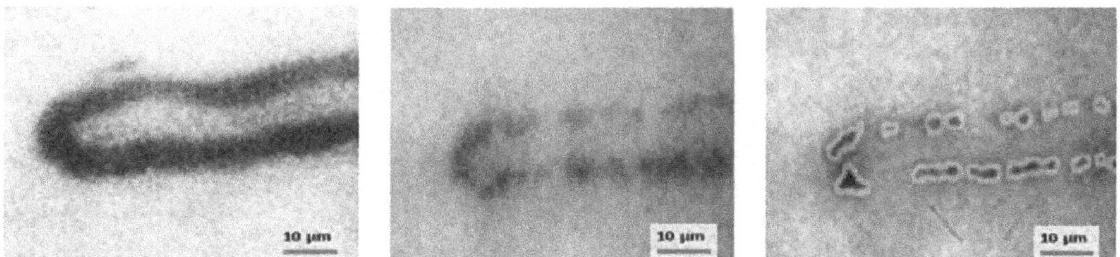

FIGURE 3.8 Visualization of RBCs in the capillary bed: normal laminar capillary blood flow in the absence of RBC aggregates – Aggregates = 0 (a, b, c); capillary blood flow in the presence of RBC aggregates – Aggregates = 1; results of image processing for the detection of aggregates (gray contours) in the capillary bed (This figure has been adapted from Ref. [71].)

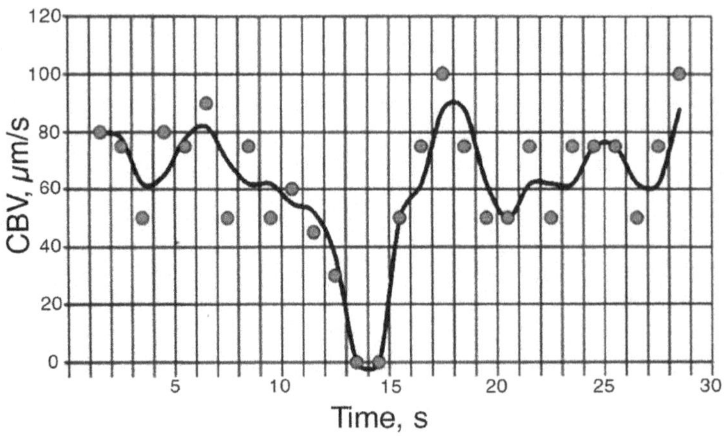

FIGURE 3.9 Typical time dependence of CBV measured *in vivo* by VDC for a patient suffering from AH and T2DM and exhibiting a stasis. (This figure has been adapted from Ref. [71].)

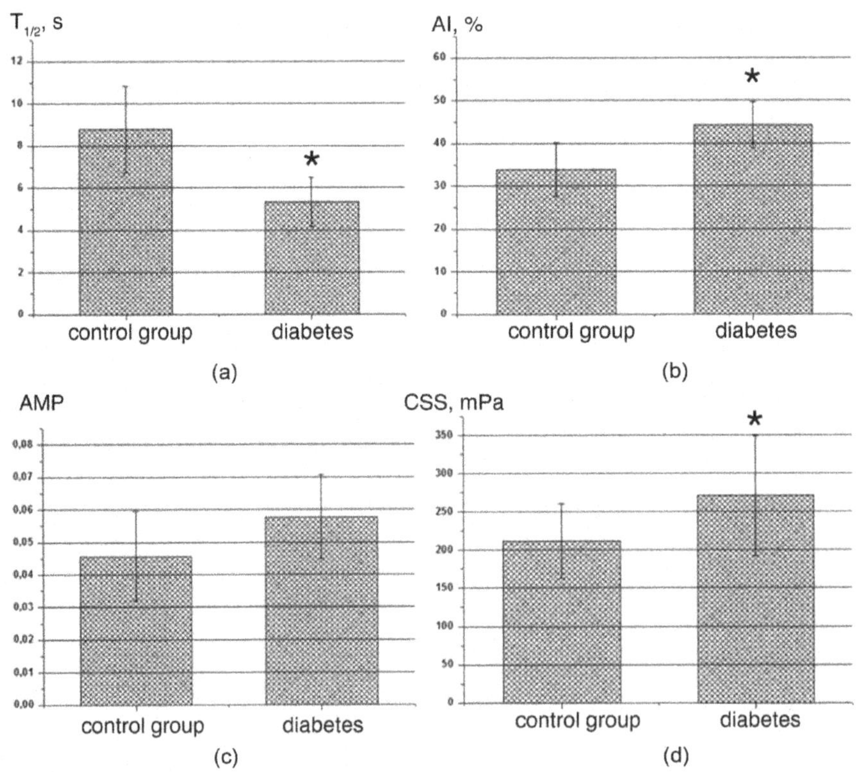

FIGURE 3.10 Blood aggregation parameters measured by diffuse light scattering using a RheoScan device in the control group ($n = 10$) and in the group of patients suffering from diabetes ($n = 10$): (a) characteristic time of formation of aggregates $T_{1/2}$. (b) RBC aggregation index, AI. (c) amplitude of aggregation, AMP. (d) critical shear stress, CSS. (*$p < 0.05$, nonparametric Mann-Whitney criterion. This figure has been adapted from Ref. [87].)

aggregation of RBCs in patients suffering from T2DM. The number of RBCs that are involved in the aggregation process in the T2DM group increases compared to the control group. This can be judged by a significant increase in the value of the AI from 33.8%±6.2% in the control group to 44.3%±5.4% in the T2DM group (Figure 3.10b). It is shown that the CSS values are larger for patients with diabetes (270.7±78.5 mPa) than in the control group (211.9±48.6 mPa) (Figure 3.10d). Also, the AMP parameter is increased in the case of diabetes, which corresponds to increased RBC aggregation. Thus, we can conclude that in T2DM, the rate of formation of aggregates, their number, and strength increase. This indicates an increased ability of RBCs to aggregate in the blood of patients suffering from T2DM. Since these properties of aggregation directly determine the blood flow in the body, RBC aggregation can lead to a violation of blood circulation and, as a result, to a deterioration in the supply of organs and tissues with oxygen and nutrients.

Thus, AP measurements at the macro level on a large ensemble of cells showed an increase in RBC aggregation in 2TDM. It is shown that diffuse light scattering and the RheoScan device functioning on its on diffuse light scattering make it possible to monitor the alterations of RBC aggregation parameters in diabetes.

The results of LT-based measurements showed (Figure 3.11) that the value of AF of RBCs in patients suffering from T2DM is 4.2±1.2 pN ($n=10$), which is higher than that in the control group, 2.7±1.5 pN ($n=10$). The difference is statistically significant ($p<0.05$). These results obtained on doublets of cells correlate with the results of the hydrodynamic strength of aggregates obtained in whole-blood samples (Figure 3.10d).

There were no significant differences between the groups in the force of disaggregation ($p>0.05$): 5.2±1.0 pN in the DM group and 5.0±1.0 pN in the control group. Interestingly, it was shown that for each of the groups, the disaggregation force statistically significantly ($p<0.05$) exceeds the AF.

One can introduce the parameter of interaction of two RBCs, as the ratio of the average values of forces during their disaggregation and aggregation:

$$R = \frac{DF}{AF}$$

This parameter is calculated for each blood sample and then averaged over the entire group. The obtained average values of R were compared between the control group and the group of diabetic patients (Figure 3.12). It is shown that the value of R is normally almost two times higher than the value measured in the group of diabetic patients, which is statistically significant ($p<0.05$): control group, 2.1±0.7, and the diabetes group, 1.3±0.1.

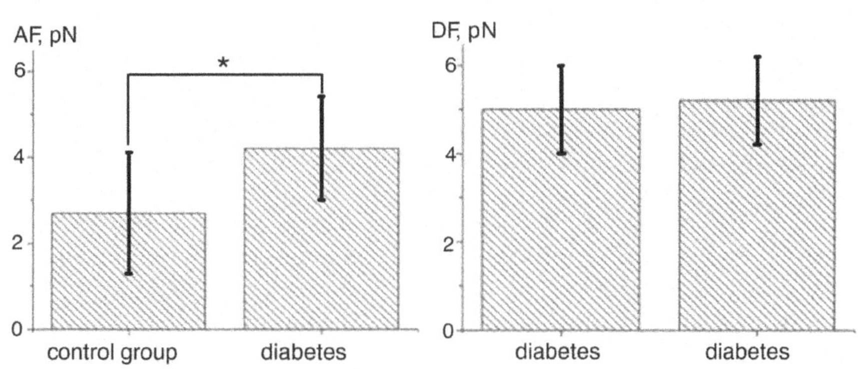

FIGURE 3.11 Interaction forces of RBCs during their (a) aggregation and (b) disaggregation in the control group ($n=10$) and the group of diabetic patients ($n=10$). For each value, the standard deviation is indicated on the diagram. (*$p<0.05$, nonparametric Mann-Whitney criterion. This figure has been adapted from Ref. [87].)

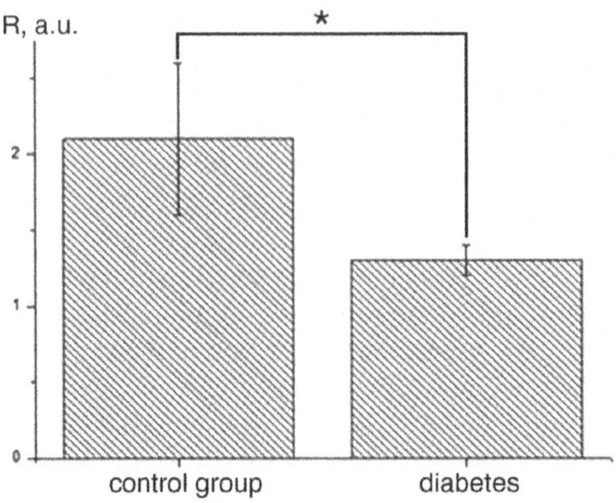

FIGURE 3.12 Comparison of the R parameter in the control group ($n=10$) and the group of diabetic patients ($n=10$). For each value, the standard deviation is indicated in the diagram. (*$p<0.05$, nonparametric Mann-Whitney criterion. This figure has been adapted from Ref. [87].)

Based on the foregoing observations, it can be assumed that evaluating the parameter R of the interaction of blood cells during their aggregation and disaggregation is promising for use as an indicator of the presence of a pathological process associated with diabetes in the body or a predisposition to it. The use of this parameter for the diagnosis and monitoring of the effectiveness of diabetes therapy and possible other pathologies, in particular, cardiovascular diseases, requires additional research and thorough verification.

To compare the *in vitro* and *in vivo* measurement techniques and assess the consistency of these methods, we presented the measured microrheological parameters for both experimental groups of patients (AH and CHD complicated with T2DM) in terms of subgroups depending on the results of measurements of the native capillary blood flow. Three parameters were obtained by VDC, which is why three types of subgroups were considered: (1) by distinguishing the absence (aggregates=0) or presence (aggregates=1) of aggregates in the capillaries through visual detection; (2) by distinguishing CBV with values lower or higher than 800 μm/s; (3) by distinguishing the absence (stasis=0) or presence (stasis=1) of stasis through visual detection. The results are shown in Figure 3.12–3.15, where the mean values of the measured parameters, standard deviations with error bars, and statistical significance (*$p<0.05$ and **$p<0.1$) are indicated on the diagrams.

The left-hand side diagram in Figure 3.13 demonstrates that the number of aggregated RBCs in the blood sample – AI, measured with a RheoScan – was significantly increased ($p=0.04$) in patients suffering from hypertension in the presence of blood aggregates detected using VDC. For patients suffering from CHD, significant differences were also obtained for this parameter ($p=0.09$). Measurements of AI in the presence and absence of the capillary blood flow stasis did not reveal significant differences for patients suffering from AH ($p=0.12$), while statistically significant differences were found for patients suffering from CHD ($p=0.04$). AI measured with a RheoScan for individuals with characteristic capillary blood flow velocity below and higher than 800 μm/s measured by VDC was significantly different in patients suffering from AH ($p=0.02$) and CHD ($p=0.096$). Thus, the tendency is that larger numbers of aggregates detected with a RheoScan were accompanied with reduced CBV determined by the VDC (Figure 3.13).

The first diagram in Figure 3.14 represents the interrelation between the presence or absence of aggregates of the nail bed and the parameter $T_{1/2}$ detected by VDC, characterizing the average or characteristic time of aggregate formation in a whole-blood sample obtained using diffuse light scattering with the aggregometer RheoScan. In patients suffering from AH, the characteristic aggregation time measured *in vitro* is significantly reduced ($p=0.03$) in the presence of aggregates *in vivo*

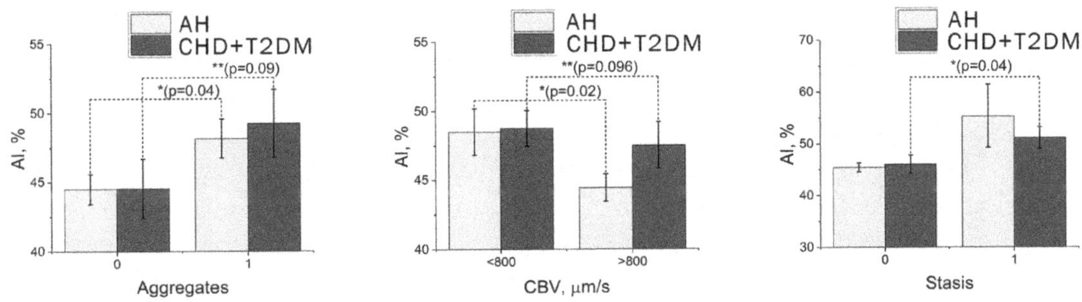

FIGURE 3.13 Aggregation indices measured *in vitro* by laser aggregometry for patients suffering from AH and those suffering from CHD+T2DM depending on the presence or absence of aggregates in the capillaries, capillary blood velocity, and the presence or absence of stasis in the blood flow detected by VDC. (This figure has been adapted from Ref. [71].)

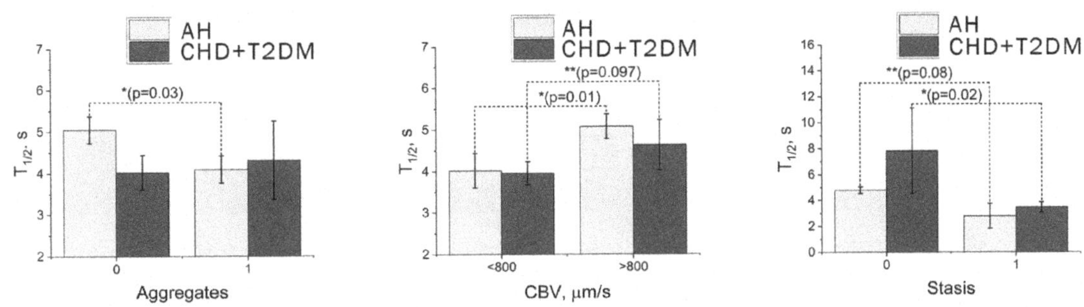

FIGURE 3.14 Characteristic times of aggregate formation measured *in vitro* by laser aggregometry for patients suffering from AH and those suffering from CHD+T2DM depending on the presence or absence of aggregates in the capillaries, capillary blood velocity, and the presence or absence of stasis in blood flow detected by VDC. (This figure has been adapted from Ref. [71].)

(more aggregates and less time for their formation). For patients suffering from CHD, there were no significant alterations of this parameter. The characteristic time, $T_{1/2}$, of aggregation measured with a RheoScan is inversely proportional to the average rate of aggregate formation. For patients suffering from AH ($p=0.01$) and CHD ($p=0.097$), significant differences in $T_{1/2}$ were found for individuals exhibiting CBV higher than 800 μm/s measured by VDC (Figure 3.14). Measurements of the $T_{1/2}$ parameter revealed significant differences in groups without and with the capillary blood flow stasis (the right part of Figure 3.14) both for patients suffering from AH ($p=0.08$) and for patients suffering from CHD ($p=0.02$).

Figure 3.15 relates the time of aggregation of a pair of cells measured with LT to the absence or presence of blood aggregates detected by VDC in patients suffering from AH and patients suffering from CHD. The former parameter is significantly reduced in hypertensive patients ($p=0.008$) and patients suffering from CHD ($p=0.01$) in the presence of blood aggregates detected by the VDC. Measurements taken with LT on individual RBCs showed significant differences in aggregation time for patients suffering from AH ($p=0.006$) and CHD ($p=0.06$) in relation to the CBV parameter obtained by VDC. That is, the higher the CBV, the slower the formation of aggregates. In the case of absence and presence of stasis (Figure 3.8c), the aggregation time measured on individual RBCs with LT revealed no significant differences for the group of hypertensive patients, whereas statistically significant differences were obtained ($p=0.03$) for the group of patients suffering from CHD. These results support the reasonable assumption that the slowing down of blood flow leads to

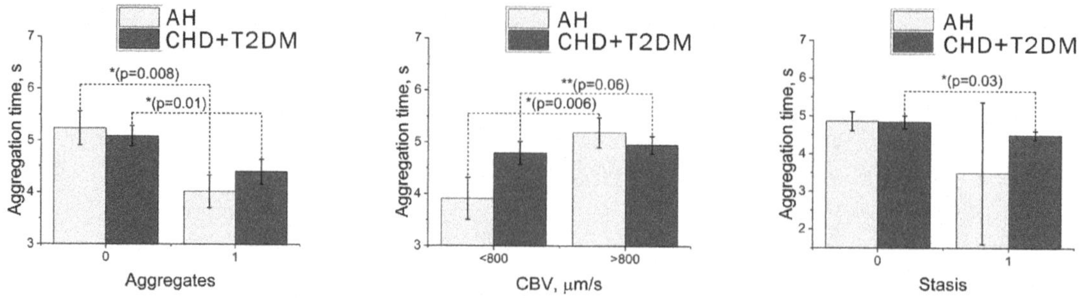

FIGURE 3.15 Aggregation times measured *in vitro* by LT for patients suffering from AH and those suffering from CHD+T2DM depending on the presence or absence of aggregates in the capillaries, capillary blood velocity, and the presence or absence of stasis in blood flow detected by VDC. (This figure has been adapted from Ref. [71].)

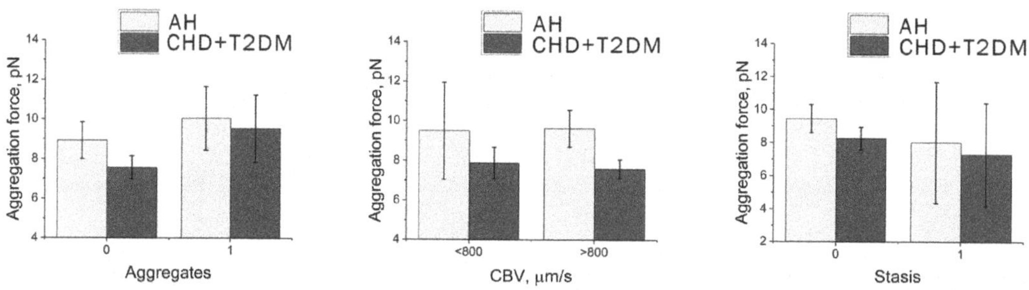

FIGURE 3.16 Aggregation forces measured *in vitro* by optical trapping for patients suffering from AH and those suffering from CHD+T2DM depending on the presence or absence of aggregates in the capillaries, capillary blood velocity, and the presence or absence of stasis in blood flow detected by VDC. (This figure has been adapted from Ref. [71].)

the formation of blood aggregates and that the presence of aggregates in the capillary bed may lead to the appearance of stasis.

Figure 3.16 shows the aggregation forces for experimental groups and subgroups. While measuring the aggregation force with LT and relating it to the microcirculation indices obtained by VDC, we found no significant differences between the values for the groups of patients suffering from AH and CHD, in the absence of blood aggregates. There is no indicated significant differences between the aggregation force and CBV for hypertensive patients and patients suffering from CHD. No significant differences for the forces were obtained in relation to the parameter of stasis measured by VDC.

The study has demonstrated that there is a high level of statistically significant differences between the information about the rheological properties of blood in the capillary bed that can be obtained with the VDC method *in vivo* and the information that can be obtained with the selected optical techniques, i.e., laser scattering aggregometry and LT, *in vitro*. The statistical results obtained for the comparative studies of blood rheological properties using the three different methods allow us to make conclusions about the applicability of these methods in clinical practice, in particular, while treating patients suffering from diabetes mellitus, cardiovascular diseases, AH, and CHD. For example, our previous results of comprehensive studies with optical trapping and laser aggregometry techniques performed in Refs [27,28] confirm the hypothesis of enhanced RBC aggregation and impairment of blood flow in patients suffering from hypertension and CHD in comparison with the normal values.

Blood rheology largely depends on the aggregation state of blood components, in particular RBCs and platelets. This work is focused on the characterization of the aggregation properties of the former cells. The activation and aggregation state of the platelets are known to affect blood rheology not only directly by raising blood viscosity and reducing its fluidity but also indirectly by influencing the aggregation of RBCs. However, this effect is beyond the scope of this work.

The need to evaluate the rheological properties of blood has increased significantly in recent years due to the widespread use of new oral anticoagulants and powerful antiplatelet agents influencing both platelets and RBCs. This application is aimed at the prevention of reocclusion of the coronary arteries because of stenting and bypass operations. With an insufficient dose of drugs, RBC aggregates continue to circulate in the blood, which greatly reduces the supply of tissues carrying oxygen and other important substances.

Another important aspect characterizing the capillary blood flow is the interaction of blood cells with the vascular wall. It should be remembered that capillaries are living tubes consisting of endothelial cells, endotheliocytes, which, in addition to enabling blood transport, play the role of a kind of paracrine organ, since nitrogen monoxide, endothelin-1, and other vasoactive substances directly and indirectly affect blood aggregation properties.

RBCs interact with endotheliocytes, glycocalyx, and blood plasma, which contain a large set of various chemical elements circulating in the bloodstream. Most of these factors are absent when conducting measurements with an optical trap and with a RheoScan aggregometer. In addition, substance stabilizers such as ethylenediaminetetraacetic acid are added to the blood sample to prevent blood from coagulating during the test.

One of the advantages of the VDC method is its non-invasiveness, in other words, the absence of any unpleasant sensation in patients during the study. Among the limitations of this method is its inability to detect *in vivo* the interaction of individual blood cells in the flow and to reveal the strength of RBC aggregates, as we do with an optical trap or a RheoScan aggregometer. However, in future, combining VDC with other types of laser microscopies and, possibly, with optical tweezers may make it possible to overcome these limitations.

3.5 SUMMARY

In this work, a series of *in vitro* and *in vivo* measurements of RBC aggregation parameters with blood samples obtained from patients suffering from T2DM and other pathologies (i.e., hypertension and CHDs) were performed. Aggregation index and the characteristic time of aggregate formation were measured using the laser aggregometry technique in whole-blood samples comprising large populations of RBCs. In addition, forces and time of aggregation on individual cell level were obtained with an optical trap. CBV, presence or absence of RBC aggregates, and stasis in blood flow were assessed *in vivo* by vital digital capillaroscopy. The relationships between the parameters measured *in vitro* demonstrate good agreement between the results obtained for patients distinguished within the subgroups in accordance with the VDC data obtained *in vivo*. Impairment of the capillary blood flow in T2DM leads to unconditional deterioration of aggregation parameters of the blood cell.

Diffuse light scattering ensures the measurement of aggregation parameters on large ensembles of cells and is also sensitive to changes in these parameters in T2DM. Moreover, in T2DM, RBCs have an increased ability to aggregate compared to the norm, while the deformability of RBCs does not change significantly. It was shown that the interaction forces of RBCs during aggregation in T2DM are significantly greater than in normal conditions. The forces arising from the separation of aggregates in T2DM do not statistically significantly differ from those of the control, while the disaggregation force is greater than the aggregation force in all groups. This statistically significant difference made it possible to introduce a new parameter for quantifying the interaction of RBCs during aggregation and disaggregation, which is promising for use in the diagnosis and monitoring of the effectiveness of therapy for both DM and other vascular pathologies.

Optical trapping systems have shown themselves to be a convenient and promising tool for conducting studies at the level of individual cells, allowing one to track the kinetics of RBC interactions during aggregation and disaggregation and measure the forces that arise during these processes with high accuracy. Alterations of RBC aggregation parameters measured *in vitro* can be used to evaluate the alterations of vital capillary blood flow parameters in the human body.

ACKNOWLEDGMENTS

This work was financially supported by the Russian Scientific Foundation (Grant No. 20-45-08004). This work was performed according to the Development Program of the Interdisciplinary Scientific and Educational School of Lomonosov Moscow State University «Photonic and Quantum technologies. Digital medicine».

REFERENCES

1. M.P. Czech, "Insulin action and resistance in obesity and type 2 diabetes," *Nat. Med.* **23**(7), 804–814 (2017).
2. V.P. Singh, A. Bali, N. Singh, and A.S. Jaggi, "Advanced glycation end products and diabetic complications," *Korean J. Physiol. Pharmacol.* **18**(1), 1–14 (2014).
3. O.A. Gerasimenko, "End products of excessive glycosylation as a potential target for "turning off" metabolic memory". *Endocrinology* №4. (2011) (in Russian).
4. D. Vashishth, "Advanced glycation end-products and bone fractures," *IBMS boneKEy* **6**(8), 268–278 (2009).
5. R. Testa, A.R. Bonfigli, F. Prattichizzo, L. La Sala, V. De Nigris, and A. Ceriello, "The "Metabolic Memory" theory and the early treatment of hyperglycemia in prevention of diabetic complications," *Nutrients* **9**(5), 437 (2017).
6. C. García-Pastor, S. Benito-Martínez, R.J. Bosch, et al., "Intracellular prostaglandin E2 contributes to hypoxia-induced proximal tubular cell death," *Sci. Rep.* **11**, 7047 (2021).
7. I. Juhan, P. Vague, M. Buonocore, J.P. Moulin, M.F. Calas, B. Vialettes, and J.J. Verdot "Effects of insulin on RBC deformability in diabetics--relationship between RBC deformability and platelet aggregation," *Scand. J. Clin. Lab. Investig. Suppl.* **156**, 159–164 (1981).
8. C. Le Dévéhat, M. Vimeux, and T. Khodabandehlou "Blood rheology in patients with diabetes mellitus,". *Clin. Hemorheol. Microcirc.* **30**, 297–300 (2004).
9. H. Schmid-Schönbein and F. Volger, "Red-cell aggregation and red-cell deformability in diabetes," *Diabetes* **25**, 897–902 (1976).
10. I. Vermes, E.T. Steinmetz, L.J. Zeyen, and E.A. van der Veen, "Rheological properties of white blood cells are changed in diabetic patients with microvascular complications," *Diabetologia* **30**(6), 434–436 (1987).
11. R.E. Wells Jr. and E.W. Merrill, "Influence of flow properties of blood upon viscosity-hematocrit relationships," *J. Clin. Investig.* **8**, 1591–1598 (1962).
12. T.W. Secomb and A.R. Pries, "Blood viscosity in microvessels: Experiment and theory," *Comptes Rendus Physique* **14**(6), 470–478 (2013).
13. I. Cicha, Y. Suzuki, N. Tateishi, and N. Maeda, "Changes of RBC aggregation in oxygenation-deoxygenation: pH dependency and cell morphology," *Am. J. Physiol. Heart Circ. Physiol.* **284**(6), H2335–H2342 (2003).
14. N. Elie, S. Sarah, R. Marc, F. Romain, L. Nathalie, G. Nicolas, G. Alexandra, A.J. Sophie, R. Céline, H.D. Marie-Dominique, S. Emeric, J. Philippe, B. Yves, and C. Philippe, "Blood rheology: Key parameters, impact on blood flow, role in sickle cell disease and effects of exercise," *Front. Physiol.* **10**, 1–12 (2019).
15. F. Skovborg, A.V. Nielsen, J. Schlichtkrull, and J. Ditzel, "Blood-viscosity in diabetic patients," *Lancet* **1**(7429), 129–131 (1966).
16. L. Dintenfass, "Blood viscosity factors in severe nondiabetic and diabetic retinopathy," *Biorheology* **14**(4), 151–7 (1977).
17. Y.I. Cho, M.P. Mooney, and D.J. Cho, "Hemorheological disorders in diabetes mellitus," *J. Diabetes Sci. Technol.* **2**(6), 1130–1138 (2006).
18. A.J. Barnes, P. Locke, T.L. Dormandy, and J.A. Dormandy, "Blood viscosity and metabolic control in diabetes mellitus," *Clin. Sci. Mol. Med.* **52**, 24–25 (1977).
19. H.G. Grigoleit, F. Lehrach, and R. Muller, "Diabetic angiopathy and blood viscosity," *Acta Diabetologica Latina* **10**, 1311–1324 (1973).

20. B. Chmiel and L. Cierpka, "Organ preservation solutions impair deformability of RBCs in vitro," *Transplant. Proc.* **35**(6), 2163–2164 (2003).
21. O. Linderkamp, J. Pöschl, and P. Ruef, "Blood cell deformation in neonates who have sepsis," *Neo Rev.* **7**, e517–e523 (2006).
22. H. Schmidt-Schönbein and P. Teitel, "In vitro assessment of "covertly" abnormal blood rheology: Critical appraisal of presently available microrheological methodology. A review focusing on diabetic retinopathy as a possible consequence of rheological occlusion," *Clin. Hemorheol. Microcirc.* **7**, 203–238 (1987).
23. J. Ditzel, "Haemorheological factors in the development of diabetic microangiopathy," *Br. J. Ophthalmol.* **51**(12), 793–803 (1967).
24. G.D. Lowe, J.M. Lowe, M.M. Drummond, S. Reith, J.J. Belch, C.M. Kesson, A. Wylie, W.S. Foulds, C.D. Forbes, A.C. MacCuish, and W.G. Manderson, "Blood viscosity in young male diabetics with and without retinopathy," *Diabetologia* **18**(5), 359–363 (1980).
25. G. Cicco and A. Pirrelli, "Red blood cell (RBC) deformability, RBC aggregability and tissue oxygenation in hypertension," *Clin. Hemorheol Microcirc.* **21**(3–4), 169–177 (1999).
26. O. Baskurt, B. Neu, and H. Meiselman, *Red Blood Cell Aggregation*, CRC Press, Boca Raton, FL (2012).
27. K. Lee, C. Wagner, and A. Priezzhev "Assessment of the "cross-bridge"-induced interaction of red blood cells by optical trapping combined with microfluidics," *J. Biomed. Optics* **22**, 091516 (2017).
28. S. Chien and K. Jan, "Red cell aggregation by macromolecules: Roles of surface adsorption and electrostatic repulsion," *J. Supramol. Struc.* **1**, 385–409 (1973).
29. S. Asakura and F. Oosawa, "On interaction between two bodies immersed in a solution of macromolecules," *J. Chem. Phys.* **22**, 1255–1256 (1954).
30. H.J. Meiselman, "Red blood cell role in RBC aggregation: 1963–1993 and beyond," *Clin. Hemorheol.* **13**, 575–592 (1993).
31. T.V. Korotaeva, N.N. Firsov, A. Bjelle, and M.A. Vishlova, "RBCs aggregation in healthy donors at native and standard hematocrit: The influence of sex, age, immunoglobulins and fibrinogen concentrations," Standardization of Parameters. *Clin. Hemorheol. Microcirc.* **36**, 335–343 (2007).
32. R. Bateman and C. Ellis, "Sepsis impairs microvascular autoregulation and delays capillary response within hypoxic capillaries," *Crit. Care* **5**, 19–389 (2015).
33. M. Martínez, A. Vayá, R. Server, A. Gilsanz, and J. Aznar, "Alterations in RBC aggregability in diabetics: The influence of plasmatic fibrinogen and phospholipids of the red blood cell membrane," *Clin. Hemorheol. Microcirc.* **18**, 253–258 (1998).
34. B. Chong-Martinez, T.A. Buchanan, R.B. Wenby, and H.J. Meiselman, "Decreased red blood cell aggregation subsequent to improved glycaemic control in type 2 diabetes mellitus," *Diabet. Med.* **20**, 301–306 (2003).
35. L. Coppola, G. Verrazzo, C. LaMarca, P. Ziccardi, A. Tirelli, and D. Giugliano, "Effect of insulin on blood rheology in non-diabetic subjects and in patients with type 2 diabetes mellitus," *Diabetes Med.* **14**, 959–963 (1997).
36. Y. Budak, H. Demirci, M. Akdogan, and D. Yavuz, "RBC membrane anionic charge in type 2 diabetic patients with retinopathy," *BMC Ophthalmol.* **4**, 10–14 (2004).
37. J. Wautier, "Blood cells and vascular cell interaction and in diabetes," *Clin. Hemorheol. Microcirc.* **25**, 49–53 (2001).
38. C. Foscat, P. Bronchi, P. Vague, M. Mirshahi, J. Saria, and S. Juhan-Vague, "The effect of plasma proteins and glycosylated fibrinogen on red blood cells aggregation in diabetes," *Clin. Hemorheol.* **11**, 630–649 (1991).
39. R. Barazzoni, E. Kiwanuka, M. Zanetti, M. Cristini, M. Vettore, and P. Tessari, "Insulin acutely increases fibrinogen production in individuals with type 2 diabetes but not in individuals without diabetes," *Diabetes* **52**, 18–51 (2003).
40. O. Ziegler, B. Guerci, S. Muller, H. Candiloros, L. Méjean, M. Donner, J.F. Stoltz, and P. Drouin, "Increased RBC aggregation in insulin-dependent diabetes mellitus and its relationship to plasma factors: A multivariate analysis," *Metabolism* **43**(9), 1182–1186 (1994).
41. K. Elishkevitz, R. Fusman, M. Koffler, I. Shapira, D. Zeltser, D. Avitzour, N. Arber, S. Berliner, and R. Rotstein, "Rheological determinants of red blood cell aggregation in diabetic patients in relation to their metabolic control," *Diabetes Med.* **19**(2), 152–6 (2002).
42. M. Mantskava, N. Momtselidze, N. Pargalava, and G. Mchedlishvili, "Hemorheological disorders in patients with type 1 or 2 diabetes mellitus and foot gangrene," *Clin. Hemorheol. Microcirc.* **35**, 307–310 (2006).

43. C. Le Devehat, T. Khodabandehlou, and M. Vimeux, "Impaired hemorheological properties in diabetic patients with lower limb arterial ischaemia," *Clin. Hemorheol.* **25**, 43 (2001).
44. C. Sun and L.L. Munn, "Influence of RBC aggregation on leukocyte margination in postcapillary expansions: A lattice Boltzmann analysis," *Phys. A* **362**(1), 191–196 (2006).
45. A. Symeonidis, G. Athanassiou, A. Psiroyannis, V. Kyriazopoulou, K. Kapatais-Zoumbos, Y. Missirlis, and N. Zoumbos, "Impairment of RBC viscoelasticity is correlated with levels of glycosylated hemoglobin in diabetic patients," *Clin. Lab. Haematol.* **23**, 103 (2001).
46. B. Schauf, U. Lang, P. Stute, S. Schneider, K. Dietz, B. Aydeniz, and D. Wallwiener, "Reduced red blood cell deformability, an indicator for high fetal or maternal risk, is found in preeclampsia and IUGR," *Hypertens. Pregnancy* **21**(2), 147–160 (2002).
47. R.M. Bauersachs, S.J. Shaw, A. Zeidler, and H.J. Meiselmann, "Red blood cell aggregation and blood viscoelasticity in poorly controlled type 2 diabetes mellitus," *Clin. Hemorheol.* **9**, 935–952 (1989).
48. A. Pérez-Martin, M. Dumortier, E. Pierrisnard, E. Raynaud, J. Mercier, and J.F. Brun, "Multivariate analysis of relationships between insulin sensitivity and blood rheology: Is plasma viscosity a marker of insulin resistance?" *Clin. Hemorheol. Microcircu.* **25**(3–4), 91–103 (2001).
49. S.H. James and A.M. Meyers, "Microangiopathic hemolytic anemia as a complication of diabetes mellitus," *Am. J. Med. Sci. Lung Transplant* **315**, 211 (1998).
50. M. Bessis and N. Mohandas, "La déformabilití des globules rouges, intérêt de sa mesure en clinique" (Red cell deformability, importance of its measurement in clinical medicine), *Schweiz. Med. Wochenschr.* **105**(47), 1568–1570 (1975) (in French).
51. H. Schmid-Schönbein, R.E. Wells, and R. Schildkraut, "Microscopy and viscometry of blood flowing under uniform shear rate (rheoscopy)," *J Appl. Physiol.* **26**, 674–684 (1969).
52. H. Zhao, X. Wang, M. Gentils, G. Cauchois, and J.F. Stoltz, "Evaluation of RBC deformability by two laser light scattering methods," *Clin. Hemorheol. Microcirc.* **20**, 241 (1999).
53. R.S. Schwartz, J.W. Madsen, A.C. Rybicki, and R.L. Nagel, "Oxidation of spectrin and deformability defects in diabetic RBCs," *Diabetes* **40**, 701 (1991).
54. N. Babu and M. Singh, "Influence of hyperglycemia on aggregation, deformability and shape parameters of RBCs," *Clin. Hemorheol. Microcirc.* **31**, 273 (2004).
55. K. Tsukada, E. Sekizuka, C. Oshio, and H. Minamitani, "Direct measurement of RBC deformability in diabetes mellitus with a transparent micro- channel capillary model and high-speed video camera system," *Microvasc. Res.* **61**, 231 (2001).
56. D. Bareford, A.E. Jennings, P.C.W. Stone, S. Baer, A.H. Barnett, and J. Stuart, "Effects of hyperglycaemia and sorbitol accumulation on RBC deformability in diabetes mellitus," *J. Clin. Pathol.* **39**, 722 (1986).
57. J.M. Leiper, G.D. Lowe, J. Anderson, P. Burns, H.N. Cohen, W.G. Manderson, C.D. Forbes, J.C. Barbenel, and A.C. MacCuish, "Effects of diabetic control and biosynthetic human insulin on blood rheology in established diabetics," *Diabetes Res.* **1**(1), 27–30 (1984).
58. M.J. Hilz, H. Marthol, and B. Neundorfer, "Diabetic somatic polyneuropathy. Pathogenesis, clinical manifestations and therapeutic concepts," *Fortschr Neurol.* **68**(6), 278–288 (2000) (in German).
59. N. Babu and M. Singh, "Influence of hyperglycemia on aggregation, deformability and shape parameters of RBCs," *Clin. Hemorheol. Microcirc.* **31**(4), 273–280 (2004).
60. H. Lawall and B. Angelkort, "Correlation between rheological parameters and RBC velocity in nailfold capillaries in patients with diabetes mellitus," *Clin. Hemorheol. Microcirc.* **20**, 41 (1999).
61. O.K. Baskurt, M. Boynard, G.C. Cokelet, P. Connes, B.M. Cooke, S. Forconi, F. Liao, M.R. Hardeman, F. Jung, H.J. Meiselman, et al., "New guidelines for hemorheological laboratory techniques," *Clin. Hemorheol. Microcirc.* **42**, 75–97 (2009).
62. M.R. Hardeman, J.G.G. Dobbe, and C. Ince, "The laser-assisted optical rotational cell analyzer (LORCA) as red blood cell aggregometer," *Clin. Hemorheol. Microcirc.* **25**(1), 1–11 (2001).
63. V.N. Lopatin, A.V. Priezzhev, A.D. Aponasenko, N.V. Shepelevich, P.V. Pozhilenkova, and I.V. Prostakova, *Methods of Light Scattering in the Analysis of Dispersed Biological Media*, Fizmatlit, Moscow (2004) (book in Russian).
64. A.V. Priezzhev, K. Lee, N.N. Firsov, J. Lademann, "Optical study of RBC aggregation in whole blood samples and single cells," Chapter 1 in *Handbook of Optical Biomedical Diagnostics*, Volume 2: Methods, V.V. Tuchin, Ed., 2nd ed., SPIE Press, Bellingham (2016).
65. S. Shin, Y. Yang, and J.S. Suh, "Measurement of RBC aggregation in a microchip-based stirring system by light transmission," *Clin. Hemorheol. Microcirc.* **41**, 197–207 (2009).
66. H. Kiesewetter, H. Radtke, R. Schneider, K. Mussler, A. Scheffler, and H. Schmid-Schonbein, "The mini RBC aggregometer: A new apparatus for the rapid quantification of the extent of RBC aggregation," *Biomed. Technol.* **27**(9), 209–213 (1982).

67. P. Steffen, C. Verdier, and C. Wagner, "Quantification of depletion-induced adhesion of red blood cells," *Phys. Rev. Lett.* **110**, 018102-1–018102-5 (2013).
68. K. Buxbaum, E. Evans, and D.E. Brooks, "Quantitation of surface affinities of red blood cells in dextran solutions and plasma," *Biochemistry* **21**, 3235–3239 (1982).
69. K. Lee, M. Kinnunen, M.D. Khokhlova, E.V. Lyubin, A.V. Priezzhev, I. Meglinski, and A. Fedyanin, "Optical tweezers study of red blood cell aggregation and disaggregation in plasma and protein solutions," *J Biomed. Opt.* **21**(3), 035001 (2016).
70. D.C. Silva, C.N. Jovino, C.A. Silva, H.P. Fernandes, M. Milton Filho, S.C. Lucena, A.M. Costa, C.L. Cesar, M.L. Barjas-Castro, B.S. Santos, and A. Fontes, "Optical tweezers as a new biomedical tool to measure zeta potential of stored red blood cells," *PLoS One* **7**(2), e31778 (2012).
71. A.E. Lugovtsov, Y.I. Gurfinkel, P.B. Ermolinskiy, A.I. Maslyanitsina, L.I. Dyachuk, and A.V. Priezzhev, "Optical assessment of alterations of microrheologic and microcirculation parameters in cardiovascular diseases," *Biomed. Opt. Express* **10**(8), 3974–3986 (2019).
72. H.B. Mann and D.R. Whitney, "On a test of whether one of two random variables is stochastically larger than the other," *Ann. Math. Stat.* **18**(1), 50–60 (1947).
73. S. Shin, Y. Yang, and J. Suh, "Measurement of erythrocyte aggregation in a microchip stirring system by light transmission," *Clin. Hemorheol. Microcirc.* **41**, 197–207 (2009).
74. S. Shin, J. Hou, and J. Suh "Measurement of cell aggregation characteristics by analysis of laser-backscattering in a microfluidic rheometry," *Korea-Australia Rheol. J.* **19**(2), 61–66 (2007).
75. P. Ermolinskiy, A. Lugovtsov, A. Maslyanitsina, A. Semenov, L. Dyachuk, and A. Priezzhev, "In vitro assessment of microrheological properties of erythrocytes in norm and pathology with optical methods," *Ser. Biomech.* **32**, 20–25 (2018).
76. P.B. Ermolinskiy, A.E. Lugovtsov, A.I. Maslyanitsina, A.N. Semenov, L.I. Dyachuk, and A.V. Priezzhev, "Interaction of erythrocytes in the process of pair aggregation in blood samples from patients with arterial hypertension and healthy donors: Measurements with laser tweezers," *J. Biomed. Photon. Eng.* **4**(3), 030303-1–030303-8 (2018).
77. A.Yu. Maklygin, A.V. Priezzhev, A.V. Karmenyan, S.Yu. Nikitin, I.S. Obolensky, A.E. Lugovtsov, and K. Li, "Measuring the force of interaction between erythrocytes in the aggregate using laser tweezers," *Quantum Electron.* **42**(6), 500–504 (2012) (in Russian).
78. K. Svoboda and S. Block, "Biological applications of optical forces," *Ann. Rev. Biophys. Biomol. Struct.* **23**, 247–285 (1994).
79. I. Krasnikov, A. Seteikin, and I. Bernhardt, "Thermal processes in red blood cells exposed to infrared laser tweezers," *J Biophoton.* **4**(3), 206–212 (2011).
80. K. Lee, A. Priezzhev, S. Shin, F. Yaya, and I. Meglinski, "Characterization of shear stress preventing red blood cells aggregation at the individual cell level: The temperature dependence," *Clin. Hemorheol. Microcirc.* **64**, 853–857 (2016).
81. K. Lee, M. Kinnunen, A.V. Danilina, V.D. Ustinov, S. Shin, I. Meglinski, and A.V. Priezzhev, "Characterization at the individual cell level and in whole blood samples of shear stress preventing red blood cells aggregation," *J. Biomech.* **3**(7), 1021–1026 (2016).
82. O. Baskurt, M. Hardeman, M. Uyuklu, P. Ulker, M. Cengiz, N. Nemeth, S. Shin, T. Alexy, and H. Meiselman "Comparison of three commercially available ektacytometers with different shearing geometries," *Biorheology* **46**, 251–264 (2009).
83. S. Shin, "Validation and application of a microfluidic ektacytometer (RheoScan-D) in measuring erythrocyte deformability" *Clin. Hemorheol. Microcirc.* **37**(4), 319–28 (2007).
84. H. Lee, S.B. Ahn, C.W. Moon, J.S. Won, and S. Shin, "Potential diagnostic hemorheological indexes for chronic kidney disease in patients with type 2 diabetes," *Front. Physiol.* **10**, 1062 (2019).
85. M. Etehad Tavakol, A. Fatemi, A. Karbalaie Emrani, and B.E. Erlandsson, "Nailfold capillaroscopy in rheumatic diseases: Which parameters should be evaluated?" *BioMed Res. Int.* **2015**, 974530 (2015).
86. Y.I. Gurfinkel, A.V. Priezzhev, M.L. Sasonko, and M.I. Kuznetzov, "Importance of image processing in digital optical capillaroscopy for early diagnostics of arterial hypertension," *BioPhotonics -2015 International Conference Proceedings*, Florence, Italy, 20–22, 1–4 (2015).
87. A.N. Semenov, A.E. Lugovtsov, K. Li, A.A. Fabrichnova, Yu.A. Kovaleva, and A.V. Priezzhev, "The use of diffuse light scattering and optical trapping methods to study the rheological properties of blood` Erythrocyte aggregation in diabetes mellitus," *Bull. Saratov Univ. New Episode Ser. Phys.* **17**(2), 85–97 (2017) (in Russian).
88. A. Maslianitsyna, P. Ermolinskiy, A. Lugovtsov, A. Pigurenko, M. Sasonko, Y. Gurfinkel, and A. Priezzhev, "Multimodal diagnostics of microrheologic alterations in blood of coronary heart disease and diabetic patients," *Diagnostics* **11**(1), 76 (2021).

4 Diagnostics of Functional Abnormalities in the Microcirculation System Using Laser Doppler Flowmetry

Irina A. Mizeva, Elena V. Potapova, and Elena V. Zharkikh

CONTENTS

4.1 CUTANEOUS CIRCULATION AS A MODEL OF GENERALIZED MICROVASCULAR FUNCTION

The key issue in developing modern diagnostic techniques lies in how to guarantee their non-invasiveness, in other words, how to conduct a study without violating tissue integrity. Since the skin is the most accessible organ of the human body, it has been hypothesized that the cutaneous circulation can be identified as a model of generalized microvascular function [1–3].

This hypothesis is supported by experimental evidence that the vascular reactivities of different segments of the vascular bed (e.g., coronary arteries, brachial artery, and cutaneous microcirculation) in healthy people correlate with those in patients with coronary artery disease, at least in terms of the endothelial function [4]. The diagnostic value of the skin microvascular endothelial function for the early detection and treatment of cardiovascular diseases was described in detail [5]. Microcirculation abnormalities were observed in different clinical events (asthma [6], obstructive

DOI: 10.1201/9781003112099-4

apnea [7], and rheumatoid arthritis [8]). Chronic inflammation is a common feature of these diseases, and the dysfunction of the blood microcirculation system correlates with the biomarkers of systemic inflammation [9,10]. In acute inflammatory conditions, sepsis peripheral microcirculation can be used as a marker of systemic microvascular dysfunction [11]. It has been established that there is a relationship between cutaneous microvascular reactivity and clinical complications, risk factors, or signs of cardiometabolic disorders [12].

Diabetes mellitus (DM) significantly influences the microcirculation in different parts of the human body: renal, eyes, skin, etc. The skin is the most accessible organ and is preferred for the microcirculation assessment [13,14]. Clinical observations show that a constant high blood sugar level leads to the damage of vessels and nerve fibers, which leads to microcirculation system abnormalities that occur at the preclinical stages of DM [15] and correlates with the stages of diabetic microangiopathy [16].

Thus, skin microcirculation studies in patients can help to describe the main pathophysiological mechanisms, or they can be used as prognostic markers or markers of response to treatment. Despite the large amount of data, this problem has not been solved to date [17] due to the substantial heterogeneity of the methods used to assess microvascular disorders, the difference in patient populations, the significant variability of physiological responses [18], and the lack of elaboration and unification of research protocols.

A comprehensive analysis of the current state of research in the field of skin microcirculation is given in Ref. [17] and a brief overview is provided here. Normal skin microcirculation is organized in two plexuses parallel to the skin surface [19,20]. The superior plexus, located in the papillary dermis, plays an important role in skin nutrition (Figure 4.1). It consists of small arterioles and venules, with an outer diameter of approximately 20 μm, with capillary loops that extend perpendicularly to the surface of the skin [21], located within 1 mm of the surface. The lower plexus is located on the skin–hypodermal border and consists of arteries and veins that exit from the underlying muscle and adipose tissue and perforate the fascia to form ascending arterioles and descending venules connected to the superficial plexus [22]. The ascending arterioles are randomly arranged, but after about every 1.5 mm, they branch out to form microvascular networks around the sweat glands and hair

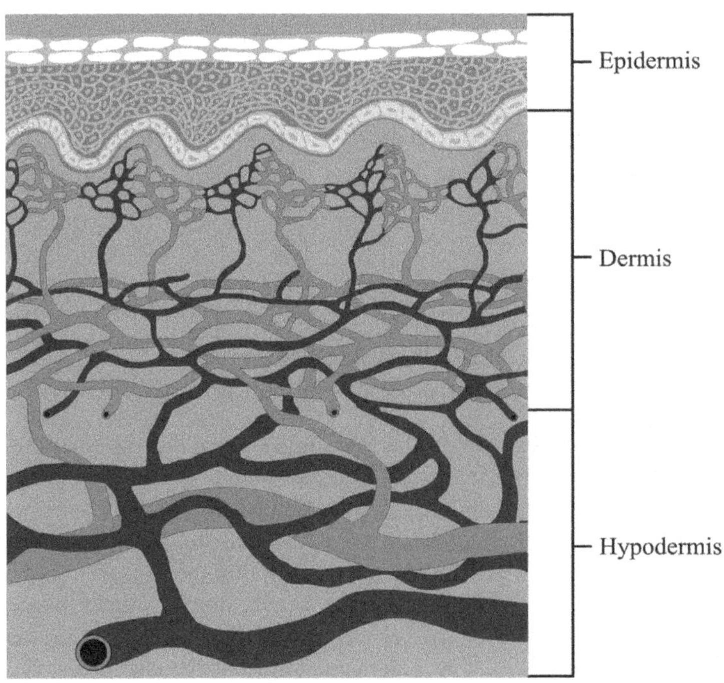

FIGURE 4.1 Architecture of the microvascular bed.

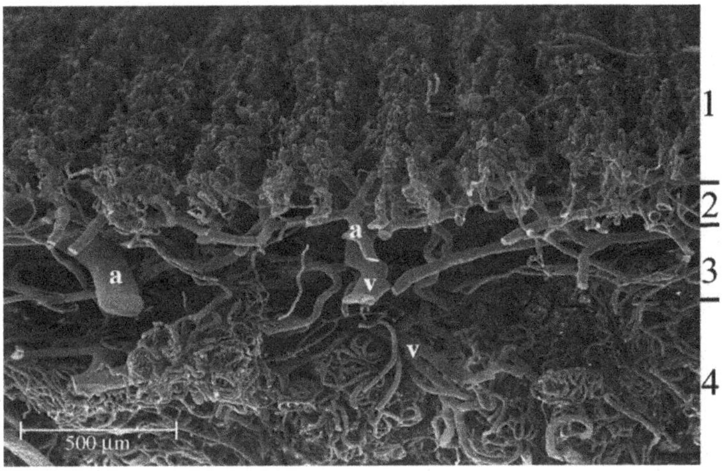

FIGURE 4.2 A corrosive preparation obtained by filling the vascular bed of the palmar surface of the skin of a surgically removed finger of an 18-year-old man with a solidifying filler. In the picture, one can distinguish the papillary layer from which the capillary loops arise (1), the subpapillary plexus (2), the reticular layer (3), and the hypodermal layer (4) together with the ascending arterioles (a) and the descending venules (v). Regular linear arrangement of capillary loops repeats the pattern of fingerprint lines. (Reproduced from Ref. [23] with the permission of John Wiley and Sons.)

follicles. Highly innervated arterioles control blood flow through papillary loops and, as a consequence, actively participate in controlling the nutritive blood flow. There are significant differences in the structure of glabrous (skin of the palms, feet, and lips) and nonglabrous skin. The glabrous skin is characterized by a lack of hair and contains a large amount of arteriovenous anastomosis (AVAs), which have a smaller surface area and lie deeper in the dermis than papillary loops (Figure 4.2).

4.2 FUNCTIONS OF THE BLOOD MICROCIRCULATION SYSTEM

4.2.1 Transport

The main function of the microcirculation is to transport blood and substances to and from tissues. When the blood moves along the capillary, a huge number of water molecules and dissolved particles diffuse in one direction and in the other through the capillary wall, ensuring constant mixing of the tissue fluid and plasma. The permeability of capillaries for various substances is different and depends on the size of their molecules and the type of capillaries themselves. One of the most important purposes of delivering oxygen to the tissues is achieved by blood transport. Cells are constantly in need of oxygen, which they receive by diffusion from the blood through the interstitial fluid. The primary function of the blood is to deliver oxygen to tissues. The circulatory system pumps oxygen-rich blood through the capillary network. The microcirculatory system is organized in such a way that all cells have at least one adjacent capillary for diffusive oxygen exchange, and red blood cells deliver oxygen to the tissues as they pass through the capillaries [24].

Furthermore, microcirculation is essential for delivering glucose, amino acids, fatty acids, and other nutrients to tissues, removing metabolic products from tissues, maintaining optimal concentrations of various ions in tissues, and releasing hormones.

4.2.2 Exchange of Nutrients between Blood and Tissue

Peripheral microvessels form extensive networks with small cross-sectional areas and large luminal surface areas. An exchange of various substances between the blood and the surrounding interstitium, the migration of white blood cells into the tissue [25–28] occurs in these vessels. In addition to the delivery of nutrients to tissues and the removal of decay products, microcirculation plays an important role in the exchange of fluid between the blood and the tissue [29,30], the delivery of hormones from the endocrine organs directly to the tissues.

4.2.3 Pressure Regulation

Arterioles have an increased contractile capacity. The Poiseuille equation states that the flow rate in the tube is proportional to the 4th degree of the radius of the tube section for an ideal fluid, and therefore changes in vasoconstriction and vasodilation substantially affect the blood flow. The largest pressure drop occurs in arterioles, for example in skeletal muscles, which are responsible for 50%–60% of the changes in vascular pressure and, as a result, for the resistivity of the vascular bed [31]. The capillary link usually accounts for no more than 15% of the pressure drop because of the changes in the number of functioning capillaries.

4.2.4 Thermoregulation

In the neutral state, the skin temperature is about 33 °C, but it varies in a wide range. The main regulator of heat exchange with the environment is papillary loops, which are located in the immediate vicinity of the dermoepidermal junction [19]. Even though the volume of capillaries is small, their surface area is large compared to other skin vessels, and the blood flow rate is quite small, which contributes to heat exchange. To a lesser extent, the vessels of the upper venous plexus, lying parallel to the surface of the skin, are involved in thermoregulation. These vessels are less efficient in terms of heat transfer than capillary loops, but they provide a useful backup mechanism when heat transfer through capillary loops is overloaded. AVAs lying deeper in the dermis are considered less effective for thermoregulation. When the body is close to thermoneutral conditions, AVAs play a large role in temperature control, as the body temperature is controlled exclusively by skin blood flow without changes in metabolism.

4.2.5 Vascular Autoregulation

In the cutaneous vascular system, the diameter of arterioles and blood pressure within them are two major factors affecting the changes in blood flow. Maintaining a constant blood flow despite an increase in blood pressure is called autoregulation of blood flow and usually occurs at pressures between 70 and 170 mm Hg. Both the metabolic state of the tissue (e.g., hypoxia) and the myogenic response prevent the blood vessel walls from excessive stretching under high pressure, thus providing constant tension in the vessel wall.

The forces exerted by the flowing blood on the vessel wall can be decomposed into two main components: the pressure normal to the vessel wall, or transmural pressure, and the shear stress applied directly parallel to the vessel surface to create a frictional force in the endothelium. Shear stress is a viscous force that the blood flow exerts on the vessel wall and is expressed in force-area unit (typically dyn/cm^2). The magnitude of the force acting on the wall can be estimated as the product of blood flow velocity and local blood viscosity divided by the transverse size of the vessel. Shear stress is inversely proportional to the third power of the vessel radius. For the high flow in a wide tube (e.g., in the aorta) and for the low flow in a narrow tube (e.g., in capillaries), this magnitude falls within the same range.

Autoregulation maintains a constant wall tension despite an increase in the transmural pressure. The sensitivity of the smooth muscle tissue to pressure-induced changes in tensile stress is known as the Bayliss effect or myogenic response [32,33]. This phenomenon involves the stretch-induced activation of non-selective cation channels in myocytes [34]. There is also a regulatory mechanism realized by means of the vegetative nervous system via the vascular wall baroreceptors, which are sensitive to the transmural pressure [35].

Cutaneous arterioles are innervated by autonomic and sensory nerve fibers, which play an important role in regulating cutaneous blood flow in response to various stimuli, such as changes in local or ambient temperature, skin inflammation, or pressure. Reflex control is mediated by sympathetic mechanisms that include noradrenergic vasoconstriction, similar to that found in skeletal muscle, and a separate, non-adrenergic vasodilator mechanism found in thick skin [36]. In the blood circulation of the skin, spontaneous sympathetic activity has a vasoconstrictor effect [17]. Sympathetic activity is involved in reflex vasoconstriction after cooling the entire body when the skin temperature drops below 33 °C. The decrease in cutaneous blood flow due to general exposure to cold may be due to two different mechanisms: reflex effects of remote cooling of the skin associated with sympathetic vasoconstriction and local cooling effects.

Axon reflexes differ from the spinal cord reflexes in that they bypass integration centers or intermediate neurons to transmit a signal. Initially, they were considered unidirectional: from the cutaneous sensory nerve endings through the afferent axonal branch to the perivascular neuroeffector terminal network. However, current evidence suggests that individual nerve endings include both a sensory transducer and an acting mediator with a bidirectional nature of the reflex.

The luminal surface of the endothelium is covered with a well-developed glycocalyx, a thick layer consisting of macromolecules (i.e., proteins, glycolipids, glycoproteins, and proteoglycans) associated with endothelial membranes. Its molecular domains provide cellular adhesion (i.e., selectins and integrins involved in immune responses to inflammation), modulate coagulation and fibrinolysis by activating and/or inhibiting thrombin, tissue factor, and plasminogen, and control fluid and metabolic transport. A feature common to all endothelial cells studied in various vascular localities and cell cultures is their ability to respond to local changes in blood flow, in particular to acute and persistent changes in hemodynamic shear stresses that deform the cell [37,38]. The interaction between blood cells and the luminal surface of the vascular endothelium makes an important contribution to the shear stress and determines the activation of molecular pathways, in particular the synthesis of nitric oxide (NO) [39,40].

Microcirculation, in various organs and tissues, is ensured by the relationship between the regulatory mechanisms. There are factors of humoral (local and hormonal) and nervous regulation. The factors of humoral regulation include hormones that have an endocrine, mainly vasoconstrictor effect, and biologically active substances that act paracrine, mainly causing vasodilation. Both hormones and biologically active substances are more characterized by a non-periodic action associated with their secretion, that is, the effect of humoral factors on the basal tone of microvessels is more of an adaptive mechanism. The exception is the periodic secretory activity of the endothelium, which to a greater extent changes the basal tone in the direction of vasodilation. The factor of nervous regulation is the activity of the autonomic nervous system. It is believed that the adrenergic sympathetic fibers innervate AVAs, thus regulating the distribution of blood flow to the network of nutritive capillaries.

In contrast to the humoral regulation, sympathetic innervation plays a role both in maintaining basal tone periodic pulsation, of 1–3 pulses per second, and in adapting to the changing conditions of the external and internal environment by affecting the walls of arterioles, precapillary sphincters and venules. The sucking force of the chest and the pumping activity of the ventricles of the heart are considered as external regulatory factors in relation to the microcirculatory bed. The integral effect of the interaction of these factors is the basal tone of microvessels, which varies under different conditions.

4.3 OSCILLATIONS OF THE VASCULAR TONE

Oscillatory regimes have been observed in numerous biological systems [41] on many time scales [42,43]. Such fluctuations allow the system to adjust to the environmental conditions. The rhythmic activity provides temporary coordination between physiological processes through synchronization and involvement and guarantees an effective hierarchy between the individual parts of the system.

Periodic fluctuations in physiological characteristics are the result of the activity of autonomous oscillators located in the body. The heart is one such autonomous oscillator and is responsible for fluctuations in several physiological variables, including blood pressure and blood flow at the heart rate and its multiples. Autonomous oscillators interact with each other in such a way that their rhythmic activity can be modulated by the action of others. For example, heart rate is modulated by the respiratory centers located in the brainstem through the vagus innervation of the sinus node, causing changes in the cardiac period synchronous with the respiratory rate, which contributes to respiratory sinus arrhythmia [44]. The activity of these autonomous oscillators disrupts the mechanisms responsible for controlling physiological variables, and the resulting reaction can also manifest as periodic activity. For example, the heart-related rhythm of integrated muscular sympathetic nervous activity is mainly due to the periodic modification of central inhibition caused by the pulse-synchronous baroreceptor nervous activity caused by fluctuations in blood pressure [45]. The oscillatory behavior of physiological variables is a consequence not only of the continuous activity of internal rhythmic sources, but also of periodic external influences. Physiological fluctuations that occur at frequencies below the heart rate (i.e., below 0.5 Hz) in a person at rest on the back) are usually referred to as variability. This term has gained popularity with the spread of the analysis of variations in the interval between heartbeats on the electrocardiogram [46]. As a result of interaction, these rhythms manifest in various signals, such as electrocardiograms, blood pressure, blood flow, and heart rate variability [47].

When analyzing the signals related to blood circulation, physiological fluctuations with a frequency below 0.15 Hz are associated with vasomotions – variations in vascular tone. Vascular smooth muscle cells exhibit different types of calcium dynamics. A static vascular tone is associated with unsynchronized calcium waves, and the strength developed depends on the number of cells involved in the process. Global calcium transients, synchronized among a large number of cells, cause the rhythmic development of the process.

Vasomotions are attracting increasing interest of those who study the properties of hemodynamics of cardiac activity and blood flow by quantifying rhythmic changes in the diameter of blood vessels [48]. Simultaneous recording of electrical and hemodynamic activities in resting people revealed the presence of oscillating activity up to frequencies of 0.01 Hz and below, associated with myogenic, neurogenic, and endothelial activities and metabolic processes [49,50]. The contribution of the vascular system to the oscillatory behavior comes from the rhythmic activity of smooth muscle cells in the vascular walls, the rhythmic activation of sympathetic nerves innervating the vessels, and the involvement of endothelial cells lining the inner layer of the vascular walls.

Although the variability of the peripheral blood flow can be interpreted as a measurement error or as a harmless side effect of the application of the homeostatic principle aimed at maintaining the relative constancy of the internal environment of the body, its deeper significance was immediately recognized [51]. The studies devoted to heart rate variability [52,53] and blood circulation in the microcirculation system [48] made it possible to quantify the variability and proved that such variability carries information about regulatory mechanisms. The analysis of spontaneous variability became more important when it became clear that an abnormal or absent response could be used as a sign of a pathological condition, and the markers of variability were very sensitive in detecting dysfunction [46].

The specific features of the mechanisms responsible for low-frequency microcirculatory fluctuations were intensively discussed [54,55]. According to Ref. [54], central mechanisms [56] predominate in the generation of low-frequency microcirculatory oscillations. On the other hand, it has been established that the central baroreflective mechanism [57] synchronizes low-frequency oscillations throughout the cardiovascular system. The methods developed made it possible to assess the activity of certain mechanisms by analyzing the energy of oscillations in certain frequency ranges. However, although the spectral approach is rather simple, it has not yet found wide applications in clinical practice. The disadvantages of the spectral approach are the low reproducibility resulting from the heterogeneity of the microvascular bed and the need for a long monitoring time [58].

Spontaneous pulsations of blood flow in microvessels attract considerable attention in the context of biomedical research [59–61]. In the frequency range of 0.005–1.6 Hz, there are six frequency ranges [62–64] the nature of which, as well as the possible diagnostic value, is discussed in detail below.

4.3.1 Pulse Wave

The frequency range of heart contractions is 0.6–2 Hz (different sources indicate a different frequency range, e.g. 0.8–1.6 Hz in [65]). Blood flow fluctuations of this frequency range in the microvascular bed occur due to the propagation of a pulse wave of increased pressure through the aorta and arteries. This wave is caused by the release of blood from the left ventricle during the contraction of the heart muscle (systole phase) [66]. The velocity of the pulse wave propagation through the vessels does not depend on the velocity of the blood flow, but is determined by the elasticity and diameter of the vessel, the thickness of its wall, and blood density [67]. An increase in pulse wave amplitude at increased or normal perfusion indicates that the increased blood flow enters the vessels on the arterial side of the microcirculation [65].

4.3.2 Respiratory Modulation of Blood Flow

Cutaneous blood flow oscillations with frequencies of 0.145–0.6 Hz are associated with respiratory activity [66,68]. Prof. A.I. Krupatkin hypothesized that the main mechanism leading to a variation in the respiratory rhythm of the microcirculation is the arteriolo-venular pressure gradient [69]. It is indicated that its decrease leads to an increase in the amplitude of oscillations in the respiratory frequency range and indicates flow stagnation in venules [69]. This is also true for cardiovascular pathology [70,71]. Dysfunction of the venular vascular bed plays an important role in the pathogenesis of arterial hypertension.

A decrease in the blood outflow leaving the microcirculatory bed is followed by an increase in the volume of blood in the venules. This increases the laser Doppler flowmetry (LDF) signal reflected by red blood cells, which in turn increases the amplitude of the respiratory wave [65].

4.3.3 Myogenic Vascular Tone Oscillations

It is known that even isolated arterioles exhibit spontaneous oscillations with a frequency of around 0.1 Hz, which occur in the cross-sectional area due to the synchronous activity of myocytes [50,72]. However, some authors suggest that 0.1 Hz oscillations can originate in the cerebrovascular tone (vasomotion) [73]. One study predicted that myogenic oscillations with a frequency close to 0.1 Hz can be weakly synchronized at different points [47], and this was demonstrated experimentally during a local thermal test [74].

Blood flow oscillations in the frequency range 0.05–0.145 Hz are currently associated with myogenic pulsations [75]; minor variations in the range limits are given in some publications. However, owing to the ambiguous physiological nature, it is reasonable to distinguish two subranges of myogenic oscillations: 0.05–0.069 Hz [76] and 0.07–0.145 Hz. The first corresponds to sensory peptidergic fibers and the second to actual myogenic fluctuations or vasomotion. Experimental evidence indicates that myogenic oscillations of different frequencies may occur in the arterioles of different diameters. The vasomotion frequency in small-diameter blood vessels is higher than that in large arterioles [77]. By analyzing myogenic oscillations (range, 0.07–0.145 Hz), the state of the oscillatory component of the muscle tone of the precapillaries regulating blood flow to the nutritive bed is evaluated.

4.3.4 Neurogenic Oscillations

The range of neurogenic oscillations (0.021–0.052 Hz and 0.02–0.046 Hz as revealed in different studies) is associated (in the spectrum of cutaneous blood flow pulsations) with the activity of the sympathetic nervous system [78]. Measurement of the spectral energy of blood flow oscillations in this frequency band makes it possible to assess the sympathetic adrenergic regulation of arterioles and anastomoses.

4.3.5 Endothelial Oscillations

There are two frequency ranges of blood flow fluctuations associated with endothelial activity: the endothelial NO-dependent frequency range (0.0095–0.021 Hz) [79] and the NO-independent range (0.005–0.0095 Hz) [62,63]. Analysis of the correlation between metabolic and microhemodynamic processes in the skin [80] revealed that improvements in oxygen uptake and glucose metabolism lead to a significant increase in the amplitude of the endothelial rhythm. The relationship of vascular endothelial oscillations of the human skin with blood oxygen saturation has been established [81]. This makes it possible to assess endothelial dysfunction by analyzing changes in the amplitudes of oscillations in the NO-dependent frequency range.

The study of the heart rate variability [82] demonstrated that the endothelial, neurogenic, myogenic, and respiratory mechanisms are synchronized with cardiac activity. Information about changes in the individual microcirculation rhythms under different physiological and pathological conditions is systematized for clinical use [69]. The regulation of vascular tone occurs due to the influence of nervous, humoral, and local factors that affect both the average and oscillatory components of the local blood flow [83]. Thus, the characteristics of blood flow fluctuations in the microcirculatory link carry information not only about the physical properties of the vessel, but also about the speed and possibility of passing biochemical reactions, the state of the nervous system, and the synthesis of substances that affect the properties of blood vessels. Physiologically, these mechanisms are configured to meet the metabolic needs of the tissue and are related to the level of metabolism [84].

4.4 MORPHOLOGICAL AND FUNCTION ABNORMALITIES CAUSED BY DIABETES MELLITUS

The pathogenesis of DM is based on the aggressive action of hyperglycemia, which leads to the development of diabetic angiopathy, in small vessels (microangiopathy) and to medium- and large-caliber vessels (macroangiopathy). The effect of hyperglycemia on the microcirculatory bed and tissue metabolic processes is associated with the activation of biochemical reactions, which leads to changes in the properties of the vessel wall: non-enzymatic glycation of proteins; polyol pathway of glucose metabolism and sorbitol accumulation; activation of protein kinase C; and increased free radical formation and oxidative stress. One of the serious complications of DM that significantly reduce the quality of life is diabetic foot syndrome, which combines a number of foot lesions due to damage to nerves and blood vessels [85]. Timely diagnosis, monitoring, and treatment of complications reduce the severity of their manifestations and potentially prevent their further development [86].

Diabetes is a disease that affects the microvascular bed in the tissues of the entire body. Pathologies of the retina, kidneys, and nervous system that accompany diabetes occur in each case due to the microcirculatory disorders in relatively small areas of the body. Nevertheless, considering the skin as an organ, we are dealing with a large area of microvessel targets affected in DM, which, if not functional, can affect the overall metabolic function. Pathology-induced vascular dysfunctions (including endothelium-dependent vasodilation disorders) manifest in cutaneous blood circulation and may reflect general systemic vascular dysfunction in magnitude and underlying mechanisms [4,87].

Since the endothelium is in direct contact with the blood, it is sensitive to changes in the chemical composition of the blood, including hyperglycemia, which increases the risk of vascular complications. Vascular endothelial dysfunction is considered as an important factor not only in the occurrence of vascular complications of DM, but also in its progression and clinical consequences. The mechanism of endothelial dysfunction is based on changes in the structure of the endothelial receptor apparatus and its sensitivity to insulin, the effect of oxidative stress, activation of endothelial cells by anti-inflammatory factors, and mitochondrial dysfunction [88]. Violation of the ability of the endothelium to respond adequately to physiological stimuli eventually leads to an imbalance between the factors that provide local processes of regulation of hemostasis, proliferation, migration of blood cells into the vascular wall, and vascular tone. There is evidence that endothelial dysfunction may precede the development of DM, and therefore, in healthy individuals who do not suffer from DM and have the first degree of kinship with a patient with DM2, there is a violation of endothelium-dependent vasodilation as well as an increase in the level of plasma markers of endothelial dysfunction [89]. The occurrence of endothelial dysfunction before the development of DM2 suggests the presence of common pathophysiological mechanisms and a causal relationship between hyperglycemia and endothelial dysfunction. The association of DM with endothelial dysfunction is one of the factors of increased risk of cardiovascular diseases in patients with diabetes. Vascular endothelial dysfunction is considered as an important factor not only in the occurrence of vascular complications of DM, but also in its progression and clinical consequences. Its leading role in endothelial damage in DM is attributed to the influence of hyperglycemia. The mechanism of endothelial dysfunction is based on changes in the structure of the endothelial receptor apparatus and its sensitivity to insulin, the effect of oxidative stress, activation of endothelial cells by anti-inflammatory factors, and mitochondrial dysfunction.

An important role in the pathogenesis of the development of microangiopathic complications of DM is played by the deterioration of the rheological properties of the blood [90]. Hyperglycemia increases plasma and whole-blood viscosity, enhances erythrocyte aggregation, reduces the ability of red blood cells to undergo a wide range of deformations, and affects hemoglobin and membrane proteins in red blood cells, which correlates with reduced membrane fluidity.

For DM, the increased tortuosity of capillaries is typical. It is assumed that such changes occur as a compensation of the increase in the diffuse surface in order that the quality of blood circulation may be improved despite a deterioration in blood rheological properties. After that, the capillaries are subjected to axial torsion (or twist) due to body movement, and the intracapillary pressure increases. Besides, a narrowing of the arterial section and a thickening of the capillary wall occur. In capillaries, the thickness of the basement membrane increases, and it is the so-called ultrastructural sign of diabetic microangiopathy [91]. As a result, the space between pericytes and endothelial cells increases and the contact between them decreases [92]. These structural changes lead to the instability of the vessel wall and explain the changes in vascular permeability. Changes in the structure of postcapillary venules are similar to those in capillaries. Patients with DM have a thickened basal membrane, an irregular luminal surface formed by microvilli protrusions on endothelial cells, and a partial increase in interendothelial cell spaces [93].

In endothelial cells of arterioles in patients with DM, microscopic studies show swelling, cytoplasmic vacuoles, bleb formation, and a decrease in contacts between endothelial cells, as well as the deposition of collagen fibers in the subendothelium and between cells, leading to a thickening of the basement membrane [94].

4.5 LOCAL HEATING TEST

The functioning of the microcirculatory bed is often assessed by analyzing the effects of stress tests: thermal [95–98], mental [99], pharmacological [100], orthostatic [101–103], respiratory [104–109], and occlusive [58,69,110,111]. External factors affect the local properties of blood flow and are used as provocative tests to determine the function of the blood microcirculation system and the state of the regulatory mechanisms [112]. These tests allow researchers to evaluate the reserves of adaptation of the microcirculatory network to external effects and can be used to identify microcirculation disorders that accompany various pathologies.

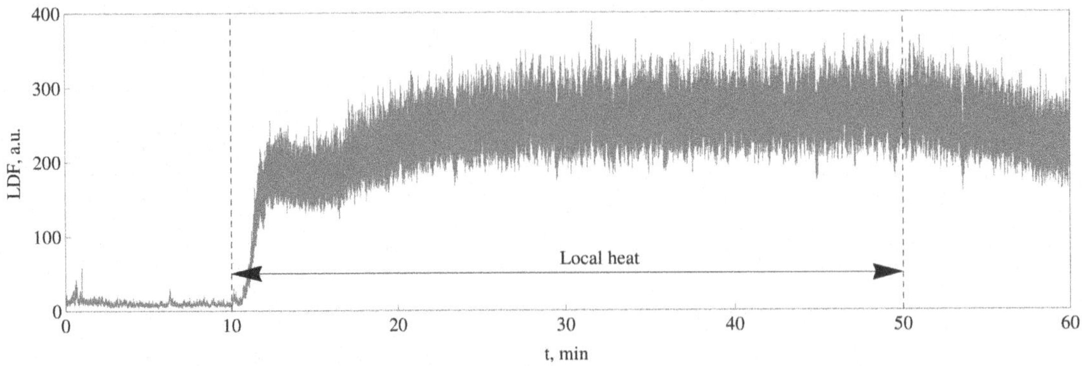

FIGURE 4.3 Typical perfusion behavior, measured by LDF, under a local heating conditions. The sample is collected on the forearm.

Being non-invasive and easy to implement, thermal tests are the most commonly used methods of diagnosing microcirculation disorders [113]. They can induce both vasodilation and vasoconstriction and thus reflect the function of the regulatory mechanisms of blood flow [55,96]. Controlled by the sympathetic vasoconstrictor nerves [114], the microvascular system is able to reduce blood flow at low temperatures. Diabetes primarily damages the unmyelinated nociceptive C-fibers, which are activated when heated above 42 °C [115]. Therefore, temperature tests are used to assess microvascular disorders in patients with DM [10,116]. At the moment, the sequence of physiological reactions that cause blood flow variation in the microcirculatory bed in response to local heating is currently described in sufficient detail [95,117,118].

The time-dependent perfusion curve of this physiological test has several phases (Figure 4.3). After the heater is turned on for a few minutes (5–6 minutes), the blood flow increases rapidly; this phase is presumably associated with the axon reflex [119,120]. Axon reflex, mediated by TRPV1 channels, activates C-fiber afferent neurons that release substance P and calcitonin gene-linked peptide (GCRP) with a modest contribution of NO [121] and may reflect both endothelial and small nerve fiber function.

Blood flow variations significantly depend on the heating mode (speed, maximum temperature, and duration of heating) and the location of the heater. Usually, the local heating test is understood as heating a small area of the skin (2–3 cm^2) to a temperature of 42–42 °C. When heating continues at the same temperature, the perfusion decreases and a local minimum appears on the curve, followed by a slow increase in perfusion, followed by a plateau phase [118]. This slow vasodilation is caused by another physiological mechanism – the local synthesis of NO [95].

Provided that the output to the stationary state reaches 10–15 minutes, the minimum time of the thermal test should be about 40 minutes. The time dependence of perfusion differs in healthy patients and in patients with pathologies [122–125].

4.6 ANALYSIS OF BLOOD FLOW OSCILLATIONS IN DIABETES MELLITUS STUDIES

Apparently, DM was one of the first diseases examined using LDF techniques [126]. Further, numerous studies have shown that the basal cutaneous blood flow is generally lower in patients with DM [127,128]. Patients with DM exhibit a reduced vasodilator response to heating tests [85]. Clinically diagnosed diabetic complications tend to decrease this response [125,129]. Microcirculation disorders in DM by means of LDF were studied at different levels of maximum heating and heating duration, yet only some studies indicated the heating rate. Many authors suggest that the LDF signal reduces substantially at the moment of maximum vasodilation caused by heating [85,116,130–133].

Despite the significant scientific interest in the amplitude–frequency analysis of the LDF signal in DM, there are very few such studies. The properties of peripheral blood flow fluctuations in diabetes and prediabetes [134], DM2 [135] and in laboratory animals [136] were extensively studied. Statistically significant differences in the parameters of blood flow oscillations in patients with DM and in controls were established even under basal conditions [128,137].

Some studies have shown a decreased sympathetic activity in the microcirculation of patients with diabetes, which manifests itself in a decrease in vasomotion at a frequency of 0.1 Hz [138]. This can be considered as a marker of the early development of sympathetic dysfunction preceding sympathetic neuropathy [139,140]. There are also studies conducted with the participation of relatives of patients with DM and with patients with impaired glucose tolerance and metabolic syndrome, showing impairment of their vascular function compared to controls [15,127]. According to the results obtained by Hu et al., the state of microcirculation gradually changes from the control group to the prediabetic and diabetic groups [134]. It has been shown [141] that increased metabolism and decreased myogenic regulation in people with DM contribute to the increased blood flow in the skin of the plantar foot. Finzhgar et al. studied the transcutaneous application of the gaseous CO_2 for improvement in microvascular function in patients with DM, wherein the microvascular function was controlled via LDF. It was shown that endothelial, neurogenic, and myogenic activities in patients before and after treatment is lower than in healthy subjects, due to endothelial dysfunction, sympathetic neuropathy, and impaired vasodilation associated with the local veno-arteriolar axon reflex [142].

Blood flow oscillations in various diseases, including DM, measured by flow-mediated skin fluorescence has been extensively studied in the Lodz laboratory of Jerze Gebicki [143]. The main functional test utilized in these studies is brachial artery occlusion and further analyses of oscillating components. The major role of myogenic vascular tone oscillations was revealed in Refs [144,145]. The authors constructed the methodology for the differentiation of diabetic foot ulcers based on the analyses of myogenic oscillations caused by the transient ischemia [146].

In a series of studies conducted by the Institute of Continuous media mechanics (Perm, Russia), skin blood volume pulsations during microcirculation were examined by high-resolution thermometry. It was taken into account the relationship between the fluctuations recorded by LDF and variations in skin temperature in the low frequency range to characterize vascular tone regulation mechanisms [147,148]. Endothelial function disorders, characteristic of DM, were detected at the impaired glucose tolerance stage [149], and a link between the manifestations of the metabolic syndrome and endothelial function was revealed [150]. It is shown that the functional disorders of the endothelium, determined by the response of the microcirculation system to the heat test, are associated with the biochemical markers of endothelial dysfunction [151] and the rheological characteristics of the blood [152].

4.7 EXPERIMENTAL STUDY

This section describes the clinical trials conducted to reveal the microvascular abnormalities associated with blood flow oscillations in DM subjects. The study involved 40 patients with DM1 and DM2 from the Endocrinology Department of the Orel Regional Clinical Hospital (Russia). All the patients were divided into groups according to the report of WHO Consultation [153]. The laboratory, clinical and anthropometric characteristics measured during the standard laboratory procedures are given in Table 4.1. The groups included subjects of slightly different age, but of close disease duration.

The optical probe was installed into the hole of the Peltier element, and the temperature was controlled by a thermistor with the accuracy of 0.1 °C. The Peltier element was mounted on the dorsal surface of the foot (Figure 4.4a) at a point located on the plateau between the 1st and 2nd metatarsals (Figure 4.4b).

Before the measurement taken about 2 hours after a meal, the volunteers were adapted for at least 10 minutes to laboratory conditions. All trials were performed with the subjects lying in the supine

TABLE 4.1
Main Characteristics of the Groups Under Study

	Controls $n=40$	DM1 $n=17$	DM2 $n=23$
Sex (M/F)	26/14	10/7	10/13
Foot temperature, °C	27±1	30±1	30±2
Age, years	39±9	35±9	50±6
Diabetes duration, years	-	14±10	7±6
Body mass index, kg/m²	23±3	25±5	35±5
Fasting glucose, mmol/l		8.1±4.7	9.2±3.5
HbA1c, %		7.9±0.8	8.8±0.9
Total cholesterol, mmol/l		4.6±0.9	5.4±0.9
Creatinine, μmol/l		88±37	74±16
Urea, mmol/l		6.1±3.4	5.7±1.8
ALT, IU/L		27±14	36±18
AST, IU/L		26±10	27±10
Systolic BP, mmHg	125±9	124±16	141±13
Diastolic BP, mmHg	80±5	78±7	86±6

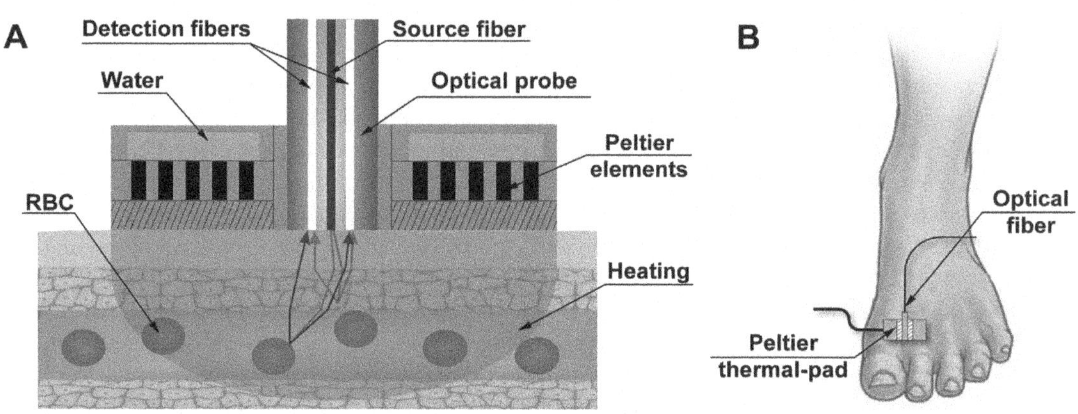

FIGURE 4.4 The scheme of the optical probe coupled with the Peltier element (A); probe location on the human lower limb (B). (Reproduced from Ref. [133] with the permission of Elsevier.)

position. In the basal state, the subjects had different skin temperatures (Table 4.1). A temperature protocol for thermal analysis included the following steps: the basic test for 4 min, cooling to 25 °C for 4 minutes, and a few local thermal tests at the temperatures of 35 °C and 42 °C for 4 and 10 minutes, respectively; in the second step, temperature conditions were unified. Thus, the measurement duration for one foot was 22 minutes. Both feet were investigated in all patients. The LDF sampling on each leg was collected continuously; to avoid remaining in the supine position for a long time, the one-leg test was conducted. An example of the collected LDF samples is presented in Figure 4.5.

The LDF signal was decomposed using a wavelet transform as:

$$W(\nu,\tau)=\nu\int_{-\infty}^{\infty} f(t)\psi^{*}\big(\nu(t-\tau)\big)dt, \tag{4.1}$$

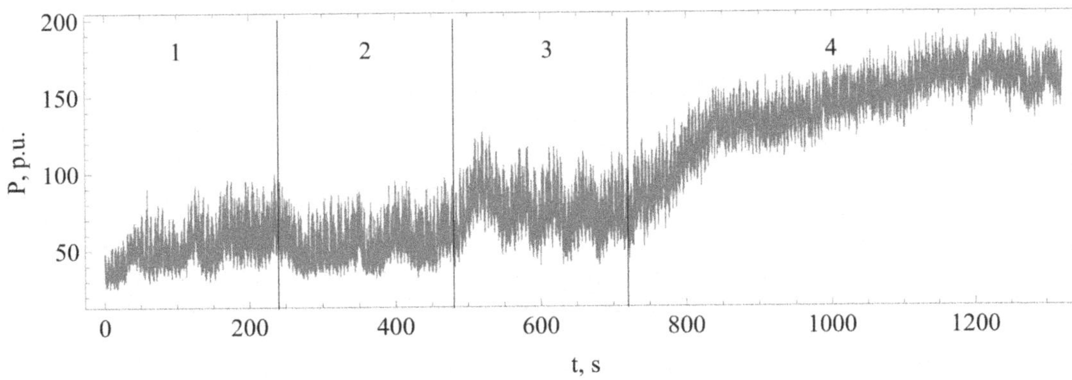

FIGURE 4.5 Typical LDF sample collected from the patients with diabetes (right foot, diabetes duration – 30 years). Numbers indicate experimental stages: 1 – basal conditions, 2 – cooling, 3 – first heating up to 35 °C, and 4 – second heating up to 42 °C. (Reproduced from Ref. [130] with the permission of Elsevier.)

where * means complex conjugation. The Morlet wavelet written in the form

$$\psi(t) = e^{2\pi i t} e^{-t^2/\sigma} \tag{4.2}$$

was used for the series expansion with the decay parameter $\sigma = 1$. Integrating the power over time gives the global wavelet spectrum

$$M(\nu) = \frac{1}{T}\int_0^T |W(\nu,\tau)|^2\, dt. \tag{4.3}$$

The wavelet coefficients for the frequency range 0.01–2 Hz with the logarithmic partitioning on 50 frequency bands were calculated. The $W(\nu,t)$ was calculated for every record, in the $M(\nu)$ calculation the cone of influence of boundaries and transition processes was excluded. The integral wavelet spectra were averaged over the group. For each frequency band, energy distribution was determined, and the results obtained for health and pathological subjects were compared. The frequency bands corresponding to different physiological mechanisms are plotted.

For reliable statistics, one should ideally include 10 cycles for each of the frequency under investigation. We have 4-minute recording for each of the first three phases, which is why reliable results can be obtained only for frequencies higher than 0.04 Hz. For lower frequencies, the results are presented to demonstrate the tendency of qualitative data only.

The Mann-Whitney test was used to compare the intergroup results, and the Wilcoxon statistical test was used to evaluate the intragroup variations. Finally, we performed sample-size estimations to obtain robust results.

Thermal tests provoke significant variations both in the average perfusion (Figure 4.7) and in its oscillation component (see the third experimental stage in Figure 4.5). The statistical analysis (Figure 4.6) demonstrates close values of perfusion (*P*, p.u.) – the parameter measured by LDF – without any significant differences in the basal state. Cooling provokes weak vasoconstriction and heating induces significant vasodilation. Note that the vasodilation dynamics varies between groups (Figure 4.6). All measurements showed a peak at the beginning of heating caused by the axon reflex. The highest rate of vasodilation is found in the control group, and this characteristic is impaired in both diabetic groups. Moreover, the heating clarified the difference between groups, and perfusion of the heated skin significantly differs in the groups of healthy and diabetes subjects, but is similar in two diabetes groups.

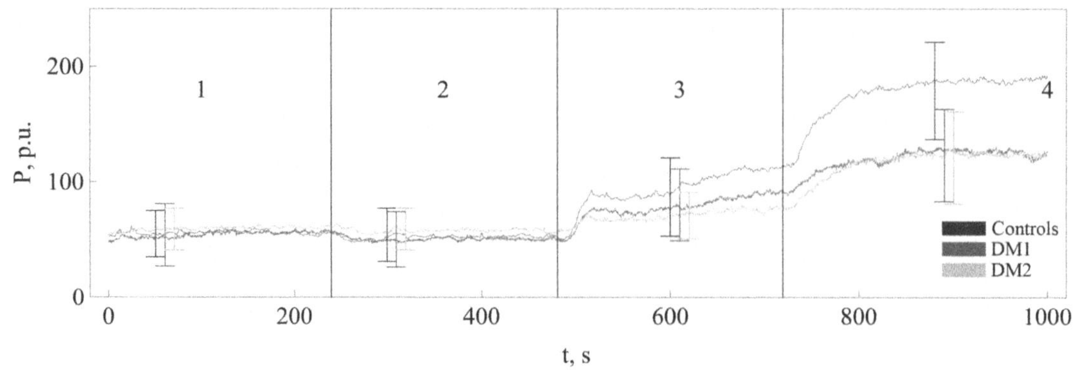

FIGURE 4.6 Dynamics of perfusion averaged over all measurements. First, we applied the moving-average filter with a window of 0.25 s and then estimated a mean value at each instant. Error bars indicate a mean standard deviation at a certain stage of the experiment. Numbers show experimental stages: 1 – native conditions, 2 – cooling, 3 – first heating up to 35 °C, and 4 – second heating up to 42 °C. (Reproduced from Ref. [133] with the permission of Elsevier.)

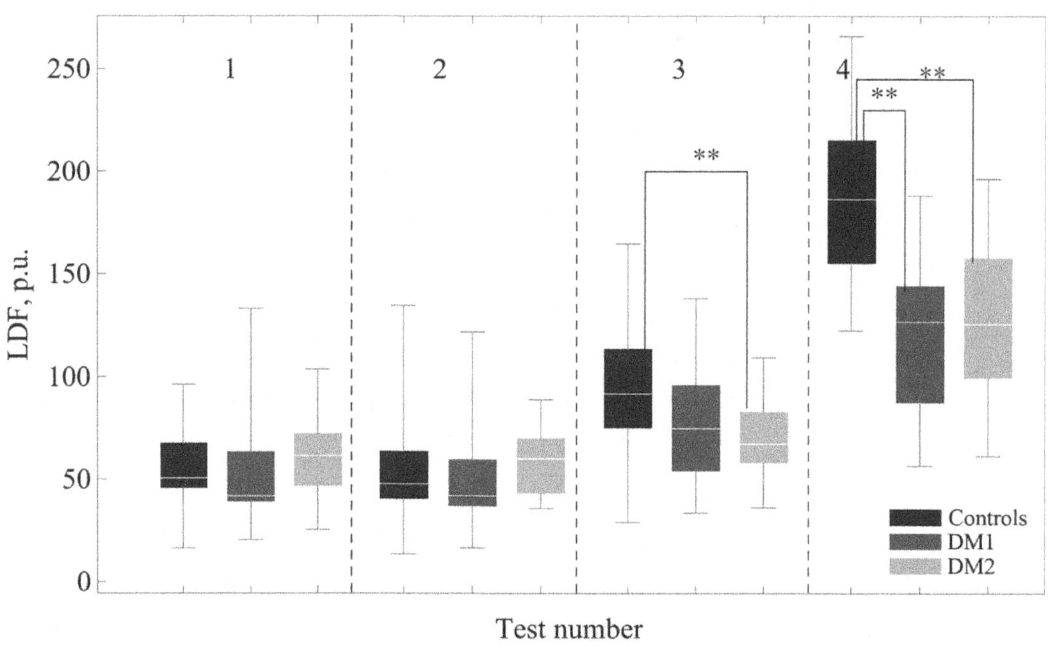

FIGURE 4.7 Box-Whisker diagram of mean perfusion during the four experimental stages: 1 – native conditions, 2 – cooling, 3 – first heating up to 35 °C, and 4 – second heating up to 42 °C. By asterisks, we mark the level of significance estimated using the Mann-Whitney test (** – p <0.01). An increase in perfusion during tests 3 and 4 was also significant. In both groups, tests provoked significant variation in perfusion (the level of significance was estimated using the Wilcoxon test, p <0.001). (Reproduced from Ref. [133] with the permission of Elsevier.)

The lowest perfusion under basal conditions was observed in the control group ($P=53\pm18$ p.u.). Patients with both types of diabetes had slightly higher perfusion, 54 ± 27 p.u. in DM1 and 58 ± 20 p.u. in DM2. Both diabetes groups had an impaired amplitude of perfusion oscillations in the frequency range 0.012–0.045 Hz (Figure 4.8) in comparison with the control group. These frequencies fall into the intervals that correspond to neurogenic and endothelial vascular tone regulation

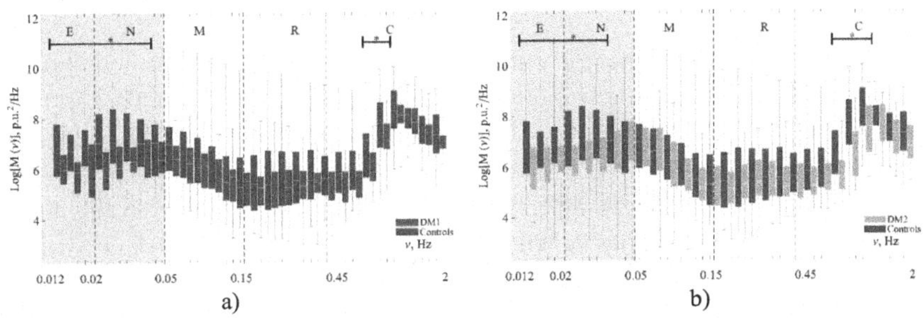

FIGURE 4.8 Averaged spectra of LDF samples in basal conditions. Thick lines in the upper parts of the plot indicate the frequency band, where $M(\nu)$ is significantly different (p <0.05). The low-frequency part of the spectra (gray shade) has insufficient statistics and is shown just to demonstrate the tendency. (Reproduced from Ref. [133] with the permission of Elsevier.)

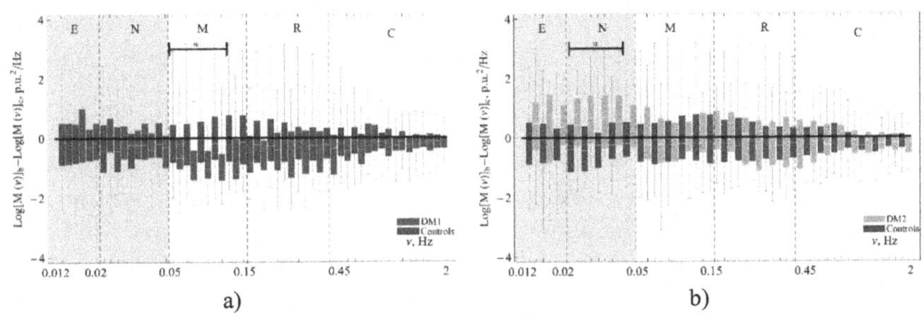

FIGURE 4.9 Variation in the spectral energy caused by cooling. For every frequency, we calculated $M(\nu)$ during cooling $\left(M(\nu)_c\right)$ and in basal conditions $\left(M(\nu)_b\right)$, then estimated their difference for every frequency $M(\nu)_c - M(\nu)_b$ and for all LDF samples. After that, we constructed the Box-Whisker diagram. The thick lines in the upper part of the plot show the frequency bands, where the variation in energy of pulsation is significant (p<0.05). the low-frequency part of the spectra (gray shade) is shown just to demonstrate the tendency similar to that of Figure 4.8. (Reproduced from Ref. [133] with the permission of Elsevier.)

mechanisms. The result is not statistically reliable due to the small data sample. Fluctuations in the range of 0.5–1 Hz were weaker in both diabetes groups as well. Moreover, the oscillations of these frequencies were significantly lower in patients with DM2 than in patients with DM1.

Local cooling-induced vasoconstriction alters the spectral properties of LDF signals. The averaged perfusion is 46±16 p.u. in the control group, 50±23 p.u. in the DM1 group, and 55±15 p.u. in the DM2 group. To study the phenomenon of spectral variation, we estimated the difference between $M(\nu)$ in basal conditions and at cooling for every subject. The results obtained are presented as a Box-Whisker plot (Figure 4.9). The spectral characteristic of LDF samples of healthy subjects only slightly varies in response to cooling.

At an exposure temperature of 25 °C, a significant reduction in oscillations was observed in the frequency band of 0.05–0.14 Hz in patients with DM1. This frequency band falls within the range associated with the myogenic activity. The spectral energy of the LDF signal of DM2 patients in this frequency band remained unchanged. The trend observed toward increased oscillations of the 0.02–0.04 Hz frequency band associated with the neurogenic activity during cooling in DM2 should be checked on larger samples. Note that the variation in the amplitude of pulsations in these frequency bands caused by local cooling is significantly different in the two diabetic groups.

Local heating up to 35 °C provoked vasodilation. The difference in averaged perfusion became significant between the examined groups; the perfusion increased to 92±28 p.u. in healthy subjects, while still remaining slightly lower at the level of 79±30 p.u. in DM1 and 67±16 p.u. in DM2.

At this stage, oscillations in the frequency band 0.05–0.45 Hz increased in the control group and in both diabetic groups (Figure 4.10). There is a sharp peak in the spectral variation diagram for the controls at a frequency close to 0.14 Hz. The variation mentioned above is significantly lower in both patient groups in comparison with the control group. The smallest reaction was observed in subjects with DM1. Together with oscillations corresponding to the myogenic activity, low-frequency oscillations increased in all three groups. The difference in the amplitude of oscillations in basal and heated states was found to be significant for all three groups. Taking into account the relationship between pulsatile and averaged components of LDF signal [154] and rising the average perfusion due to heating, it is difficult to explain this result. On the other hand, averaged perfusion in the groups with diabetes is statistically similar ($p>0.05$), but it is worth noting here that the amplitude of oscillations in the endothelial frequency band in the pathological groups varies in a more weakly manner. Hence, we can conclude that the mechanisms involved in the low-frequency modulation of the cutaneous blood flow are strongly related to DM1.

The next stage of the experiment was prolonged heating up to 42 °C. Higher temperature provokes stronger vasodilation, so the mean P rose up to 190±27 p.u. in the control group. The vasodilation response was impaired in both patient groups (128±38 p.u. in DM1 and 122±38 p.u. in DM2). Local heating up to 42 °C increased all frequency bands (Figure 4.11). Note that the amplitude of oscillations is lower in both diabetes groups in comparison with controls, and it is significant for high frequency pulsations of patients with DM2.

Even in basal conditions, all subjects had a similar level of perfusion, which was slightly higher in patients with DM2. These results match those of previous studies [86,116]. Moreover, the analysis of blood flow oscillations revealed a significant difference in the microhemodynamic parameters of healthy and pathological subjects. The amplitude of 1 Hz oscillations was lower in both DM groups than in the controls in basal conditions. Since local microvascular tone regulation mechanisms are not involved in the modulation of cardiac activity, this difference is related to the morphological abnormalities of the microvascular system in DM. The lowest energy of 1 Hz pulsations was observed in patients with DM2, and this energy was slightly higher in the DM1 group. However, both values are markedly lower compared to that of the control group. The cardiac stroke volume [155] in patients with DM is higher than in healthy ones, and therefore one can conclude that cardiac wave is dumped stronger by the cardiovascular system in patients with DM than in healthy ones. Metabolic syndrome, insulin resistance, impaired glucose tolerance, and accumulation of advanced glycation end products are positively correlated with increased arterial stiffness [156]. Therefore, it is suggested that the difference in blood flow oscillations associated with cardiac activity indirectly characterizes the elastic properties of vessels and indicate the increased arterial stiffness [157] of patients with DM. The results obtained are consistent with [158] where statistically significant differences were found in the contribution of relative energy in the cardiac frequency band to the total blood flow between the

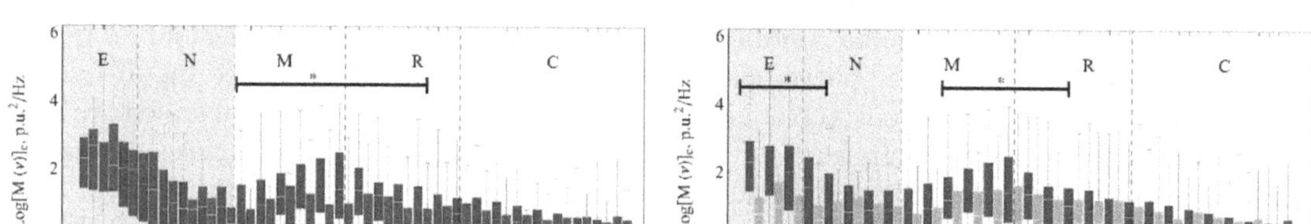

FIGURE 4.10 Variation in the spectral energy $\left(M(\nu)_h - M(\nu)_b\right)$ caused by heating up to 35 °C.The plot algorithm is similar to the one used in Figure 4.9. The low-frequency part of the spectra (gray shade) is shown just to demonstrate the tendency similar to Figure 4.8. (Reproduced from Ref. [133] with the permission of Elsevier.)

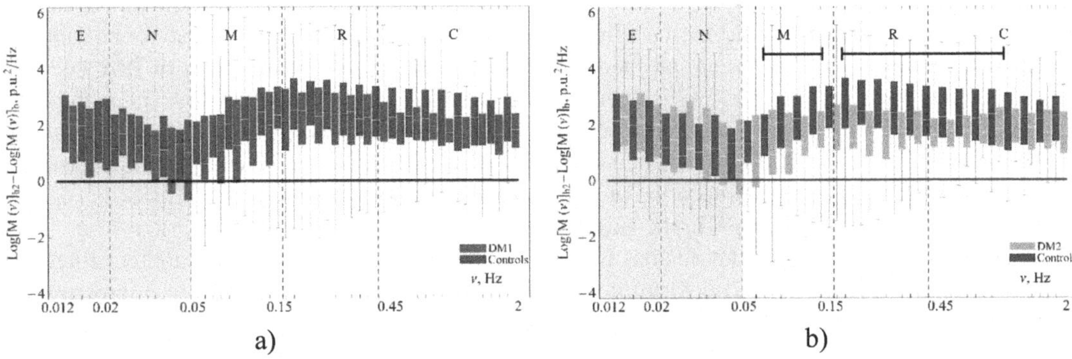

FIGURE 4.11 Variation of the spectral energy $\left(M(\nu)_h - M(\nu)_b\right)$ caused by heating up to 42 °C. The plot algorithm is similar to the one used in Figure 4.9. The low-frequency part of the spectra (gray shade) is shown just to demonstrate the tendency similar to that of Figure 4.8. (Reproduced from Ref. [133] with the permission of Elsevier.)

control group and subjects with DM2. The authors advanced a similar assumption that these changes are due to an increase in vascular resistance caused by vasoconstriction.

The response of smooth vascular muscles to sympathetic system stimulation during local cooling [120] provokes vasoconstriction. The spectral characteristics of LDF signals in controls are practically not disturbed by cooling; small variations are observed at the left end of the neurogenic frequency band. Patients with DM1 demonstrated a decrease in myogenic oscillations caused by local cooling.

Local mild heating initiates the sequence of reflexes, which leads to vasodilation [95]. At the beginning of heating, we observed a local peak on the perfusion–time curve associated with the axon reflex; vasodilation and its rate were impaired in patients with DM in comparison with controls. Further, after the local minimum on the perfusion–time curve, one can see a repeated increase in perfusion associated with the release of NO [95]. Perfusion at this stage was impaired in patients with DM, as in Refs [159,160]. Having the highest perfusion in basal conditions, patients with DM1 have the lowest one under local heating conditions. In Ref. [161], such a behavior was interpreted as a low reserve of the microcirculation system in pathological conditions. The LDF record during this test is nonstationary, and its slow variations are related to the axon reflex peak in the first part of the test and to endothelial activity in its second part. For this reason, as far as signal length is short for the analyses of slow perfusion oscillations, the discussion of the low-frequency part of the spectra is dropped from the consideration. We revealed an increase in myogenic activity in all groups; its highest variation, which was observed in the control group, was accompanied by a sharp peak at a frequency close to 0.14 Hz. Similar behavior observed in Ref. [55] was associated with high pre-capillary pressure and the stretching of arterioles, causing myogenic oscillations. Patients with DM had a lower response of myogenic activity to heating.

Prolonged heating induces NO-mediated vasodilation, which is lower in patients with DM and is associated with endothelial dysfunction [162]. Endothelial dysfunction is characterized by the decreased NO production and bioavailability, the increased production of vasoconstrictors (endothelin-1), the high level of oxidative stress and the process of angiogenesis, which are typical for diabetes. Long-term exposure to high temperature increases perfusion and variations in its spectral properties. The spectra of LDF signals for healthy volunteers vary weakly in the neurogenic frequency band compared to another frequency bands. This is indicative of a key role of endothelial activity in vasodilation at this stage. In both diabetes groups, the vasomotions were impaired in comparison with controls.

4.8 SUMMARY

Blood microcirculation performs several important physiological functions in the body. The study of these functions and timely detection of blood regulation abnormalities yields valuable diagnostic information. LDF has shown considerable potential as a tool for evaluating the cutaneous circulation. This method was introduced over 30 years ago and has undergone a continual development since.

It offers substantial advantages over other methods. Recent studies have shown that it is both highly sensitive and responsive to local blood perfusion and is also versatile and easy to use for continuous subject monitoring. One of the acute problems of modern medicine is the diagnosis of DM, because it has a serious effect on microvessels and nerve endings, leading to microangiopathy and neuropathy and even to the development of purulent-necrotic tissue damage. The use of LDF makes it possible to detect early symptoms of diabetes. The LDF method has found wide applications in diabetology, where it is used to evaluate the reactions of skin microcirculation to provocative tests or to register oscillatory hemodynamic changes in the skin.

Controlling blood flow oscillations has been a topic of primary interest to researchers since the late 1990s as an opportunity to perform *in vivo* studies of different mechanisms of microvascular tone regulation. This chapter gives a brief description of the past and current studies on oscillations of human peripheral blood flow by LDF using wavelet analysis. However, it should be mentioned that although wavelet analysis is a promising methodology for the study of microcirculatory function, it has not been extensively used for this purpose. The wavelet spectra of LDF signals are frequently undergone an additional post-processing procedure, such as the calculation of the average density in the frequency band [163]. The study described here uses the data-processing technique proposed in Ref. [164], which allows one to avoid signal post processing, to compare the raw spectra of signals, and to determine frequency bands having significantly different characteristics. The results of the study demonstrate the ability of LDF to provide useful information for disease classification in a noninvasive manner even in basal conditions.

The proposed approach has clinical benefits as a method for diagnostics and monitoring of microvascular disorders of different etiologies, and it can be used in the detection of microcirculation system disorders. In this chapter, we reviewed studies devoted to the detection of functional abnormalities of microcirculation in DM. Today this method is used for endothelial dysfunction studies [165–167] and obesity [168]. The areal of LDF usage is much broader; in Ref. [169], microvessel reactivity as a biomarker of cardiovascular diseases in COVID-19 is estimated, and brain blood flow is studied in Ref. [170].

Blood flow oscillations in microvessels are intensively studied for various pathological conditions [171,172] and treatment control [173,174]. It is important to note that the approach developed can be applied to the blood flow oscillation recorded by other optical or temperature techniques [175].

ACKNOWLEDGMENTS

IM gratefully acknowledges the financial support provided by the Ministry of Science and High Education of Russia (theme No. AAAA-A19-119012290101-5).

REFERENCES

1. M. Rossi, A. Carpi, F. Galetta, F. Franzoni, and G. Santoro, "The investigation of skin blood flowmotion: A new approach to study the microcirculatory impairment in vascular diseases?," *Biomed. Pharmacother.* **60**(8), 437–442 (2006).
2. J. Stewart, A. Kohen, D. Brouder, et al., "Noninvasive interrogation of microvasculature for signs of endothelial dysfunction in patients with chronic renal failure," *Am. J. Physiol. Circ. Physiol.* **287**(6), H2687–H2696 (2004).
3. M. Rossi, S. Taddei, A. Fabbri, et al., "Cutaneous vasodilation to acetylcholine in patients with essential hypertension," *J. Cardiovasc. Pharmacol.* **29**(3), 406–411 (1997).
4. L. A. Holowatz, C. S. Thompson-Torgerson, and W. L. Kenney, "The human cutaneous circulation as a model of generalized microvascular function," *J. Appl. Physiol.* **105**(1), 370–372 (2008).
5. M. Hellmann, M. Roustit, and J.-L. Cracowski, "Skin microvascular endothelial function as a biomarker in cardiovascular diseases?," *Pharmacol. Rep.* **67**(4), 803–810 (2015).
6. I. V. Tikhonova, N. I. Kosyakova, A. V. Tankanag, and N. K. Chemeris, "Oscillations of skin microvascular blood flow in patients with asthma," *Microcirculation* **23**(1), 33–43 (2016).

7. A. A. Tahrani, A. Ali, N. T. Raymond, et al., "Obstructive sleep apnea and diabetic neuropathy," *Am. J. Respir. Crit. Care Med.* **186**(5), 434–441 (2012).
8. E. Arosio, S. De Marchi, A. Rigoni, et al., "Forearm haemodynamics, arterial stiffness and microcirculatory reactivity in rheumatoid arthritis," *J. Hypertens.* **25**(6), 1273–1278 (2007).
9. T. Dimitroulas, J. Hodson, A. Sandoo, J. Smith, and G. D. Kitas, "Endothelial injury in rheumatoid arthritis: A crosstalk between dimethylarginines and systemic inflammation," *Arthritis Res. Ther.* **19**(1), 32 (2017).
10. A. Parshakov, N. Zubareva, S. Podtaev, and P. Frick, "Local heating test for detection of microcirculation abnormalities in patients with diabetes-related foot complications," *Adv. Skin Wound Care* **30**(4), 158–166 (2017).
11. C. Ince, E. C. Boerma, M. Cecconi, et al., "Second consensus on the assessment of sublingual microcirculation in critically ill patients: Results from a task force of the European Society of Intensive Care Medicine," *Intensive Care Med.* **44**(3), 281–299 (2018).
12. A. Stirban, "Microvascular dysfunction in the context of diabetic neuropathy," *Curr. Diab. Rep.* **14**(11), 1–9 (2014).
13. M. Roustit, S. Blaise, C. Millet, and J. L. Cracowski, "Reproducibility and methodological issues of skin post-occlusive and thermal hyperemia assessed by single-point laser Doppler flowmetry," *Microvasc. Res.* **79**(2), 102–108 (2010).
14. I. Eleftheriadou, A. Tentolouris, P. Grigoropoulou, et al., "The association of diabetic microvascular and macrovascular disease with cutaneous circulation in patients with type 2 diabetes mellitus," *J. Diabetes Complicat.* **33**(2), 165–170 (2019).
15. A. E. Caballero, S. Arora, R. Saouaf, et al., "Microvascular and macrovascular reactivity is reduced in subjects at risk for type 2 diabetes.," *Diabetes* **48**(9), 1856–1862 (1999).
16. S. Tehrani, K. Bergen, L. Azizi, and G. Jörneskog, "Skin microvascular reactivity correlates to clinical microangiopathy in type 1 diabetes: A pilot study," *Diabetes Vasc. Dis. Res.* **17**(3) (2020). doi: 10.1177/1479164120928303.
17. J.-L. Cracowski and M. Roustit, "Human skin microcirculation," *Compr. Physiol.* **10**(3), 1105–1154 (2020).
18. D. Balaz, A. Komornikova, P. Kruzliak, et al., "Regional differences of vasodilatation and vasomotion response to local heating in human cutaneous microcirculation," *Vasa - Eur. J. Vasc. Med.* **44**(6), 458–465 (2015).
19. J. M. Johnson, C. T. Minson, and D. L. Kellogg Jr., "Cutaneous vasodilator and vasoconstrictor mechanisms in temperature regulation," *Compr. Physiol.* **4**(1), 33–89 (2014).
20. I. M. Braverman, "The cutaneous microcirculation," *J. Investig. Dermatol. Symp. Proc.* **5**(1), 3–9 (2000).
21. A. Yen and I. M. Braverman, "Ultrastructure of the human dermal microcirculation: The horizontal plexus of the papillary dermis," *J. Invest. Dermatol.* **66**(3), 131–142 (1976).
22. I. M. Braverman, "Ultrastructure and organization of the cutaneous microvasculature in normal and pathologic states," *J. Invest. Dermatol.* **93**(2 SUPPL.), S2–S9 (1989).
23. S. Sangiorgi, A. Manelli, T. Congiu, et al., "Microvascularization of the human digit as studied by corrosion casting," *J. Anat.* **204**(2), 123–131 (2004).
24. R. N. Pittman, "Oxygen transport in the microcirculation and its regulation," *Microcirculation* **20**(2), 117–137 (2013).
25. E. M. Renkin, "Control of microcirculation and blood-tissue exchange," in Renkin, E. M., Michel C.C., eds. *Handbook of Physiology.* The Cardiovascular System, Microcirculation, pp. 627–687. American Physiological Society, Bethesda, MD (1984).
26. M. L. Smith, D. S. Long, E. R. Damiano, and K. Ley, "Near-wall μ-PIV reveals a hydrodynamically relevant endothelial surface layer in venules in vivo," *Biophys. J.* **85**(1), 637–645 (2003).
27. K. Suzuki, T. Yamamoto, T. Usui, et al., "Expression of hyaluronan synthase in intraocular proliferative diseases: Regulation of expression in human vascular endothelial cells by transforming growth factor-β," *Jpn. J. Ophthalmol.* **47**(6), 557–564 (2003).
28. J. Scallan, V. H. Huxley, and R. J. Korthuis, "Capillary fluid exchange: Regulation, functions, and pathology," *Colloq. Ser. Integr. Syst. Physiol. From Mol. Funct.* **2**(1), 1–94 (2010).
29. T. W. Secomb, R. Hsu, and A. R. Pries, "Effect of the endothelial surface layer on transmission of fluid shear stress to endothelial cells," *Biorheology* **38**, 143–150 (2001).
30. T. W. Secomb, R. Hsu, and A. R. Pries, "Motion of red blood cells in a capillary with an endothelial surface layer: Effect of flow velocity," *Am. J. Physiol. Circ. Physiol.* **281**(2), H629–H636 (2001).
31. K. Fronek and B. W. Zweifach, "Microvascular pressure distribution in skeletal muscle and the effect of vasodilation," *Am. J. Physiol. Content* **228**(3), 791–796 (1975).

32. W. M. Bayliss, "On the local reactions of the arterial wall to changes of internal pressure," *J. Physiol.* **28**(3), 220–231 (1902).
33. S. A. Regirer and N. K. Shadrina, "A simple model of a vessel with a wall sensitive to mechanical stimuli," *Biophysics (Oxf)* **47**, 908–913 (2002).
34. T. Voets and B. Nilius, "TRPCs, GPCRs and the Bayliss effect," *EMBO J.* **28**(1), 4–5 (2009).
35. M. S. Olufsen, J. T. Ottesen, H. T. Tran, et al., "Blood pressure and blood flow variation during postural change from sitting to standing: Model development and validation," *J. Appl. Physiol.* **99**(4), 1523–1537 (2005).
36. G. J. Hodges and J. M. Johnson, "Adrenergic control of the human cutaneous circulation," *Appl. Physiol. Nutr. Metab.* **34**(5), 829–839 (2009).
37. P. F. Davies, "Flow-mediated endothelial mechanotransduction," *Physiol. Rev.* **75**(3), 519–560 (1995).
38. T. M. Griffith, "Endothelial control of vascular tone by nitric oxide and gap junctions: A haemodynamic perspective," *Biorheology* **39**, 307–318 (2002).
39. S. Weinbaum, X. Zhang, Y. Han, H. Vink, and S. C. Cowin, "Mechanotransduction and flow across the endothelial glycocalyx," *Proc. Natl. Acad. Sci.* **100**(13), 7988–7995 (2003).
40. S. Forconi and T. Gori, "Endothelium and hemorheology," *Clin. Hemorheol. Microcirc.* **53**, 3–10 (2013).
41. T. Penzel, A. Porta, A. Stefanovska, and N. Wessel, "Recent advances in physiological oscillations," *Physiol. Meas.* **38**(5), E1–E7 (2017).
42. L. Glass and M. C. Mackey, *From Clocks to Chaos: The Rhythms of Life*, Princeton University Press, Princeton, NJ (1988).
43. G. Buzsáki and A. Draguhn, "Neuronal oscillations in cortical networks," *Science.* **304**(5679), 1926–1929 (2004).
44. D. L. Eckberg, "Point:Counterpoint: Respiratory sinus arrhythmia is due to a central mechanism vs. respiratory sinus arrhythmia is due to the baroreflex mechanism," *J. Appl. Physiol.* **106**(5), 1740–1742 (2009).
45. S. M. Barman, P. J. Fadel, W. Vongpatanasin, R. G. Victor, and G. L. Gebber, "Basis for the cardiac-related rhythm in muscle sympathetic nerve activity of humans," *Am. J. Physiol. Circ. Physiol.* **284**(2), H584–H597 (2003).
46. A. J. Camm, M. Malik, J. T. Bigger, et al., "Heart rate variability: Standards of measurement, physiological interpretation and clinical use. Task Force of the European Society of Cardiology and the North American Society of Pacing and Electrophysiology" (1996).
47. A. Stefanovska and M. Hožič, "Spatial synchronization in the human cardiovascular system," *Prog. Theor. Phys. Suppl.* **139**, 270–282 (2000).
48. A. Colantuoni, S. Bertuglia, and M. Intaglietta, "Quantitation of rhythmic diameter changes in arterial microcirculation," *Am. J. Physiol. Circ. Physiol.* **246**(4), H508–H517 (1984).
49. Y. Shiogai, A. Stefanovska, and P. V. E. McClintock, "Nonlinear dynamics of cardiovascular ageing," *Phys. Rep.* **488**(2), 51–110 (2010).
50. C. Aalkjær, D. Boedtkjer, and V. Matchkov, "Vasomotion – what is currently thought?," *Acta Physiol.* **202**(3), 253–269 (2011).
51. G. Billman, "Heart rate variability – A historical perspective," *Front. Physiol.* **2**, 86 (2011).
52. S. Akselrod, D. Gordon, J. B. Madwed, et al., "Hemodynamic regulation: Investigation by spectral analysis," *Am. J. Physiol. Circ. Physiol.* **249**(4), H867–H875 (1985).
53. S. Akselrod, D. Gordon, F. A. Ubel, et al., "Power spectrum analysis of heart rate fluctuation: A quantitative probe of beat-to-beat cardiovascular control," *Science.* **213**(4504), 220–222 (1981).
54. A. V. Tankanag, A. A. Grinevich, T. V. Kirilina, et al., "Wavelet phase coherence analysis of the skin blood flow oscillations in human," *Microvasc. Res.* **95**, 53–59 (2014).
55. L. W. Sheppard, V. Vuksanović, P. V. E. McClintock, and A. Stefanovska, "Oscillatory dynamics of vasoconstriction and vasodilation identified by time-localized phase coherence," *Phys. Med. Biol.* **56**(12), 3583 (2011).
56. E. G. Salerud, T. Tenland, G. E. Nilsson, and P. A. Oberg, "Rhythmical variations in human skin blood flow.," *Int. J. Microcirc. Clin. Exp.* **2**(2), 91–102 (1983).
57. D. D. Heistad, F. M. Abboud, A. L. Mark, and P. G. Schmid, "Interaction of thermal and baroreceptor reflexes in man.," *J. Appl. Physiol.* **35**(5), 581–586 (1973).
58. M. Roustit and J.-L. Cracowski, "Non-invasive assessment of skin microvascular function in humans: An insight into methods," *Microcirculation* **19**(1), 47–64 (2012).
59. A. Stefanovska, "Physics of the human cardiovascular system," *Contemp. Phys.* **40**(1), 31–55 (1999).
60. M. Rossi, A. Carpi, F. Galetta, F. Franzoni, and G. Santoro, "Skin vasomotion investigation: A useful tool for clinical evaluation of microvascular endothelial function?," *Biomed. Pharmacother.* **62**(8), 541–545 (2008).

61. J. A. Schmidt, G. A. Breit, P. Borgström, and M. Intaglietta, "Induced periodic hemodynamics in skeletal muscle of anesthetized rabbits, studied with multiple laser doppler flow probes," *Int. J. Microcirc.* **15**(1), 28–36 (1995).
62. P. Kvandal, S. A. Landsverk, A. Bernjak, et al., "Low-frequency oscillations of the laser Doppler perfusion signal in human skin," *Microvasc. Res.* **72**(3), 120–127 (2006).
63. H. D. Kvernmo, A. Stefanovska, K. A. Kirkebøen, and K. Kvernebo, "Oscillations in the human cutaneous blood perfusion signal modified by endothelium-dependent and endothelium-independent vasodilators," *Microvasc. Res.* **57**(3), 298–309 (1999).
64. H. D. Kvernmo, A. Stefanovska, M. Bracic, K. A. Kirkebøen, and K. Kvernebo, "Spectral analysis of the laser doppler perfusion signal in human skin before and after exercise," *Microvasc. Res.* **56**(3), 173–182 (1998).
65. A. I. Krupatkin, "Oscillatory processes in the diagnosis of the state of microvascular-tissue systems," *Hum. Physiol.* **44**(5), 581–591 (2018).
66. A. I. Krupatkin, "Cardiac and respiratory oscillations of the blood flow in microvessels of the human skin," *Hum. Physiol.* **34**(3), 323–329 (2008).
67. T. Pedley, *The Fluid Mechanics of Large Blood Vessels (Cambridge Monographs on Mechanics)*, Cambridge University Press, Cambridge (1980).
68. A. Bollinger, A. Yanar, U. Hoffmann, and U. K. Franzeck, "Is high-frequency flux motion due to respiration or to vasomotion activity? 1," in *Vasomotion and Flow Motion* **20**, pp. 52–58. Karger Publishers, Basel (1993).
69. A. I. Krupatkin and V. V. Sidorov, *Handbook of Laser Doppler Flowmetry of Blood Microcirculation (in Russian)*, Knizhnyj dom "LIBROKOM," Moscow (2013).
70. A. A. Fedorovich, A. N. Rogoza, S. B. Gorieva, and T. S. Pavlova, "Correlation between function of venular part of blood microcirculation and diurnal rhythm of blood pressure in normal feature and under arterial hypertension," *Kardiol. Vestn.* **3**(2), 21–31 (2008).
71. A. Fedorovich, "The functional state of regulatory mechanisms of the microcirculatory blood flow in normal conditions and in arterial hypertension according to laser doppler flowmetry," *J. Hypertens.* **28**, e178 (2010).
72. C. Aalkjær and H. Nilsson, "Vasomotion: Cellular background for the oscillator and for the synchronization of smooth muscle cells," *Br. J. Pharmacol.* **144**(5), 605–616 (2005).
73. V. A. Shvartz, A. S. Karavaev, E. I. Borovkova, et al., "Investigation of statistical characteristics of interaction between the low-frequency oscillations in heart rate variability and photoplethysmographic waveform variability in healthy subjects and myocardial infarction patients," *Russ. Open Med. J.* **5**(2), 203 (2016).
74. F. Liao and Y.-K. Jan, "Enhanced phase synchronization of blood flow oscillations between heated and adjacent non-heated sacral skin," *Med. Biol. Eng. Comput.* **50**(10), 1059–1070 (2012).
75. G. J. Hodges and A. T. Del Pozzi, "Noninvasive examination of endothelial, sympathetic, and myogenic contributions to regional differences in the human cutaneous microcirculation," *Microvasc. Res.* **93**, 87–91 (2014).
76. A. I. Krupatkin, "Blood flow oscillations at a frequency of about 0.1 Hz in skin microvessels do not reflect the sympathetic regulation of their tone," *Hum. Physiol.* **35**(2), 183–191 (2009).
77. A. Colantuoni, S. Bertuglia, G. Coppini, and L. Donato, "Superposition of arteriolar vasomotion waves and regulation of blood flow in skeletal muscle microcirculation," in *Oxygen Transport to Tissue XII*, pp. 549–558. Springer, Boston, MA (1990).
78. S. A. Landsverk, P. Kvandal, T. Kjelstrup, et al., "Human skin microcirculation after brachial plexus block evaluated by wavelet transform of the laser doppler flowmetry signal," *Anesthesiology* **105**(3), 478–484 (2006).
79. A. Bernjak and A. Stefanovska, "Importance of wavelet analysis in laser Doppler flowmetry time series," in *2007 29th Annual International Conference of the IEEE Engineering in Medicine and Biology Society*, pp. 4064–4067, IEEE (2007).
80. A. A. Fedorovich, "Non-invasive evaluation of vasomotor and metabolic functions of microvascular endothelium in human skin," *Microvasc. Res.* **84**(1), 86–93 (2012).
81. C. E. Thorn, H. Kyte, D. W. Slaff, and A. C. Shore, "An association between vasomotion and oxygen extraction," *Am. J. Physiol. Circ. Physiol.* **301**(2), H442–H449 (2011).
82. M. B. Lotric, A. Stefanovska, D. Stajer, and V. Urbancic-Rovan, "Spectral components of heart rate variability determined by wavelet analysis," *Physiol. Meas.* **21**(4), 441–457 (2000).
83. A. I. Krupatkin, "Dynamic oscillatory circuit of regulation of capillary hemodynamics," *Hum. Physiol.* **33**(5), 595–602 (2007).
84. M. Jacob, D. Chappell, and B. F. Becker, "Regulation of blood flow and volume exchange across the microcirculation," *Crit. Care* **20**(1), 319 (2016).

85. D. Fuchs, P. P. Dupon, L. A. Schaap, and R. Draijer, "The association between diabetes and dermal microvascular dysfunction non-invasively assessed by laser Doppler with local thermal hyperemia: A systematic review with meta-analysis," *Cardiovasc. Diabetol.* **16**(1), 11 (2017).
86. J. C. Schramm, T. Dinh, and A. Veves, "Microvascular changes in the diabetic foot," *Int. J. Low. Extrem. Wounds* **5**(3), 149–159 (2006).
87. E. J. Barrett, Z. Liu, M. Khamaisi, et al., "Diabetic microvascular disease: An endocrine society scientific statement," *J. Clin. Endocrinol. Metab.* **102**(12), 4343–4410 (2017).
88. C. E. Tabit, W. B. Chung, N. M. Hamburg, and J. A. Vita, "Endothelial dysfunction in diabetes mellitus: Molecular mechanisms and clinical implications," *Rev. Endocr. Metab. Disord.* **11**(1), 61–74 (2010).
89. B. M. Balletshofer, et al., "Endothelial dysfunction is detectable in young normotensive first-degree relatives of subjects with type 2 diabetes in association with insulin resistance," *Circulation* **101**(15), 1780–1784 (2000).
90. Y. I. Cho, M. P. Mooney, and D. J. Cho, "Hemorheological disorders in diabetes mellitus," *J. Diabetes Sci. Technol.* **2**(6), 1130–1138 (2008).
91. Ö. Aagenaes and H. Moe, "Light- and electron-microscopic study of skin capillaries of diabetics," *Diabetes* **10**(4), 253–259 (1961).
92. I. M. Braverman, J. Sibley, and A. Keh, "Ultrastructural analysis of the endothelial-pericyte relationship in diabetic cutaneous vessels," *J. Invest. Dermatol.* **95**(2), 147–153 (1990).
93. A. C. Shore, "The microvasculature in type 1 diabetes," *Semin Vasc. Med.* **2**(01), 9–20 (2002).
94. M. E. Salem, A.-A. A. Ismael, A. Salem, and T. Salem, "Ultrastructural changes in peripheral arteries and nerves in diabetic ischemic lower limbs, by electron microscope," *Alexandria J. Med.* **53**(4), 373–379 (2017).
95. J. M. Johnson and D. L. Kellogg, "Local thermal control of the human cutaneous circulation," *J. Appl. Physiol.* **109**(4), 1229–1238 (2010).
96. N. Charkoudian, "Skin blood flow in adult human thermoregulation: How it works, when it does not, and why," *Mayo Clin. Proc.* **78**(5), 603–612 (2003).
97. D. L. Kellogg, "In vivo mechanisms of cutaneous vasodilation and vasoconstriction in humans during thermoregulatory challenges," *J. Appl. Physiol.* **100**(5), 1709–1718 (2006).
98. Y. Isii, K. Matsukawa, H. Tsuchimochi, and T. Nakamoto, "Iced-water hand immersion causes a reflex decrease in skin temperature in the contralateral hand," *J. Physiol. Sci.*, **57**, 241–248 (2007).
99. C. Lemne, U. de Faire, and B. Fagrell, "Mental stress induces different reactions in nutritional and thermoregulatory human skin microcirculation: A study in borderline hypertensives and normotensives," *J. Hum. Hypertens.* **8**(8), 559–563 (1994).
100. V. O. Soldatov, T. N. Malorodova, T. I. Balamutova, et al., "Endothelial dysfunction: Comparative evaluation of ultrasound dopplerography, laser dopplerflowmetry and direct monitoring of arterial pressure for conducting pharmacological tests in rats," *Res. Results Pharmacol.* **4**(1), 73 (2018).
101. J. Drescher, A. Diedrich, A. N. Lebedev, et al., "Forehead skin microcirculation during tilt table testing and lower body negative pressure," *J. Gravit. Physiol.* **2**(1), P11–2 (1995).
102. I. V. Tikhonova, A. A. Grinevich, I. E. Guseva, and A. V. Tankanag, "Effect of orthostasis on the regulation of skin blood flow in upper and lower extremities in human," *Microcirculation* **28**(1), e12655 (2021).
103. M. A. Skedina, A. A. Kovaleva, and N. V. Degterenkova, "Investigation of cerebral circulation and peripheral microcirculation to passive postural orthostatic test," *Reg. Blood Circ. Microcirc.* **17**(3), 115–119 (2018).
104. H. N. Mayrovitz and E. E. Groseclose, "Inspiration-induced vascular responses in finger dorsum skin," *Microvasc. Res.* **63**(2), 227–232 (2002).
105. S. B. Wilson, P. E. Jennings, and J. J. F. Belch, "Detection of microvascular impairment in type I diabetics by laser Doppler flowmetry," *Clin. Physiol.* **12**(2), 195–208 (1992).
106. N. C. Abbot, J. S. Beck, S. B. Wilson, and F. Khan, "Vasomotor reflexes in the fingertip skin of patients with type 1 diabetes mellitus and leprosy," *Clin. Auton. Res.* **3**(3), 189–193 (1993).
107. H. Wollersheim, H. Droste, J. Reyenga, and T. Thien, "Laser Doppler evaluation of skin vasomotor reflexes during sympathetic stimulation in normals and in patients with primary Raynaud's phenomenon," *Int. J. Microcirc. Clin. Exp.* **10**(1), 33–42 (1991).
108. R. C. Littleford, F. Khan, and J. J. F. Belch, "Impaired skin vasomotor reflexes in patients with erythromelalgia," *Clin. Sci.* **96**(5), 507–512 (1999).
109. N. Charkoudian, "Mechanisms and modifiers of reflex induced cutaneous vasodilation and vasoconstriction in humans," *J. Appl. Physiol.* **109**(4), 1221–1228 (2010).

110. M. Roustit and J.-L. Cracowski, "Assessment of endothelial and neurovascular function in human skin microcirculation," *Trends Pharmacol. Sci.* **34**(7), 373–384 (2013).
111. A. A. Sagaidachnyi, "Reactive hyperemia test: Methods of analysis, mechanisms of reaction and prospects," *Reg. Blood Circ. Microcirc.* **17**(3), 5–22 (2018).
112. A. V. Stankevich, A. A. Akhapkina, and I. A. Tikhomirova, "Functional tests in evaluation of spare capacity of athletes' blood flow," *Jaroslavskij Pedagog. Vestn.* **3**(4), 190–194 (2013).
113. K. A. Roberts, T. van Gent, N. D. Hopkins, et al., "Reproducibility of four frequently used local heating protocols to assess cutaneous microvascular function," *Microvasc. Res.* **112**, 65–71 (2017).
114. P. E. Pergola, D. L. Kellogg, J. M. Johnson, W. A. Kosiba, and D. E. Solomon, "Role of sympathetic nerves in the vascular effects of local temperature in human forearm skin," *Am. J. Physiol. Circ. Physiol.* **265**(3), H785–H792 (1993).
115. M. Campero, T. K. Baumann, H. Bostock, and J. L. Ochoa, "Human cutaneous C fibres activated by cooling, heating and menthol," *J. Physiol.* **587**(23), 5633–5652 (2009).
116. Y.-K. Jan, S. Shen, R. D. Foreman, and W. J. Ennis, "Skin blood flow response to locally applied mechanical and thermal stresses in the diabetic foot," *Microvasc. Res.* **89**, 40–46 (2013).
117. Z. Marcinkevics, U. Rubins, A. Aglinska, A. Caica, and A. Grabovskis, "Remote photoplethysmography for skin perfusion monitoring using narrowband illumination," *Proc. SPIE* **11073**, 1107312 (2019).
118. M. Ciplak, A. Pasche, A. Heim, et al., "The vasodilatory response of skin microcirculation to local heating is subject to desensitization," *Microcirculation* **16**(3), 265–275 (2009).
119. C.-S. Huang, S.-F. Wang, and Y.-F. Tsai, "Axon reflex-related hyperemia induced by short local heating is reproducible," *Microvasc. Res.* **84**(3), 351–355 (2012).
120. D. P. Stephens, N. Charkoudian, J. M. Benevento, J. M. Johnson, and J. L. Saumet, "The influence of topical capsaicin on the local thermal control of skin blood flow in humans," *Am. J. Physiol. Integr. Comp. Physiol.* **281**(3), R894–R901 (2001).
121. P. Marche, S. Dubois, P. Abraham, et al., "Neurovascular microcirculatory vasodilation mediated by C-fibers and Transient receptor potential vanilloid-type-1 channels (TRPV 1) is impaired in type 1 diabetes," *Sci. Rep.* **7**(1), 44322 (2017).
122. A. Bandini, S. Orlandi, C. Manfredi, et al., "Modelling of thermal hyperemia in the skin of type 2 diabetic patients," *J. Healthc. Eng.* **4**, 729867 (2013).
123. A. Bandini, S. Orlandi, C. Manfredi, et al., "Effect of local blood flow in thermal regulation in diabetic patient," *Microvasc. Res.* **88**, 42–47 (2013).
124. I. Fredriksson, M. Larsson, F. H. Nyström, et al., "Reduced arteriovenous shunting capacity after local heating and redistribution of baseline skin blood flow in type 2 diabetes assessed with velocity-resolved quantitative laser Doppler flowmetry," *Diabetes* **59**(7), 1578–1584 (2010).
125. S. Arora, P. Smakowski, R. G. Frykberg, et al., "Differences in foot and forearm skin microcirculation in diabetic patients with and without neuropathy," *Diabetes Care* **21**(8), 1339–1344 (1998).
126. M. Oimomi, S. Nishimoto, S. Matsumoto, et al., "Evaluation of periflux blood flow measurement in diabetic patients with autonomic neuropathy," *Diabetes Res. Clin. Pract.* **1**(2), 81–85 (1985).
127. W. Guillaume, O. Philippe, D. Frédéric, et al., "Metabolic syndrome individuals with and without type 2 diabetes mellitus present generalized vascular dysfunction," *Arterioscler. Thromb. Vasc. Biol.* **35**(4), 1022–1029 (2015).
128. V. Urbancic-Rovan, A. Stefanovska, A. Bernjak, K. Ažman-Juvan, and A. Kocijančič, "Skin blood flow in the upper and lower extremities of diabetic patients with and without autonomic neuropathy," *J. Vasc. Res.* **41**(6), 535–545 (2004).
129. B. A. Brooks, S. V. McLennan, S. M. Twigg, and D. K. Yue, "Detection and characterisation of microcirculatory abnormalities in the skin of diabetic patients with microvascular complications," *Diabetes Vasc. Dis. Res.* **5**(1), 30–35 (2008).
130. A. S. Shah, Z. Gao, L. M. Dolan, et al., "Assessing endothelial dysfunction in adolescents and young adults with type 1 diabetes mellitus using a non-invasive heat stimulus," *Pediatr. Diabetes* **16**(6), 434–440 (2015).
131. V. V. Dremin, V. V. Sidorov, A. I. Krupatkin, et al., "The blood perfusion and NADH/FAD content combined analysis in patients with diabetes foot," *Proc. SPIE* **9698**, 969810 (2016).
132. V. V. Dremin, E. A. Zherebtsov, V. V. Sidorov, et al., "Multimodal optical measurement for study of lower limb tissue viability in patients with diabetes mellitus," *J. Biomed. Opt.* **22**(8), 085003 (2017).
133. I. Mizeva, E. Zharkikh, V. Dremin, et al., "Spectral analysis of the blood flow in the foot microvascular bed during thermal testing in patients with diabetes mellitus," *Microvasc. Res.* **120**, 13–20 (2018).

134. H.-F. Hu, H. Hsiu, C.-J. Sung, and C.-H. Lee, "Combining laser-Doppler flowmetry measurements with spectral analysis to study different microcirculatory effects in human prediabetic and diabetic subjects," *Lasers Med. Sci.* **32**(2), 327–334 (2017).
135. D. Balaz, A. Komornikova, P. Sabaka, et al., "Changes in vasomotion–effect of hyperbaric oxygen in patients with diabetes Type 2," *Undersea Hyperb. Med. J. Undersea Hyperb. Med. Soc. Inc.* **43**(2), 123–134 (2016).
136. X. T. Tigno, B. C. Hansen, S. Nawang, R. Shamekh, and A. M. Albano, "Vasomotion becomes less random as diabetes progresses in monkeys," *Microcirculation* **18**(6), 429–439 (2011).
137. E. V. Zharkikh, I. A. Mizeva, I. I. Makovik, et al., "Blood flow oscillations as a signature of microvascular abnormalities," *Proc. SPIE* **10685**, 106854C (2018).
138. K. B. Stansberry, H. R. Peppard, L. M. Babyak, et al., "Primary nociceptive afferents mediate the blood flow dysfunction in non-glabrous (hairy) skin of type 2 diabetes: A new model for the pathogenesis of microvascular dysfunction.," *Diabetes Care* **22**(9), 1549–1554 (1999).
139. L. Bernardi, A. Radaelli, P. L. Solda, et al., "Autonomic control of skin microvessels: Assessment by power spectrum of photoplethysmographic waves," *Clin. Sci.* **90**(5), 345–355 (1996).
140. M. F. Meyer, C. J. Rose, J.-O. Hülsmann, H. Schatz, and M. Pfohl, "Impaired 0.1-Hz vasomotion assessed by laser Doppler anemometry as an early index of peripheral sympathetic neuropathy in diabetes," *Microvasc. Res.* **65**(2), 88–95 (2003).
141. Y.-K. Jan, F. Liao, G. L. Y. Cheing, et al., "Differences in skin blood flow oscillations between the plantar and dorsal foot in people with diabetes mellitus and peripheral neuropathy," *Microvasc. Res.* **122**, 45–51 (2019).
142. M. Finžgar, H. B. Frangež, K. Cankar, and I. Frangež, "Transcutaneous application of the gaseous CO_2 for improvement of the microvascular function in patients with diabetic foot ulcers," *Microvasc. Res.* **133**, 104100 (2021).
143. J. Katarzynska, T. Cholewinski, L. Sieron, A. Marcinek, and J. Gebicki, "Flowmotion monitored by flow mediated skin fluorescence (FMSF): A tool for characterization of microcirculatory status," *Front. Physiol.* **11**, 702 (2020).
144. J. Gebicki, A. Marcinek, and J. Zielinski, "Assessment of microcirculatory status based on stimulation of myogenic oscillations by transient ischemia: From health to disease," *Vasc. Health Risk Manag.* **17**, 33–36 (2021).
145. J. Gebicki, J. Katarzynska, and A. Marcinek, "Can efficient stimulation of myogenic microcirculatory oscillations by transient ischemia predict low incidence of COVID-19 infection?," *Respir. Physiol. Neurobiol.* **286**, 103618 (2021).
146. A. Los-Stegienta, J. Katarzynska, A. Borkowska, et al., "Differentiation of diabetic foot ulcers based on stimulation of myogenic oscillations by transient ischemia," *Vasc. Health Risk Manag.* **17**, 145–152 (2021).
147. S. Podtaev, M. Morozov, and P. Frick, "Wavelet-based correlations of skin temperature and blood flow oscillations," *Cardiovasc. Eng.* **8**(3), 185–189 (2008).
148. P. Frick, I. Mizeva, and S. Podtaev, "Skin temperature variations as a tracer of microvessel tone," *Biomed. Signal Process. Control* **21**, 1–7 (2015).
149. E. Smirnova, S. Podtaev, I. Mizeva, and E. Loran, "Assessment of endothelial dysfunction in patients with impaired glucose tolerance during a cold pressor test," *Diabetes Vasc. Dis. Res.* **10**(6), 489–497 (2013).
150. E. Smirnova, E. Loran, S. Shulkina, and S. Podtaev, "The relationship of endothelial dysfunction and metabolic manifestations of obesity," *Reg. Blood Circ. Microcirc.* **19**, 53–59 (2020).
151. E. Smirnova, S. Shulkina, E. Loran, S. Podtaev, and N. Antonova, "Relationship between skin blood flow regulation mechanisms and vascular endothelial growth factor in patients with metabolic syndrome," *Clin. Hemorheol. Microcirc.* **70**, 129–142 (2018).
152. N. Antonova, K. Tsiberkin, S. Podtaev, et al., "Comparative study between microvascular tone regulation and rheological properties of blood in patients with type 2 diabetes mellitus," *Clin. Hemorheol. Microcirc.* **64**, 837–844 (2016).
153. K. Alberti and P. Zimmet, "Definition, diagnosis and classification of diabetes mellitus and its complications. Part 1: Diagnosis and classification of diabetes mellitus. Provisional report of a WHO consultation," *Diabet. Med.* **15**(7), 539–553 (1998).
154. I. Mizeva, P. Frick, and S. Podtaev, "Relationship of oscillating and average components of laser Doppler flowmetry signal," *J. Biomed. Opt.* **21**(8), 1–10 (2016).
155. R. B. Devereux, M. J. Roman, M. Paranicas, et al., "Impact of diabetes on cardiac structure and function," *Circulation* **101**(19), 2271–2276 (2000).

156. S. J. Zieman, V. Melenovsky, and D. A. Kass, "Mechanisms, pathophysiology, and therapy of arterial stiffness," *Arterioscler. Thromb. Vasc. Biol.* **25**(5), 932–943 (2005).
157. M. Jaiswal, E. M. Urbina, R. P. Wadwa, et al., "Reduced heart rate variability is associated with increased arterial stiffness in youth with type 1 diabetes," *Diabetes Care* **36**(8), 2351–2358 (2013).
158. C. Lal and S. N. Unni, "Correlation analysis of laser Doppler flowmetry signals: A potential non-invasive tool to assess microcirculatory changes in diabetes mellitus," *Med. Biol. Eng. Comput.* **53**(6), 557–566 (2015).
159. M. J. Stevens, E. L. Feldman, and D. A. Greene, "The aetiology of diabetic neuropathy: The combined roles of metabolic and vascular defects," *Diabet. Med.* **12**(7), 566–579 (1995).
160. M. J. Stevens, J. Dananberg, E. L. Feldman, et al., "The linked roles of nitric oxide, aldose reductase and, (Na+, K+)-ATPase in the slowing of nerve conduction in the streptozotocin diabetic rat.," *J. Clin. Invest.* **94**(2), 853–859 (1994).
161. I. A. Mizeva, "Phase coherence of 0.1 Hz microvascular tone oscillations during the local heating," *IOP Conf. Ser. Mater. Sci. Eng.* **208**, 12027 (2017).
162. Y. Shi and P. M. Vanhoutte, "Macro- and microvascular endothelial dysfunction in diabetes," *J. Diabetes* **9**(5), 434–449 (2017).
163. A. Stefanovska, M. Bracic, and H. D. Kvernmo, "Wavelet analysis of oscillations in the peripheral blood circulation measured by laser Doppler technique," *IEEE Trans. Biomed. Eng.* **46**(10), 1230–1239 (1999).
164. I. Mizeva, I. Makovik, A. Dunaev, A. Krupatkin, and I. Meglinski, "Analysis of skin blood microflow oscillations in patients with rheumatic diseases," *J. Biomed. Opt.* **22**(7), 070501 (2017).
165. N. V. Malyuzhinskaya, K. V. Stepanenko, and E. I. Volchansky, "Assessment of the functional state of the microvasculature in children with diabetes mellitus type 1," *Med. Her. South Russ.* **11**(2), 71–80 (2020).
166. A. P. Vasil'ev, N. N. Strel'tsova, I. S. Bessonov, and A. V. Korotkikh, "State of microcirculation in patients with atherosclerosis and diabetes mellitus after limb revascularization," *Angiol. Sosud. Khir.* **26**(1), 22–29 (2020).
167. S. Wang, Y. Liu, H. Zhang, et al., "Effects of an SGLT2 inhibitor on cognition in diabetes involving amelioration of deep cortical cerebral blood flow autoregulation and pericyte function," *Alzheimer's Dement.* **16**(S2), e037056 (2020).
168. E. Fusco, M. Pesce, V. Bianchi, et al., "Preclinical vascular alterations in obese adolescents detected by Laser-Doppler Flowmetry technique," *Nutr. Metab. Cardiovasc. Dis.* **30**(2), 306–312 (2020).
169. P. A. Glazkova, D. A. Kulikov, A. A. Glazkov, et al., "Reactivity of skin microcirculation as a biomarker of cardiovascular events. Pilot study," *Clin. Hemorheol. Microcirc.* **78**, 247–257 (2021).
170. L. A. Allen, M. Terashvili, A. Gifford, and J. H. Lombard, "Evaluation of cerebral blood flow autoregulation in the rat using laser doppler flowmetry," *JoVE* **155**, e60540 (2020).
171. I. Mizeva, V. Dremin, E. Potapova, et al., "Wavelet analysis of the temporal dynamics of the laser speckle contrast in human skin," *IEEE Trans. Biomed. Eng.* **67**(7), 1882–1889 (2020).
172. G. Lancaster, A. Stefanovska, M. Pesce, et al., "Dynamic markers based on blood perfusion fluctuations for selecting skin melanocytic lesions for biopsy," *Sci. Rep.* **5**(1), 12825 (2015).
173. X. Hou, X. He, X. Zhang, et al., "Using laser Doppler flowmetry with wavelet analysis to study skin blood flow regulations after cupping therapy," *Ski. Res. Technol.*, **27**, 1–7 (2020).
174. E. V. Zharkikh, Y. I. Loktionova, I. O. Kozlov, et al., "Wearable laser Doppler flowmetry for the analysis of microcirculatory changes during intravenous infusion in patients with diabetes mellitus," *Proc. SPIE* **11363**, 113631K (2020).
175. A. Sagaidachnyi, A. Fomin, D. Usanov, and A. Skripal, "Real-time technique for conversion of skin temperature into skin blood flow: Human skin as a low-pass filter for thermal waves," *Comput. Methods Biomech. Biomed. Eng.* **22**(12), 1009–1019 (2019).

5 Wearable Sensors for Blood Perfusion Monitoring in Patients with Diabetes Mellitus

Evgenii A. Zherebtsov, Elena V. Zharkikh, Yulia I. Loktionova, Angelina I. Zherebtsova, Viktor V. Sidorov, Alexander I. Krupatkin, and Andrey V. Dunaev

CONTENTS

5.1 INTRODUCTION

The system of microvasculature regulation in the skin, which plays a key role in a number of pathological conditions (arterial hypertension, coronary heart disease, diabetes mellitus (DM), etc.), is involved in the pathological process at the earliest stages, leading to further complications that are severe and difficult to treat. It is known that functional disorders of the microcirculatory system manifest themselves earlier than morphological changes in the structure of the microvascular network. The search for markers of the preclinical stage of microvascular complications in patients with DM is a highly promising research area for potential early diagnostics. The identification of early markers of endothelial damage can become the basis for finding methods to prevent vascular

DOI: 10.1201/9781003112099-5

complications of DM. Assessment of the microcirculation may conveniently be performed in the skin because of its ease of access. The cutaneous blood flow can be evaluated using various optical diagnostic methods [1], of which laser speckle, video capillaroscopy, optical coherence tomography, and laser Doppler flowmetry (LDF) are most frequently used [2–4].

Wearable electronic diagnostic devices are currently undergoing rapid development. The main reason for this interest has to do with the possibility of daily monitoring, which promises a new level of diagnostic quality. The emergence on the market of ultra-compact, energy-efficient semiconductor lasers, miniature optical spectrometers, and energy-dense lithium batteries allows the development of new sensors for recording blood microcirculation and monitoring endogenous fluorescence parameters [5,6].

This opens new perspectives for innovative approaches in diagnostics with wearable medical electronic devices. Fine analysis of the structure and rhythm of capillary blood flow in several areas of the patient's skin simultaneously recorded by a group of wearable devices provides fundamentally new information about the distributed dynamics of blood microcirculation parameters of the whole organism. Monitoring of endogenous fluorescence with excitation at wavelengths of 375 nm and 450 nm has proved to provide information on the dynamics of mitochondrial activity and oxidative phosphorylation in skin cells, which in some cases may be associated with oxidative stress, as well as a marker of the excess accumulation of glycated protein structures [7,8]. The combined monitoring of blood microcirculation and skin fluorescence parameters, blood pressure parameters, and blood glucose dynamics has great potential to be a groundbreaking diagnostic solution for monitoring microcirculatory disorders and is of special interest in predicting the course and onset of diabetic complications.

5.2 MICROCIRCULATORY TISSUE SYSTEM OF THE SKIN AND DIAGNOSTIC METHODS

Microcirculatory tissue system (MTS), the functional element of an organ according to A.M. Chernukh, is a structural and functional complex composed of specialized parenchyma cells, cells of the connective tissue suspended in a non-cellular matrix, blood and lymphatic microvessels (Figure 5.1), and fiber nerve endings, which is combined into a single system by regulatory mechanisms. The MTS involves both morphological and functional substrates. From a structural point of view, it is advantageous to use a four-compartment (4C) model for studying the MTS – blood microvessels (blood compartment), interstitial space (non-cellular component of connective tissue), lymphatic microvessels (lymph compartment), and tissue cells (Figure 5.1). The functional substrate of the MTS is provided by regulatory mechanisms.

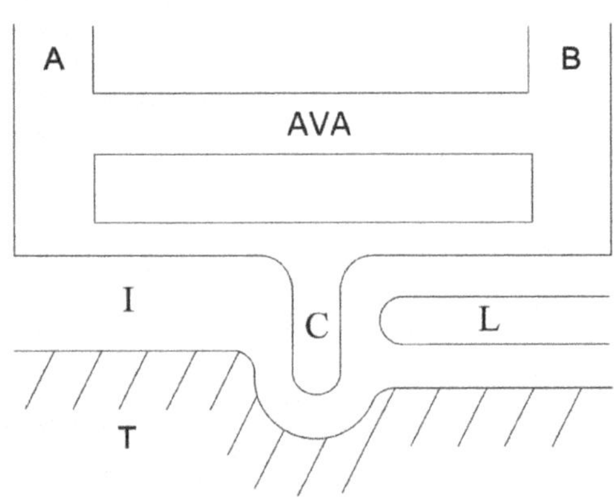

FIGURE 5.1 Schematic representation of the microcirculatory tissue system. A, arteriole; B, venule; AVA, arterio-venular anastomosis; C, capillary; I, interstitial space; L, lymphatic capillary; T, tissue cells.

TABLE 5.1
The Morphological Characteristics of Blood Vessels

Vessels	Internal Diameter (Range Limits in Parentheses), μm	Presence of Smooth Muscle Cells	Synonyms
Arterioles	20–35 (17–70), large arterioles up to 150 μm (in some studies)	+, one continuous line	Arcade arteriole, afferent arteriole
Precapillary arteriole	12–15 (7–20)	+, one discontinuous line	Terminal arteriole*, metarteriole, precapillary, arterial capillary
Precapillary sphincter	10–12	+, a few cells along a diameter of the arteriole	–
Capillary	7–8 (4–20)	–	Arterial, venous, true, main (magistral), sinusoidal capillary
Postcapillary venule	15–20 (10–30)	–	Postcapillary, collecting venule, venule, non-muscle venule
Venule	30–50 (25–100)	+, occurred in venules more than 300 μm in diameter**	Efferent venule, muscular venule
Arteriovenous anastomosis (AVA)	25–30 (15–100)	+	Arteriovenous anastomosis, arteriovenous shunt, semi-shunt

* Some authors use the term “terminal arterioles” to denote the final branching of arterioles, whose branches are metarterioles.

** A continuous layer of smooth muscle cells occurs in small veins and is absent in venous vessels less than 300 μm in diameter.

5.2.1 Microcirculation of Blood: Structure and Regulation

The morphological characteristics of blood vessels included in the MTS are given in Table 5.1.

Main directions in the regulation of peripheral microcirculation are as follows: regulation of volumetric blood flow through organs and tissues (function of arteries and arterioles); regulation of transcapillary exchange; and regulation of blood distribution (function of the venous bed).

Regulation in smooth muscle vessels is concerned with the diameter of these vessels and, as a result, with the volumetric blood flow. The contractile state of smooth muscle cells establishes the vessel diameter of a required size, which ensures active tonic tension for a long time, i.e., to maintain proper vascular tone (Table 5.2). Small arteries and arterioles play a major role in maintaining resistance to blood flow.

At the level of non-muscle metabolic microvessels (capillaries), the main objects for microcirculation regulation are the exchange surface area determined by the number of perfused capillaries and the exchange processes (mass transfer of water- and fat-soluble substances) occurred in the capillary wall.

5.2.2 Lymph Microcirculation: Structure and Regulation

The blood capillaries are responsible for delivering fluids to tissues, and the venules are known as the channels for carrying proteins. About 5%–10% of the capillary-venule filtrate is transported from the tissue to the lymph, and about 2–4 L of lymph per day is returned to the circulation. The lymphatic capillaries (LC) are blind vascular tubes with diameters ranging from 20 to 200 μm or, more often, from 10 to 60 μm (for comparison, blood capillaries are about 4–8 μm in diameter). In the English scientific literature, these capillaries are called initial, or terminal, lymphatics. When

TABLE 5.2
The Main Directions of Regulation of the Microvascular Tone under Physiological Conditions

Vessel	Sympathetic Regulation	Endothelium-Dependent Vasodilation	Myogenic (1), Metabolic (2) Regulation
Arteriole	++ / +	++(NO)	+
Arteriola	++	++(NO,EDHF)	++ (1)
Metarteriola	+	+(EDHF,NO?)	++(2)
Precapillary sphincter	–	+(EDHF,NO?)	+++(2)
True capillary	–	–	–
Muscle less venula	–	–	–
Small vein	+	+/–	+/–
Arterio-venular anastomosis	+++	+ (NO)	–

The number "+" shows the degree of regulation. NO, nitric oxide; EDHF, endothelial hyperpolarizing factor. Although precapillary sphincters contain myocytes, they do not have nervous synaptic regulation. Non-synaptic nervous regulation is retained even at the capillary level. NO-dependent vasodilation is most pronounced in small arteries and large arterioles. (1) and (2), basic regulation types.

the hydrostatic pressure of the interstitium is higher than that in the LC, interendothelial junctions are extended, peculiar pores (primary valves) of about 2 μm diameter are formed, and resorption occurs. When the pressure equilibrates, the endothelial cells close together, and the fluid ceases to flow into the lymphatic capillaries.

After lymph leaks out of the LC, it enters the lymphatic postcapillary (LP), which has a basement membrane, valves that prevent retrograde lymph flow, and single myocytes, including those in the valve zone. The next section of the lymphatic bed consists of lymphatic vessels (LV), which have a three-layer wall with smooth muscles (contractile or collecting lymphatics). The part between two valves of the LV is called a lymphangion. Each lymphangion has its own pacemaker located near valve areas. Actually, the activity produced by the valves is the major driving force of lymph (the intrinsic mechanism of lymph flow). The mechanism underlying the pacemaker activity is associated with increased intravascular pressure and stretching of vascular smooth muscles. The lymphatic system acts as an extensive drainage network, which is not attached to the heart and serves to transport and absorb proteins and their complexes with other substance lymph from the intercellular spaces between the cells of tissues [9,10].

5.2.3 Peculiarities of Peripheral Blood Circulation in Diabetes Mellitus

DM is one of the major public health challenges of the 21st century. According to the International Diabetes Federation, more than 400 million people are living with diabetes in the world, and about 90% of them are individuals with type 2 DM. DM is an endocrine disease resulting from the absolute or relative deficiency of pancreatic insulin production. There are several risks and complications associated with diabetes, including microangiopathy and/or macroangiopathy [11].

The microcirculation abnormalities, induced by incorrect functioning of MTSs, are the major cause of such serious diabetic complications, such as retinopathy, nephropathy, and polyneuropathy. From a morphological viewpoint, microvascular disease or dysfunction precedes the development of microangiopathy. There is also an opinion that the evolution of cardiovascular disorders is a risk factor for the progression of microangiopathy.

Of particular interest are the results supporting the fact that similar cardiovascular disorders, yet in a less volume, can in fact manifest themselves during the stage of prediabetes. Even before the development of diabetes, prediabetes is associated with increased risks of vascular disease, nephropathy, neuropathy, cancer, and dementia. Among patients with prediabetes, factors associated with insulin resistance, including increased oxidative stress, inflammation, impaired vascular fibrinolytic function, and dyslipidemia, contribute to microvascular and macrovascular diseases [11,12].

Diabetic macroangiopathy. DM is characterized by an increased stiffness of the arterial wall. Elevated glycemia is a major determinant of both arterial stiffness and carotid intimal media thickness. Increased peripheral vascular resistance predominantly leads to the development of arterial hypertension in DM.

Chronic hyperglycemia is known to be associated with the buildup of advanced glycation end-products (AGEs), which leads to arteriosclerosis. Endothelial dysfunction plays a primary role in the development of atherosclerosis and hypertension in DM. The development of oxidative stress that results from excessive reactive oxygen species (ROS) production and the inactivation or suppression of nitric oxide synthase not only due to hyperglycemia and but also due to the variability of blood glucose concentration are among the most common mechanisms of endothelial dysfunction [13].

Diabetic microangiopathy and the status of the microcirculatory tissue system (MTS). The course of diseases associated with tissue disorders in DM is chronic and progressive. Microvascular dysfunctions, especially in the long-term diabetes-related health problems, are characterized by a marked loss of microvessels (including capillaries), a decrease in tissue vascular density, an increase in capillary perfusion heterogeneity, vascular wall remodeling, basement membrane thickening, reduced NO bioavailability, activation of pathways for arachidonic acid metabolism, an increase in the myogenic tone of microvessels, endothelial dysfunction, and a reduction in the adaptive capabilities of microvascular networks.

Capillary bed depletion in patients with DM can be functional (a loss of perfused capillaries) or structural (a loss of vessels in the tissue) [14]. This suggestion is of particular importance for the bodies' most essential organs, namely, the heart. In DM, the depletion of the myocardial microvascular bed is not only functional but also structural later. This can trigger refractory or microvascular angina; coronary arteries remain relatively intact [15,16].

Remodeling of the vascular wall of small arteries and arterioles in response to hypertension differs in patients with or without type 2 diabetes. The total amount of wall tissue remains unchanged in patients without DM, whereas in patients with diabetes, hypertrophic remodeling appears as a result of changes in the myogenic response of vascular wall [17,18].

The severity of hypertrophy increases with diabetes duration, e.g., retinal microvascular abnormalities were recorded in patients with diabetes duration more than over 60 months [19].

Diabetes-induced endothelial dysfunction is also characterized by increased vascular permeability to albumin [20]. T. Deckert formulated a hypothesis, called the Steno hypothesis after the name Steno Diabetes Center, that vascular endothelial dysfunction is an early, key characteristic of the development of albuminuria in diabetes [21,22].

The progression of increased endothelial permeability is thought to be associated with cathepsin S. Cathepsin S secreted by the invading macrophages activates protease-activated receptor-2 on endothelial cells, resulting in the increase in microvascular permeability manifesting as albuminuria; administration of inhibitors of either cathepsin S or protease-activated receptor-2 prevented ultrastructural and functional endothelial abnormalities and attenuated albuminuria and glomerulosclerosis [22,23].

An increase in blood glucose disrupts the balance of NO produced in the vascular wall. NO-generating responses of endothelial cells to bradykinin or a calcium ionophore were diminished by elevated glucose levels. According to Goligorsky M.S, Chen J., Brodsky S. (2001), every hyperglycemic episode is associated with NO scavenging, sharp reduction in NO bioavailability, and impairment of NO-dependent endothelial functions, such as deterrence of leukocyte infiltration, platelet aggregation, imbalance in vasoreactive regulators, and reduced response to angiogenic cues (VEGF) [24].

A major component of the blood vessel, an extracellular matrix, also undergoes modifications in DM – the ratio of collagen to elastin increases due to inflammatory changes and fibrosis.

It should be noted that diabetes-induced microcirculation changes occur due to tissue metabolic changes, namely, due to increased oxidative metabolism. Impaired oxidation, reduced mitochondrial contents, lowered rates of oxidative phosphorylation, and excessive ROS production have been reported [25].

Diagnostics. Recent studies show that sensors displaying cutaneous microcirculation alteration in DM have a promising potential for predicting cardiac emergencies and, first, myocardial infarction, since structural microcirculatory disorders precede their development [26].

Analysis of the cutaneous hyperemic response confirmed the relationship between functional microvascular disorders and glycemic control. The severity of glycemic control (percentage decrease in HbA1c over a 12-month period) significantly correlated with the percentage increase in the level of cutaneous hyperemic response [27–29].

A link between the risk factors for coronary heart disease and the microcirculatory dysfunction was established via the results illustrating the functional state of skin microvessels (the number of perfused capillaries and endothelium-dependent vasodilation) [30].

Insulin-induced capillary recruitment and microvascular dilation, measured by LDF during iontophoresis of acetylcholine, correlate with insulin sensitivity and are reduced in obese compared to lean women at prediabetes [31].

The state of cutaneous microvascular function at diabetes does not correlate with the type and duration of diabetes. This fact indicates that diabetes *per se* causes the microvascular dysfunction to occur [32].

The totality of the evidence submitted above demonstrates that the noninvasive methods are very promising techniques to study the functional state of microcirculation at DM [33].

5.2.4 Laser Doppler Flowmetry of the Blood Microcirculation

The name of the method "laser Doppler flowmetry" reflects its basic ideas. For diagnostic purposes, the probing tissue area is exposed to laser radiation, and the laser beam reflected by the tissue is analyzed based on the extraction of the signal frequency (proportional to the particle velocity in the microvasculature) from the Doppler shift. This makes it possible to register microvascular changes in the blood or lymph flow using the flowmetry technique [10].

The flowmetry result can be expressed as:

$$Im = K \times N \times V_{av}, \tag{5.1}$$

where

Im – microcirculation index (signal amplitude in volts);

K – the coefficient of proportionality ($K = 1$);

N – the number of scattering particles in the probed tissue;

V_{av} – average flow velocity of scattering particles in the probed tissue.

The main scattering particles in blood microvessels are RBCs, and in microlymph vessels – those scattering particles in the interstitial spaces during lymph formation.

In English-language publications, the blood flow parameter measured by LDF is known as red (blood) cell flux, blood flux (flow), and volume flux. In 1992, the European Laser Doppler User Group (London) recommended the term "laser Doppler perfusion" to describe the output signal defined as the product of the linear velocity of red blood cells and their concentration.

The signal amplitude, which is measured in arbitrary units (AU), is proportional to product (5.1).

Active factors (tone-forming mechanisms directly affecting microvessels) regulating vascular lumen formation are as follows: endothelial, myogenic, and neurogenic. These factors modulate the blood flow by the vascular wall movement and are realized through its muscular-tonic component. The executive object or "target" of active control factors is the muscular component of the vascular wall. Under physiological conditions, the targets of neurogenic regulation are arterioles and arterio-venous anastomoses, the myogenic regulation component in its pure form is localized to precapillaries and sphincters, and the endothelial vessel diameter regulation predominantly affects the most proximal vessels (small arteries, large arterioles).

Passive factors (mechanisms that occur outside the microcirculation system) are the arterial pulse wave and the suction action caused by vein valves (respiratory pump). These oscillations are transmitted (together with the blood flow) into the probed area, since the microcirculatory bed, which is a component of the general circulatory system, is topographically located between arteries and veins.

The influence of active and passive factors on the blood flow leads to a change in the rate and concentration of the erythrocyte flow. These changes cause perfusion modulation and are recorded as a complex oscillatory process.

Active factors create transverse fluctuations in the blood flow due to alternating contraction and relaxation of vascular muscles (alternating episodes of vasoconstriction and vasodilation). Passive factors organize longitudinal fluctuations that manifest themselves as periodic changes in pressure and blood volume in the vessel. In the arterioles, the nature of these changes is determined by the pulse wave, in the venules – by the respiratory pump oscillations.

The variable $\delta I_m(t)$ provides valuable information about the blood flow modulation. The decoding, analysis, and interpretation of this component allow diagnosing the state of vascular tone and mechanisms of regulation of blood flow in the microvasculature. As soon as the constant component of the LDF signal I_m characterizes the amount of perfusion, the variable $\delta I_m(t)$ indicates mechanisms that regulate perfusion. Therefore, to diagnose the functional state of the microvasculature, it is essential to analyze both components. The oscillatory process recorded in the LDF graph occurs due to the superposition of oscillations induced by active and passive factors (Figure 5.2).

The blood flow in the microcirculatory bed is not stable, and hence, it is variable. The parameter σ (Figure 5.2) is the average perfusion fluctuation relative to the average value of the blood flow M, calculated by the formula for mean deviation, and it has a dimension in AU. The parameter σ characterizes the temporal variability of perfusion and reflects the average modulation of blood flow occurred due to the active and passive factors mentioned above.

The higher the σ value, the deeper the microcirculation modulation. In the analysis of the LDF results, it is reasonable to use the ratio of σ to M; the coefficient of variation $Kv=\sigma/M \cdot 100\%$.

The Kv value expresses the tension of the regulatory systems of the microvasculature.

The study of the microcirculation oscillation amplitudes within specific frequency ranges makes it possible to assess the functional state of some perfusion control mechanisms (Table 5.3).

Blood flow oscillations are of hemodynamic significance. An increase in the amplitudes of oscillations in the active tone-forming frequency ranges is accompanied by a decrease in the effective resistance to blood flow.

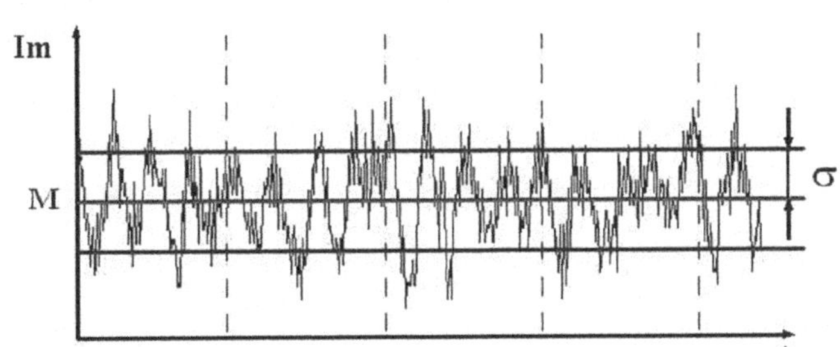

FIGURE 5.2 LDF graph, M, arithmetic average of the blood microcirculation index – I_m,; and σ, mean deviation of the microcirculation index.

TABLE 5.3
The Frequency Ranges of Blood Flow Oscillations in Skin Microvessels Calculated using Wavelet Analysis

Frequency Range		Mean Peak Frequency			
Hz	n /min	Hz	n/min	Physiological Nature	Diagnostic Value
0.005–0.0095	0.3–0.6	0.007	0.4	Endothelial	Assessment of the endothelial dysfunction by studying the changes in oscillation amplitudes at approximately 0.007 and 0.01 Hz (NO-dependent range). Oscillations with frequencies of approximately 0.01 Hz predominate (especially, in a combination with a high-amplitude pulse rhythm), which indicates dilatation of small arteries and large arterioles.
0.0095–0.02	0.6–1.2	0.01	0.6	Endothelium-dependent nitric oxide	
0.02–0.046	1.2–2.8	0.03	1.8	Neurogenic sympathetic adrenergic	Assessment of the oscillatory component of sympathetic adrenergic regulation of arterioles and AVAs, the amplitudes of which predominate among active oscillations, indicating that the ergotropic direction in the regulation of microhemocirculatory-tissue systems predominates. An increase in the amplitudes of neurogenic oscillations and a decrease in the amplitude of myogenic oscillations in the skin zones with ABA indicate a decrease in resistance and the activation of non-nutritive (shunt) blood flow pathways.
0.047–0.069	2.8–4.1	0.06	3.6	Sensory and peptidergic (including, hyperthermia, myogenic autoregulation)	The presence of high-amplitude oscillations in this range indicates the activation of sensory peptidergic fibers. Their identification in the spectrum and a significant decrease in the M value indicate the activation of the myogenic tone of arterioles. These oscillations are of great importance for the diagnosis of neurogenic inflammation, sensory-sympathetic mating, and sympathetic pain dependence, e.g., in complex regional pain syndrome. Note that sensory peptidergic fibers secreting neuropeptides (substance P, calcitonin-gene-related peptide, neurokinin A) are the main component of nervous tissue trophism.
0.07–0.145	4.2–8.7	0.1	6	Self-myogenic (vasomotions)	Assessment of the state of the oscillatory component of the muscle tone of precapillaries regulating blood flow to the nutritional bed.
0.16–0.18	9.6–10.8	0.17	10	Cholinergic, parasympathetic	These oscillations reflect the influence of central trophotropic mechanisms (including parasympathetic centers). Their occurrence in the spectrum of blood flow oscillations in the area of skin devoid of parasympathetic innervation (e.g., skin with ABA) indicates a decrease in the ergotropic central regulatory component and a shift in the central regulation in the trophotropic direction.

(Continued)

TABLE 5.3 (*Continued*)
The Frequency Ranges of Blood Flow Oscillations in Skin Microvessels Calculated using Wavelet Analysis

Frequency Range		Mean Peak Frequency			
Hz	n /min	Hz	n/min	Physiological Nature	Diagnostic Value
0.2–0.4	12–24	0.3	18	Passive respiratory	An increase in the amplitude of the respiratory wave indicates a decrease in microcirculatory pressure and/or a deterioration in venous outflow. A deterioration of the outflow of blood from the microcirculatory bed increases the number of red blood cells, and this is accompanied by an increase in the amplitude of the respiratory wave. The respiratory wave does not directly reflect the blood flow of the venous parts of the capillaries and venules; it is primarily associated with the respiratory modulation of blood flow.
0.8–1.6	48–96	1	60	Passive cardiac	An increase in the amplitude of the pulse wave with increased or normal values of the average perfusion M indicates an increase in the inflow of arterial blood into the microcirculatory bed.

To compare the amplitudes of oscillations recorded in different subjects obtained under various environmental conditions, the normalized oscillation amplitudes should be taken into consideration. In the literature devoted to LDF, there are two options for normalization of amplitudes: one is M-normalization and the other σ-normalization.

1. Determination of the contribution of the oscillation amplitude of a certain frequency range relative to the average modulation of blood flow is estimated by the formula A/σ (where A is the amplitude of oscillations, and σ is the root-mean-square deviation of perfusion oscillations, the average modulation of blood flow).
2. Determination of the contribution of the amplitude of oscillations relative to the M value is estimated by the formula A/M. The A/M value characterizes the tension regulation by individual active regulatory factors or the modulation of blood flow by passive (pulse, respiratory) mechanisms.

Analysis of the wavelet spectrum shows a pronounced increase in the oscillation amplitudes associated with endothelial and/or sympathetic activity ranges, which is indicative of a predominance of arteriole blood flow, arterio-venular anastomoses (AVAs), and a relative decrease in nutritional blood flow. On the contrary, an increase in the oscillation amplitudes or in resonant myogenic oscillations counts in favor of the activation of capillary perfusion.

There are examples of amplitude-frequency blood flow oscillation spectra at hyperemia and venous stasis, which are shown in Figures 5.3 and 5.4.

Assessment of blood flow shunting. The total value of the shunting component (SC) is calculated by the formula $SC = SC1 + SC2$, where $SC1$ is the shunting component associated with the differences in the tone in the microvessels of the nutritive and non-nutritive blood flow pathways directly within the microvascular bed, and $SC2$ is the shunting component associated with the differences in the perfusion of microvessels and larger vascular segments (arteries, venules, and veins) in cases of

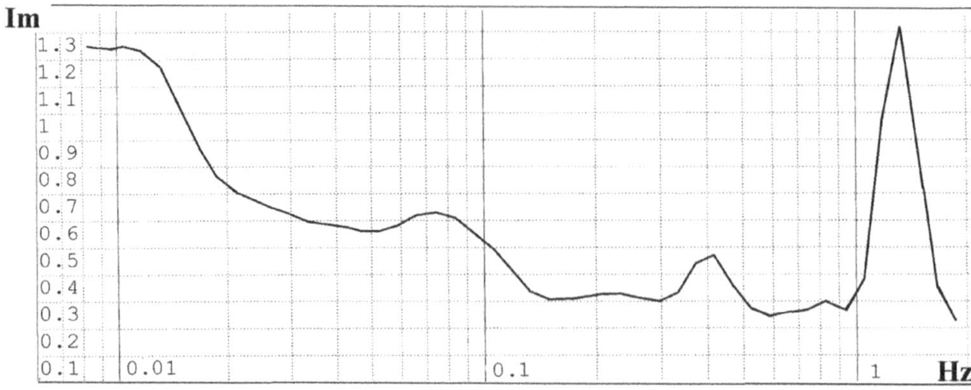

FIGURE 5.3 Wavelet spectrum of blood flow oscillations in the skin of the palmar surface of the second finger. First day after desympathization (thoracoscopic clipping of the Th3 ganglion), perfusion units (vertical axis), and oscillation frequency (Hz) (horizontal axis). A pronounced increase in the amplitudes of the endothelial rhythm with a peak frequency of 0.01 Hz and a cardiac (pulse) rhythm with a peak frequency of 1.3 Hz. The activity of oscillations of sympathetic genesis is not shown.

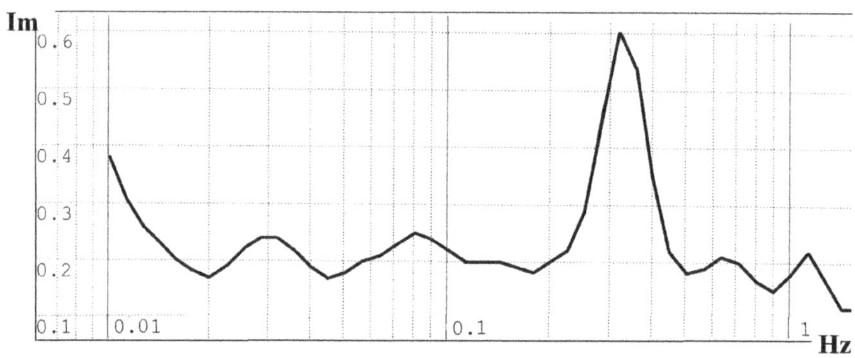

FIGURE 5.4 Wavelet spectrum of blood flow fluctuations in the skin of the palmar surface of the 5th finger. One month after damage to the ulnar neurovascular bundle, perfusion units (vertical axis), and oscillation frequency (Hz) – horizontal axis. A pronounced resonant increase in the amplitude of the respiratory rhythm with a peak frequency of 0.31 Hz, venular stasis.

arterial hyperemia or venous stasis. It should be noted here that the proportion of nutritional perfusion in the probed tissue also decreases in this case.

Calculation of SC in the areas with arterio-venular anastomoses.

In humans, AVAs are numerous in the skin of the palmar and plantar surfaces of the fingers and toes, the auricles, the tip of the nose, lips, and forehead.

I. $SC1 = A_n/A_m$, where A_n and A_m are, respectively, the maximum amplitudes of oscillations of the sympathetic adrenergic and myogenic frequency ranges. In the areas with AVA, the SC value represents the ratio of the shunt and non-shunt (nutritive) blood flows, i.e., it shows the degree to which the former exceeds the latter. At $SC1 = 1$, the shunt and non-shunt (nutritive) blood flows are 50% and 50%.

II. $SC2 = A_c(r)/A_m(sp)$, where $A_c(r)$ is the dominant amplitude of oscillations among the cardiac- and respiratory-rhythm oscillations, and $A_m(sp)$ is the dominant amplitude of oscillations among the myogenic and sensor peptidergic oscillations. *SC2* is taken into consideration only when its value ≥ 1, i.e., when the dominant amplitude in the passive ranges is equal to or exceeds the $A_m(sp)$ value. These states may correspond to arterial hyperemia (A_c dominance) or to venous congestion (A_r dominance).

Calculation of SC in the areas without arterio-venular anastomoses.

The most commonly involved areas include the skin of the forearm and dorsal surfaces of hands, feet, fingers, and toes.

$SC1 = A_{max}/A_m$, where A_{max} is the maximum averaged amplitude of oscillations, which dominates in magnitude among all active tone-forming frequency ranges – 0.005–0.145 Hz. For instance, if oscillations of the endothelial rhythm A_e dominate, then its value is used; if A_m dominates, then its value is used; if A_n dominates, then its value is used. If in the spectrum $A_{max} = A_{sp}$, then $SC1 = A_{sp}/A_m$. In the skin areas without AVA, the *SC1* value reflects the proportion of the shunt blood flow in the main vessels in the total hemodynamics of the area under study. In these areas, *SC1* is not less than 1. If $SC1 = 1$, then, during the LDF recording, the entire blood flow enters the nutritive bed. If *SC1* is greater than 1, then the difference from 1 shows to what extent the total blood flow exceeds the nutritive one.

Calculation of SC2 is similar to that carried out for the areas with AVA.

Several options are available to shunt blood flow in the microcirculatory bed. However, the achievement of nutritional perfusion in microhemodynamics is not an end in itself. The nutritional blood flow typically matches the metabolic needs of tissues, and therefore, the maintenance of shunting is an important mechanism protecting capillaries and tissues from different venous disorders, e.g., the excess buildup of fluid around tissues (called edema).

Particular attention should be paid to one more aspect of functional shunting – the possibility of redistribution of perfusion deep inside the tissue or organ. For the skin, this is a reduction in the blood supply of superficial microvessels due to a decrease in the M value, for example, in the acute stage of inflammation.

Diagnostics of the state of perfusion is via shunt and nutritional pathways. Calculation of the SC value makes it possible to assess perfusion by examining the nutritional and shunt pathways in the microvascular networks. In the areas with AVAs, the value of nutritive perfusion (M_{nutr}) is calculated by the formula $M_{nutr} = M/(1+SC_l)$. For zones without AVA, $M_{nutr} = M/SC$. Accordingly, the value of shunt perfusion (M_{shunt}) is estimated by the formula $M_{shunt} = M - M_{nutr}$. All the perfusion indices (*M*, M_{nutr}, and M_{shunt}.) are measured with AU.

Assessment of the tone of microvessels. The oscillatory component of the microvascular tone is determined by the formula σ/A, where σ is the standard deviation from *M* and *A* is the value of the amplitude of perfusion oscillations in the corresponding active tone-forming frequency range (endothelial, sympathetic neurogenic, myogenic, etc.). The stationary component of tone (MVT) of resistive vessels (small arteries and arterioles) at the entrance to the studied microvascular network is estimated by the formula $MVT = AP_{avg}/M$ (mmHg/au) (AP_{avg} – arterial pressure). This factor determined for the skin is identical to that used in the international journals for skin vascular resistance (SVR). The stationary component of the tone of the precapillaries and precapillary sphincters, as well as AVA, is not calculated, because the AP_{avg} values are not measured separately in these segments of the microvascular network using noninvasive methods.

It would also be interesting to consider the significance of the amplitude of cardiac pulse oscillations (A_c) in the study of resistance to blood flow – the general tone of resistive vessels (small arteries, arterioles), as well as in the study of intravascular resistance or rheological component. The fluctuations of the heart rate are directed along the blood flow, and their genesis is primarily associated with changes in the arterial pressure gradient. A_n increase in the value of AP_{avg}/A_c indicates an increase in vascular and intravascular resistance, and its decrease says otherwise.

The main topographic localization sites of the components of regulation of the tone of microvessels of the skin with AVAs are shown in Figure 5.5. As one can see, the oscillatory component of myogenic tone occurs in all muscle-containing microvessels, but, in the zone of precapillary sphincters and precapillaries, it is realized in a relatively "pure" form.

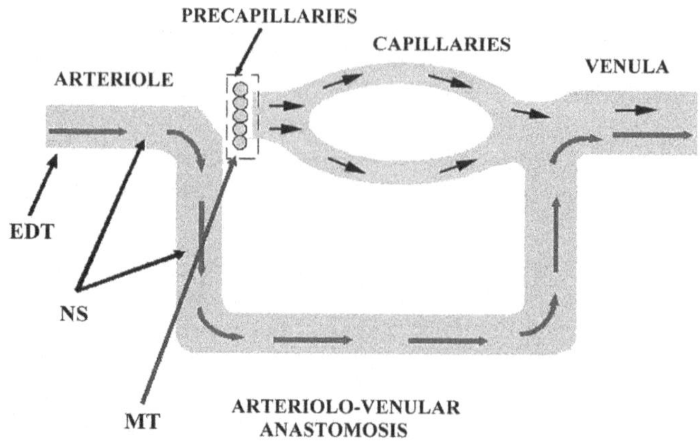

FIGURE 5.5 Localization of regulation of neurogenic sympathetic (NS), myogenic (MT), and endothelium-dependent (EDT) tone components.

TABLE 5.4
LDF Parameters at Blood Circulatory Disorders

	Arterial Hyperemia				
Parameter	**Non-nutritive**	**Nutritive**	**Venous Stasis**	**Ischemia**	**Stasis**
M, AU	↑	↑	= or ↑	↓	↓
Proportion of M_{nutr}	↓	↑	± **	± (usually ↓)	-----*
Proportion of M_{shunt}	↑	↓	±	±	-----*
A_c	↑↑ or ↑	↑	=	frequently ↓	±, ↓ (in ischemic stasis)
A_r	=	=	↑↑	↓	±, ↑ (in venous stasis)
A_c/A_r	↑ (more than 1)	↑ (more than 1)	↓ (≤ 1)	± (more than 1)	± (about 1)
Specific features of oscillations of the active tone-forming range	A_e increases	The amplitude of oscillations of the overall myogenic range dominates	A_e decreases	In chronic ischemia, the amplitudes of oscillations decrease	Sharp suppression of oscillations

* – Shunting component is not determined due to the absence of oscillations; ↑ – increase, ↓ – decrease, = – unchanged, ± – minor deviations in comparison with control, ** – in chronic venous stasis, the density and perfusion of capillaries are decreased.

5.2.4.1 Typical Disorders of Peripheral Blood Circulation

Typical disorders of peripheral blood circulation include arterial (active) hyperemia, venous hyperemia (congestion, passive hyperemia), ischemia, and stasis (Table 5.4).

The M value and the state of oscillatory processes associated with the arterial flow (A_c) and respiratory modulation of venular outflow (A_r) are of primary importance in diagnosing typical circulatory disorders. It is therefore advisable to make an overall (rather than isolated) assessment of one of the parameters.

The options shown in Table 5.4 correspond to "pure" and more often "acute" cases of circulatory disorders, yet, in clinical practice, the following mixed forms may take place: mixed arterio-venous hyperemia (congestive hyperemic form and congestive ischemic form).

The mixed arterio-venous hyperemia is transitional from arterial hyperemia to stagnation, e.g., in the inflammatory foci, and the congestive ischemic form – from venous stasis to stasis. In the case of a congestive hyperemic form, an increase in the inflow is accompanied by a decrease in the blood outflow along the venules, but the A_c/A_r amplitude ratio decreases; the remaining LDF indicators are intermediate between arterial hyperemia and congestion. In the case of a congestive ischemic form, an increase in pressure in the venules causes a reduction in pressure in the arterioles. The wavelet spectrum is characterized by a combination of a distinct decrease in M, an increase in A_r, and a decrease in A_c; the remaining LDF indicators are intermediate between stagnation and ischemia.

In differential LDF diagnostics, the following algorithm is used. At the first stage of the algorithm, the direction of the A_c/A_r ratio is determined.

I. $A_c/A_r > 1$. In these cases, an increase in the value of M makes it possible to diagnose arterial hyperemia, and a decrease in the M value is an indicative of ischemia.
II. $A_c/A_r \leq 1$, which is typical of congestive forms. If there is no decrease in the M value, and A_c is within the normal range, then the venous stasis is diagnosed. If the M and A_c values increase, then a congestive hyperemic form is diagnosed. If the M_c value is reduced, then a congestive ischemic form is diagnosed (more often, A_c is also reduced, but not in all cases).

5.2.4.2 Functional Classification of Microcirculatory Disorders

Analysis of microcirculatory disorders includes the following steps:

(1) the presence and definition of the type of microcirculatory disorders, (2) the severity of hemodynamic disorders, and (3) their compensation. This analysis shows the most promise in clinical trials.

Typical circulatory disorders in the microcirculatory bed: arterial hyperemia (non-nutritive, nutritive), venous congestion, ischemia, stasis, mixed – congestive hyperemic and congestive ischemic forms.

Severity of hemodynamic disorders:

- no disorders – perfusion parameters, oscillation amplitudes, and shunting indices correspond to the mean values of the control group with a deviation of $\pm 20\%$;
- moderate disorders – the increase or decrease in these parameters is 21%–50%, as compared to the mean values of the control group;
- pronounced disorders – the increase or decrease in these parameters is 51% or even more, as compared to the mean values of the control group.

5.2.4.3 Compensation of Hemodynamic Disorders

Blood flow is compensated if the perfused tissue and its metabolism are not affected by disorders. Certainly, it is unreasonable in this case to talk about the compensation or decompensation of blood flow only by evaluating the severity of vascular disorders. Therefore, it is advisable to assess the degree of compensation along with the functional parameters of the state of tissue metabolism (laser-induced fluorescence spectroscopy data – Section 5.2.9). It is also essential to consider one more aspect associated with the circulation compensation analysis – assessment of the compensation for the main blood flow deficiency in the vascularization zone of a large arterial vessel. In this case, hemodynamic parameters and microvascular perfusion values have to be analyzed jointly. This approach is effective in studying, for example, an occlusive arterial lesion in the low extremity, and it helps to establish the severity of ischemic disorders.

Function tests. For the purposes of the study of MTSs, the following function tests can be performed:

1. Respiratory test (breath-holding test)
2. Postural test

3. Local cold test
4. Local thermal test
5. Occlusion test
6. Pharmacological, including iontophoretic tests
7. Electrostimulation test.

Test methodologies and the interpretation of test results have been previously described in detail [10].

5.2.5 Assessment of the Adaptive Processes in the Microcirculatory Bed

Oscillatory processes are an integral component of adaptation and formation of the functional states of the microcirculatory bed. Among the functional states of microhemocirculatory-tissue systems, adaptive states (corresponding to the physiological norm), maladaptive (hyper- and hypoadaptive) states, and the failure of adaptation can be distinguished. Adaptation deficiency is the most common cause of almost all diseases. A certain level of angioadaptation is an integral component of human health, whereas the maladaptive microcircular network alterations are involved in the pathogenesis of many diseases. The key characteristics that define the state of adaptive processes are the coefficient of variation of blood flow and the state of the information mode in the wavelet spectrum of blood flow oscillation (multistable or resonance) (Table 5.5).

5.2.6 Laser Doppler Flowmetry of Lymph Microcirculation

LDF of lymph microcirculation [2] is implemented within a range of human cutaneous lymph flow velocities from 5 to 30 μm/s. The record of the lymph flow index ($I_{ml} = K \cdot N_{ras} \cdot V_{avg}$) on the pad of the third finger is given in Figure 5.6 as an example. The lymph flow index is estimated in AU.

TABLE 5.5
Key Diagnostic Criteria for the Functional States of the Microhemocirculatory Bed

Functional States	Values of Indices
Adaptive states	Kv limits - from –25% to+30%, as compared to the mean control values
	Hyperadaptive States
Moderate functional tension	Kv limits - from+31% to+50%, as compared to the mean control values
Pronounced functional tension	Kv limits - from+51% and more, as compared to the mean control values
Hypertension	The presence of resonant oscillations in one of the frequency ranges of the wavelet spectrum
Pronounced hypertension	Resonant oscillations with frequency capture of one or several frequency ranges of the wavelet spectrum; no oscillations are registered in other (one or several) frequency ranges.
	Hypoadaptive States
Reducing the tension of regulatory systems	Kv limits - from –26% to –50% %, as compared to the mean control values
Depletion of regulatory systems	Kv limits- from –51% and more, as compared to the mean control values Cases of a twofold (or more) simultaneous reduction of M and σ
	Adaptation Failure
Adaptation failure	A sharp depletion of regulatory systems, no oscillations are registered in any of the frequency ranges of the wavelet spectrum Resonant oscillations with frequency capture of the neighboring ranges of the wavelet spectrum; no oscillations are registered in other frequency ranges (LDF recording of the sinusoidal type)

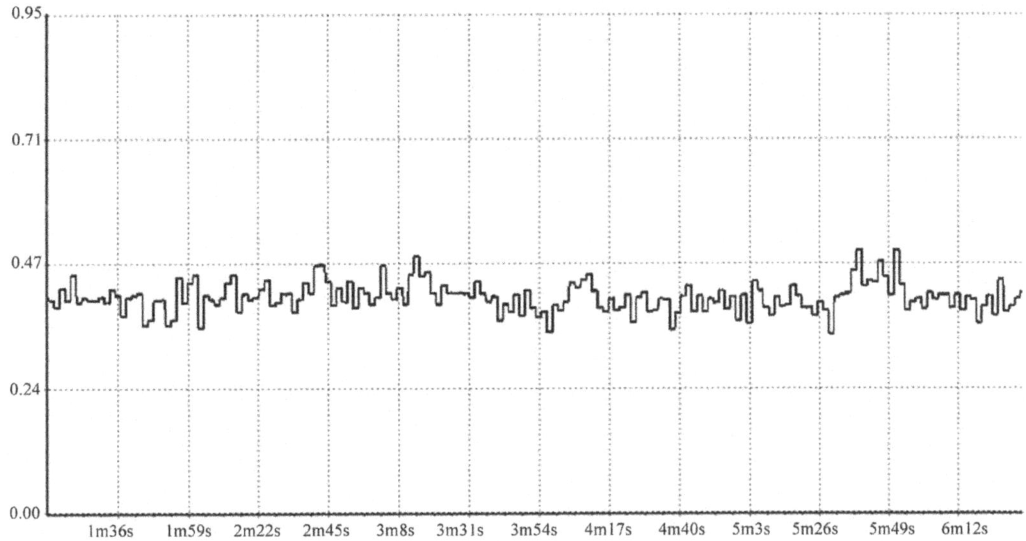

FIGURE 5.6 Record of the lymph flow index on the pad of the third finger; the y-axis – I_{ml} (ab. units), and the x-axis – time.

TABLE 5.6
Clinical and Physiological Interpretation of Changes in the A_l/M_l Index (A_l – Amplitude of Pacemaker Oscillations, ab. units, and M_l – Average Values Microlymphocirculation, ab. units)

A_l/M_l Index	No Edema	Presence of Edema
Within the limits of control values	Normal phase activity	–
Reduction	Reduction in the activity of phase oscillations	Low activity of phase oscillations
Increase	Increase in the activity of phase oscillations	Protective activation of phase oscillations

The average values of the lymph flow index and the variable are M_l=0.41 ab. units and σ_l=0.03 ab. units, respectively. For diagnostics, the calculated parameters which are similar to those from Section 5.2.9 are used.

The amplitude-frequency analysis in combination with wavelet transform makes it possible to identify the following frequency ranges of lymph flow oscillations in human skin: endothelial oscillations (0.005–0.015 Hz), pacemaker phase oscillations (0.016–0.042 Hz), myogenic oscillations (0. 05–0.145 Hz), and respiratory oscillations (0.2–0.4 Hz). Due to the high hydrodynamic resistance of the lymph nodes, respiratory rhythms can hardly be expected to penetrate into peripheral LV and microvessels, and this causes the genesis of respiratory oscillations (transmission of respiratory rhythms to thin-walled LV of adjacent venules and veins) to occur. In cases of severe edema, the activity of pacemaker oscillations is weak or not observed in the amplitude-frequency spectrum (Table 5.6).

The limits of control values are± 20%, as compared to mean values.

5.2.7 Joint Research of Blood and Lymph Flows

The lymphatic system provides additional, collateral to the veins, drainage of organs and simultaneous filtering of excess tissue fluids. From a diagnostic point of view, a joint study of microcirculation of blood flow and lymph flow is the most effective way to obtain relevant results.

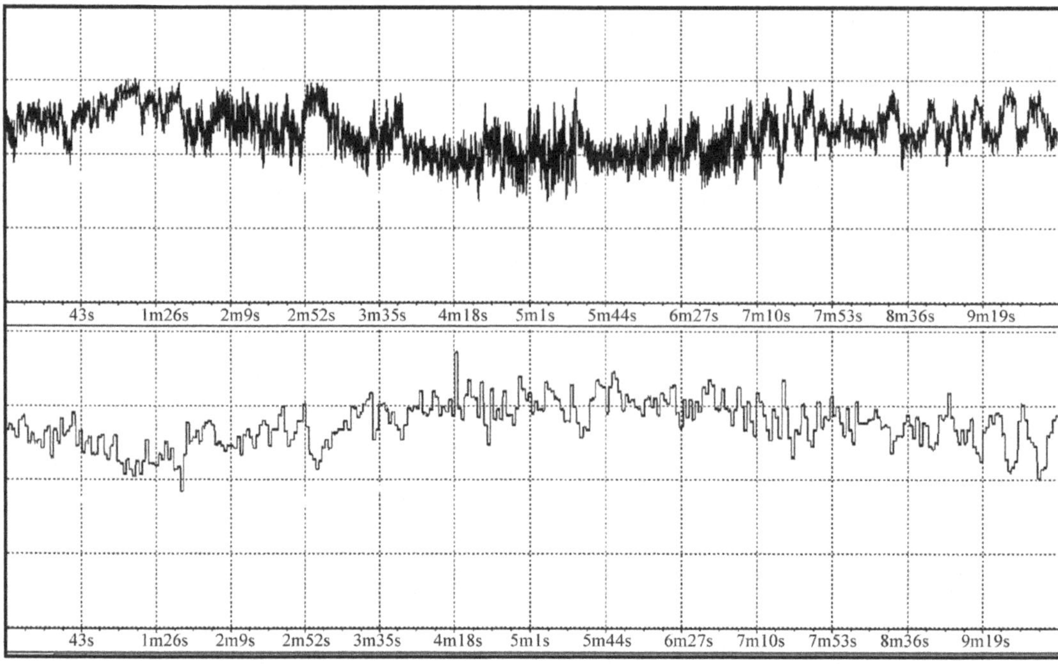

FIGURE 5.7 Example of simultaneous recording of microcirculation of blood flow (upper graph) and lymph flow (lower graph). Arrows show the antiphase oscillations of blood flow and lymph flow.

Figure 5.7 shows the "compensation effect": the dynamics of microcirculatory changes in blood flow and lymph flow in antiphase – an increase in blood flow microcirculation causes a decrease in lymph flow microcirculation, and vice versa.

Such simultaneous recordings of blood flow and lymph flow make it possible to reveal this effect, which is an indicator of the absence of disturbances in the physiological connection between these MTS compartments.

5.2.8 Laser-Induced Fluorescent Spectroscopy: Diagnostics of Oxidative Metabolism

Cellular metabolism, being an integral part of the MTS, is diagnosed using laser-induced fluorescence spectroscopy (LIFS). Oxidative metabolism is evaluated through autofluorescence imaging of such coenzymes as reduced nicotinamide adenine dinucleotide (NADH) and oxidized flavin adenine dinucleotide (FAD) contained in tissue cells [10,34]. NADH transfers electrons to molecular oxygen, and FAD acts as an electron acceptor. However, there are peculiarities associated with positioning of coenzymes in the cell – FAD is contained strictly in the mitochondria, and NADH – in both the mitochondria and the cytoplasm. It is generally considered that the NADH fluorescence intensity in the cytoplasm plays a minimal role in optical diagnostics. The fluorescence intensity of NADH is excited by radiation of a wavelength of 365 nm, taking its fluorescent spectrum in the range of 460–470 nm. To excite FAD, radiation of a wavelength of 450 nm is applied, and the fluorescence spectrum of FAD is in a 510–520 nm range.

Analysis of the dynamics of metabolism has revealed an increase in the amplitude of NADH fluorescence accompanied by the progression of pathology [35]. In DM, due to a decrease in the number of beta cells, the utilization of NADH decreases, and the fluorescence amplitude increases [34]. Thus, an increase in the concentration of NADH and FAD in tissue can be indicative of reduced oxidative metabolism.

For practical applications of laser fluorescence spectroscopy as a diagnostic indicator of the changes in the state of oxidative metabolism, it is essential to assess the dynamics of the fluorescence amplitude of the coenzyme NADH or FAD. This amplitude is normalized to the intensity of radiation sources needed to excite fluorescence [2,35].

The study of the MTS in relation to blood flow, lymph flow, and cell metabolism shows that diagnostics is most effective with the simultaneous registration of these MTS parts in one tissue volume.

According to Fick's principle, the activation of tissue oxidative metabolism is associated with an increase in blood flow to microvessels (dilatation of arterioles) and/or with an increase in the surface of diffusion and transcapillary metabolism – the number and area of capillaries – an increase in M_b or M_{nutr} indices. For the "compensation effect", an increase in perfusion is accompanied by a decrease in microlymph flow – M_l decreases.

The complex diagnostic oxidative metabolism index (OMI) caused by the state of the microvasculature of blood flow and the activity of oxidative metabolism can be represented by the following expression:

$$OMI = K \cdot M_{nutr} / Af_c, \tag{5.2}$$

where K is the coefficient of proportionality, M_{nutr} is the nutritive blood flow, and Af_c is the fluorescence amplitude of the coenzyme NADH or FAD, normalized to the intensity of radiation that excites fluorescence.

Microcirculation is spatially heterogeneous, and therefore, it is reasonable to assess the general state of the MTS by analyzing a set of data obtained in several study areas and with the aid of a distributed control system – flowmeters with a simultaneous control of the fluorescence of biomarkers of oxidative metabolism. The use of wearable compact wireless devices offers new opportunities for studying human MTSs [36]. They can be applied to identify the progression of pathology in symmetric areas even in the case of right- or left-sided asymmetry of diagnostic indicators [37]. To assess the general condition of the MTS, at least six sensors should be positioned in symmetric areas: two on the forehead in the vascularization zones of the superciliary arteries [36,38], two on the upper, and two on the lower extremities.

LDF of microhemo- and lymphocirculation and laser fluorescence spectroscopy are the main diagnostic techniques to evaluate the state of MTSs [10,33]. These methods are noninvasive and have no contraindications to their use. Skin and mucous membranes are most commonly as an object of analysis. For assessing the state of microcirculation of internal organs, it is advantageous to make use of the skin Zakharyin-Ged zones that are most suitable. Intraoperative methods are applicable to the analysis of any biological tissue surfaces.

5.3 MODERN TECHNIQUES FOR BLOOD MICROCIRCULATION MEASUREMENTS WITH WEARABLE INSTRUMENTATION

Timely diagnostics of the system of skin microvasculature regulation, which plays a key role in a number of pathological conditions (arterial hypertension, coronary heart disease, DM, etc.), is involved in the pathological process at the earliest stages, leading to further severe and difficult-to-treat complications. The discovery of early markers of endothelial damage can become the basis for finding methods of preventing vascular complications of DM. In this regard, optical measurements of blood flow dynamics have been demonstrated to be promising methods for *in vivo* diagnostics of the microvascular blood flow parameters. In modern diagnostics, there are several commonly used optical noninvasive methods to diagnose the functional state of skin microcirculation. One of these methods is LDF.

Since the early 1990s, dynamic light scattering (DLS) and LDF measurements have become an object of wide research and industrial interest in the field of life sciences. These methods are based on optical noninvasive sensing of tissue using laser light, and further analysis of the scattered radiation partially reflected by the moving red blood cells. A great advantage of the LDF technique is its ability to measure blood flow in a local area of tissue with excellent temporal resolution [39]. At present, there is a surge of interest in wearable, electronic diagnostic devices. The main reason for such interest is the possibility of daily monitoring promising a new quality of diagnosis. Round-the-clock monitoring, even of such a simple indicator as heart rate, allows the acquisition of new information about the state of the whole body, while blood perfusion measurements promise a great improvement in such kinds of diagnostics. An additional synergistic effect may be obtained by annotation of physiological data by location data as well as information about current physical activity (e.g., accelerometer data).

LDF has a long history of clinical use for the diagnosis of pathological changes in the microcirculation, including studies related to DM [40]. However, the use of conventional LDF devices imposes some restrictions on the experimenter, since light transmission along an optical fiber is quite sensitive to artifacts caused by the movement of the studied person, which affects the quality of the recorded data. For this reason, there is currently a need to develop new wireless and wearable solutions for LDF.

Some developments have already taken place in this direction. In 2010, Kimura et al. presented a wearable version of an LDF device [41]. The principle of the device is based on that of a laser diode and a photodiode, integrated in a single miniaturized package. Such a structure allows one to reduce the optical power loss of the device and miniaturize it. The authors used this sensor to assess various physical conditions of the examinee, such as dehydration [42], alcohol consumption [43], and physical activity [44], as well as in the diagnosis of microcirculatory disorders associated with systemic sclerosis [45].

Another development of a wearable version of LDF has been made in a Russian-British collaboration. The device has already been used to study the differences in perfusion between smokers and non-smokers [46]. In the present chapter, we will review the structural organization of the device and describe examples of its application in the clinical care of DM.

At present, the LDF technique has shown promise for applications in the monitoring of cardiovascular parameters, and in sports and rehabilitation medicine. At the same time, conventional blood flowmeters face several challenges, not only due to their large physical dimensions, but also due to their considerable difficulties in their application for blood perfusion monitoring.

Experimental studies were conducted using wearable LDF monitors "AMT-LAZMA 1" (Aston Medical Technology Ltd., UK) for the analysis of blood microcirculation [47]. The system comprises one or more wearable devices with an integrated LDF sensor and skin thermometer, and a wireless data acquisition module. Every wearable sensor in the system uses a VCSEL chip as a single-mode laser source to implement fiber-free direct illumination of tissue. The devices implement identical channels for recording blood perfusion and allow simultaneous measurements at several points of the body. The fiber-free solution and direct illumination of tissue by the laser diode make it possible to avoid fiber-coupling losses, as well as to decrease the movement artifacts, which are common in fiber-based LDF monitors (Figure 5.8).

The synchronization of recordings from different parts of the body makes possible the study of the synchronization of the skin blood flow under different conditions. Figure 5.9 shows the changes in the LDF signal, recorded in a test measurement on the fingertips and wrists, during the breath-holding test. The black solid line on the graphs corresponds to the average value of perfusion in two symmetrical measurement areas, and the area around the black solid line is the range of signal variations. As can be seen from the graphs presented, blood tissue perfusion is higher and more synchronous when measured on fingertips compared to measurements performed on the wrists. The breath-holding test causes a short-term vasoconstriction, which is observed as a drop in perfusion in both fingers and wrists.

As can be observed from the graphs, despite the more extensive range of perfusion measurements in the wrists, under the provocative stimuli, the microcirculatory blood flow in symmetrical areas of the body shows a rather synchronous response.

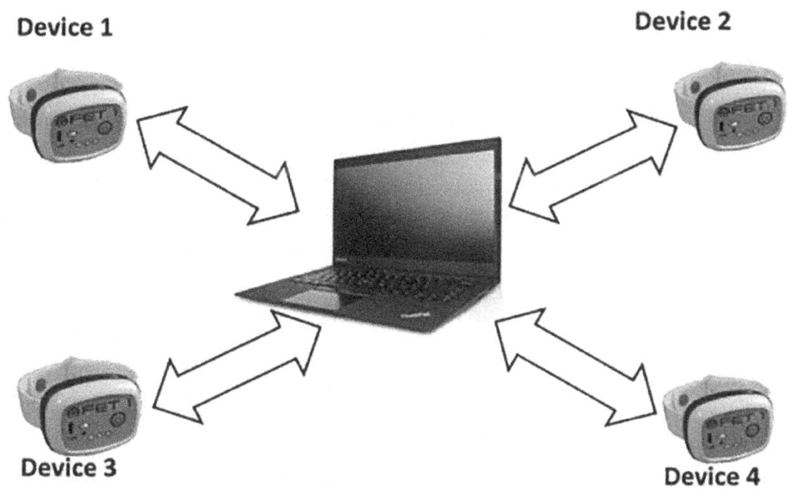

FIGURE 5.8 Wearable sensor system for multipoint measurements of blood perfusion. (Reprinted with permission from [47] © SPIE.)

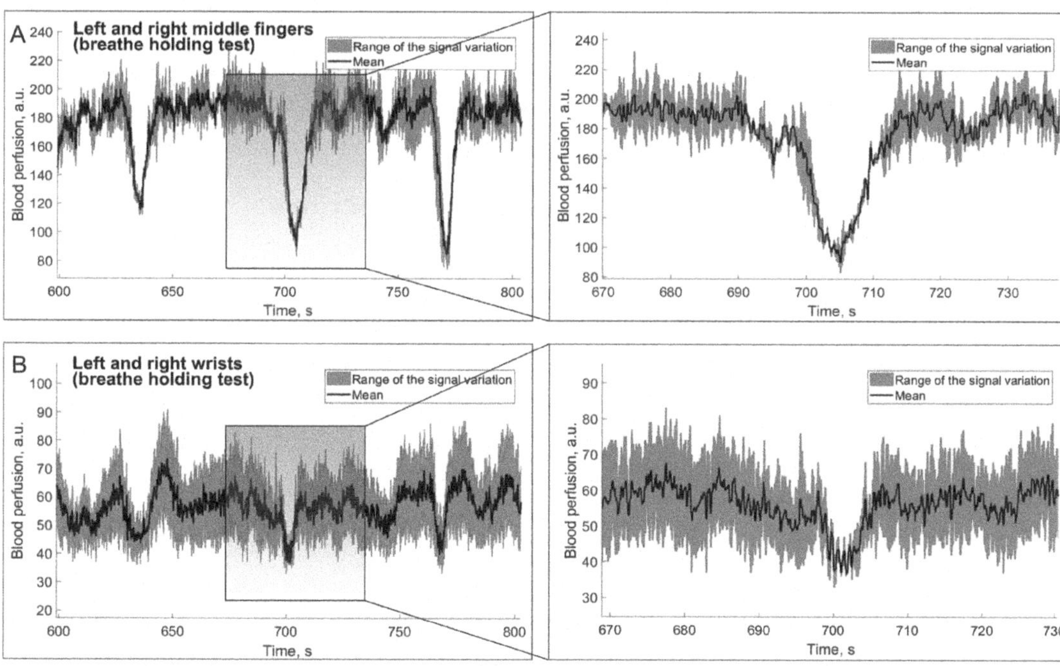

FIGURE 5.9 Test measurements with series of breath-holding tests: (a) represents measurements from fingers; (b) represents measurements from wrists. (Reprinted with permission from [47] © SPIE.)

5.4 PATHOLOGICAL CHANGES IN MICROCIRCULATION SYSTEM DURING DIABETES MELLITUS

Microcirculation disorders manifest themselves in all parts of the body and affect the functioning of various organs, including kidneys, eyes, cardiovascular system, and skin. This significantly reduces patients' quality of life and may lead to their full disability. Chronic hyperglycemia and insulin resistance in DM cause increased vascular permeability and disruption of vascular tone, causing structural and functional changes in capillaries and arterioles. The earliest, usually reversible, manifestation of these diseases is the development of microcirculatory dysfunction due to endothelial damage, and excessive expression of certain adhesion molecules and other factors.

The novel wearable sensor system for multipoint recordings of blood perfusion applied to measurements in healthy volunteers and patients with type 2 diabetes has demonstrated a good quality of recorded blood perfusion signals from areas of skin with different levels of microvascular bed density. The experiments conducted in Ref. [47] showed that the implementation of the blood perfusion sensor as a fiber-free wireless wearable device is a very convenient solution for application in point-of-care testing. Measurements in groups of different ages allowed for registration of age-specific changes in blood perfusion, as well as changes that can be associated with the development of diabetes. The wearable implementation of LDF can become a truly new diagnostic interface to monitor cardiovascular parameters, which could be of interest for diagnostics of conditions associated with microvascular disorders. The study involved 37 healthy volunteers and 18 patients with type 2 diabetes. Volunteers with cardiovascular and other serious chronic diseases that affect the blood circulation system, or with alcohol or drug dependence, were excluded from the studies. Before starting the measurements, each volunteer gave a voluntary informed written consent to participate in the experiment and passed a questionnaire to identify possible health problems. The healthy participants were divided into two groups according to their age: 16 volunteers with the age of 19.6±0.6 (group 1) and 21 volunteers with age 53.2±11.4 (group 2). The LDF signal was recorded using the blood perfusion measurement system on wrists and the fingertips of the middle fingers of both hands.

Studies were conducted in a sitting position, in a state of physical and mental rest, not earlier than 2 hours after eating. The volunteer's hands were placed on the table at the level of the heart. The blood perfusion was recorded for 10 minutes, while the sensors were attached to the palmar surface of the distal phalanx of the third fingers and both wrists.

Figure 5.10 shows the scatter plots of parameters for the studied groups.

Studies have shown that the average perfusion differs between healthy volunteers of different age groups and between healthy volunteers of the younger age group and patients with DM. There was no statistically significant difference between parameters of the older control group and patients. On the fingertips, the highest level of perfusion is observed in the second group of healthy volunteers, and the lowest values are recorded for the first group. This result might be due to structural changes in microcirculation during aging, including an increase in the total parallel vascular length [48].

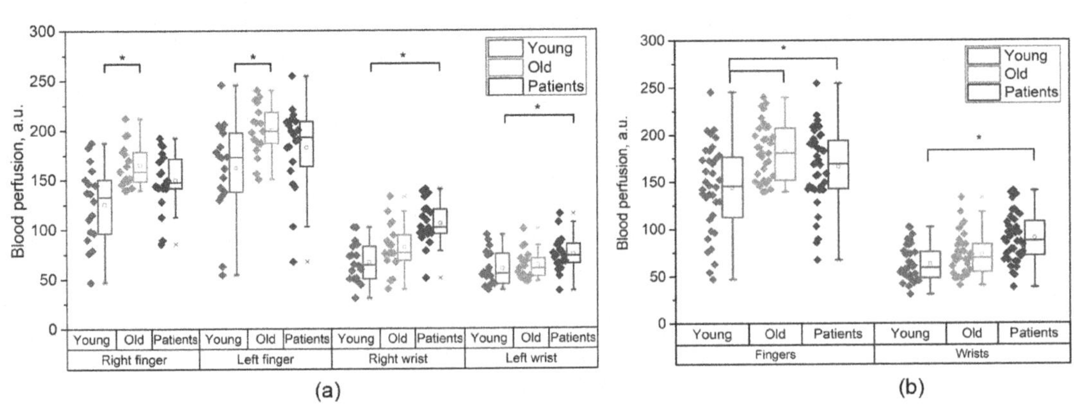

FIGURE 5.10 Measurements of mean blood perfusion in patients with type 2 diabetes: (a) separate analysis for left and right sides of body; (b) data from symmetrical points combined together. Young healthy volunteers (age of 19.6±0.6); middle-aged healthy volunteers (age of 52.6±10.2); middle-aged patients with type 2 diabetes (age of 53.2±11.4). (Reprinted with permission from [47] © SPIE.)

When measuring on the wrists, patients with diabetes had the highest level of perfusion, and the lowest were the first group of volunteers.

An increase in perfusion in patients with diabetes under basal conditions has been described in previous works by other authors in connection with the effect of diabetic neuropathy on the blood flow [49]. It is interesting to note that when measuring on the wrists, the average level of perfusion in the two distinct groups of healthy volunteers did not show the statistically significant differences found in similar measurements on the fingertips. This result may be due to the lower amplitude of the signal recorded on the wrists compared to records made on the fingers. It should be noted that when comparing the parameters of individual limbs, the difference between the level of perfusion between healthy volunteers of the younger age group and patients ceased to be statistically significant.

The LDF method allows not only to detect disorders in the microcirculatory system, including in the early stages, but also to determine the degree of their severity [50–53]. However, the complexity of differentiating the states of the microcirculatory bed is complicated by its variability during life [54]. Therefore, it is necessary to establish the boundaries of reference values for people belonging to different age groups.

The work [55] is devoted to the study of nutritive and shunt blood flow in the lower limbs of healthy volunteers and patients with type 2 DM. The experimental studies were carried out using four wearable LDF monitors for blood microcirculation analysis. The devices were attached to the inner parts of the upper thirds of the shins and the palmar surface of the big toes. In these areas, there are two types of skin: skin with AVAs in the area of the big toes and skin without AVA in the area of the shins [56].

The study of microcirculation in the big toes included three groups of volunteers, presented in Table 5.7: the first and second groups consisted of conditionally healthy volunteers of different ages, while the third group comprised patients with type 2 DM.

The study of microcirculation in the shins included two groups of volunteers, presented in Table 5.8: the first group consisted of older healthy volunteers, while the second group was made up of patients with type 2 DM.

The index of microcirculation (Im) was recorded for 10 minutes in the supine position. To assess the state of the microcirculatory system and obtain quantified diagnostic information, the bypass index (*BI*), nutritional, and shunt blood flow parameters were evaluated [57]. The total value of the *BI* was obtained by adding two components: BI_1 – the shunting index associated with differences in tone in the microvessels of the nutritive and non-nutritive blood flow pathways directly within the

TABLE 5.7
Cohort Characteristics for Measurements in Toes

	1st Group (Healthy Volunteers)	2nd Group (Healthy Volunteers)	3rd Group (Patients with Type 2 DM)
Numbers of volunteers	17 volunteers	10 volunteers	43 patients
Average age	21.7 ± 1.4 years	51.8 ± 14.4 years	56.6 ± 11.8 years

TABLE 5.8
Cohort Characteristics for Measurements in Shins

	1st Group (Healthy Volunteers)	2nd Group (Patients with Type 2 DM)
Numbers of volunteers	9 volunteers	16 patients
Average age	52.3 ± 10.7 years	57.2 ± 9.1 years

microcirculatory bed; BI_2 – the shunting index associated with differences in perfusion of microvessels and larger vascular segments (arteries, venules, and veins) in cases of arterial hyperemia or venous stagnation [58]. For zones with AVA (e.g., big toes) and without AVA (e.g., shins), *BI1* was calculated using formulas 5.3 and 5.4, respectively:

$$BI_1 = A_n / A_m, \tag{5.3}$$

$$BI_1 = A_{max} / A_m, \tag{5.4}$$

where A_n, A_m, and A_{max} – maximum amplitudes of neurogenic (*n*), myogenic (*m*), and active oscillations, respectively.

BI_2 was calculated in the same way for zones with and without AVA:

$$BI_2 = A_{c(r)} / A_m, \tag{5.5}$$

where $A_{c(r)}$ – the dominant amplitude of passive oscillations: cardiac (*c*) or respiratory (*r*).

By the calculation of *BI*, it is possible to assess the perfusion of nutritive and shunt pathways in microvascular networks. The value of nutritive perfusion (M_{nutr}) was calculated by the expressions (5.6) and (5.7) for areas with and without AVA, respectively:

$$M_{nutr} = Im/(1 + BI), \tag{5.6}$$

$$M_{nutr} = Im/BI. \tag{5.7}$$

Accordingly, the value of shunt perfusion (M_{shunt}) is estimated by the following expression:

$$M_{shunt} = Im - M_{nutr}. \tag{5.8}$$

A statistically significant difference was found in the bypass index, nutritive and shunt blood flow between older volunteers and patients with type 2 diabetes (Figure 5.11a, b). The rate of bypass index in patients is higher than in older volunteers, while nutritional and shunt blood flow are lower. In the big toes, a statistically significant difference in shunt blood flow was also found between younger volunteers and patients (Figure 5.11c).

The results obtained may indicate that DM and aging have a similar effect on the microcirculation system, but DM is an accelerated model of age-related changes.

Such calculated parameters as nutritive and shunt blood flow provide new diagnostic information about the microcirculation, which also contributes to the establishment of more accurate limits of the norm and the severity of violations of the microcirculatory system.

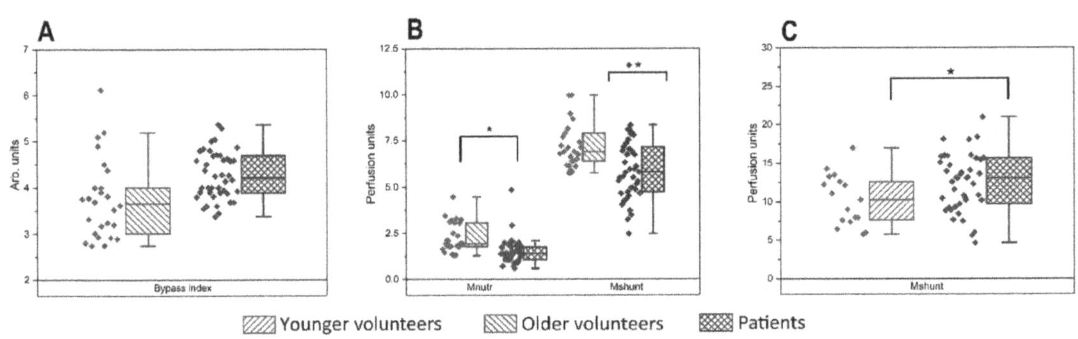

FIGURE 5.11 Results of blood flow analysis in shins (a, b) and in fingers (c) of younger volunteers, older volunteers, and patients: (a) bypass index; (b) nutritive and shunt blood flow; (c) shunt blood flow.*The significance of the difference between the values was confirmed with $p < 0.05$ according to the Mann-Whitney test. (Adapted with permission from [55] © SPIE.)

5.5 AGE-RELATED CHANGES

Aging has a prominent impact on the parameters of the microcirculation not only in patients with diabetes but also in healthy individuals [38]. In previous studies of age-related changes in the microcirculatory bed, it was revealed that the trophic supply of tissues and organs was significantly deteriorated. It has been shown that the aging of the microcirculatory bed affects the pathogenesis of various diseases [54,59,60]. The quantitative characterization of age-related changes in the microcirculatory bed can be used to identify disorders at early stages. Previously, in studies of age-related changes in the microcirculatory bed by using noninvasive diagnostic methods, various functional tests, such as thermal and occlusive tests, were used [61,62]. Among them, there are a few pieces of research devoted to the study of basal signals recorded by LDF [63,64].

Wearable sensors for LDF measurements have been successfully used for the characterization of such changes. Experimental studies were conducted using two wearable LDF monitors "AMT-LAZMA 1" (Aston Medical Technology Ltd., UK) for the analysis of blood microcirculation [65]. The studies were conducted in accordance with the principles set out in the Helsinki Declaration of 2013 by the World Medical Association. The study involved 36 healthy volunteers without cardiovascular or other serious chronic diseases that affect the blood circulation system; volunteers with alcohol or drug dependence were also excluded. The study participants were divided into two groups according to their age: 18 volunteers were under the age of 20 (first group: average age of 19.4±0.6 years) and 18 volunteers were over 40 (second group: average age of 52.6±10.2 years). The measurements were conducted in a sitting position, in a state of physical and mental rest with the volunteer's hands placed on the table at the level of heart. The skin blood perfusion was recorded for 10 minutes, while the devices were arranged as follows, as shown in Figure 5.12; sensors were attached to the palmar surface of the distal phalanx of the third fingers of both hands and the dorsal surface of the forearms without applying any pressure on the study area.

The selected areas represent two main skin types: glabrous and non-glabrous skin. Glabrous skin mostly covers the palms, soles, and face. This type of skin is primarily involved in the mechanisms of thermoregulation of the body and contains a large number of arteriovenous anastomoses (AVA). Sympathetic regulation is the dominant mechanism involved in blood flow in glabrous skin [56].

Non-glabrous or "hairy" skin covers almost the entire surface of the human body. It contains only a few AVAs [66].

Figure 5.13 shows a comparison of the index of microcirculation in the fingers (a) and wrists (b) between the two age groups. The volunteers in the older group had a higher blood perfusion level in their fingers in comparison with volunteers in the younger group (18.23±2.97 p.u. vs 14.26±4.68 p.u.). A similar situation was observed with the higher perfusion in older volunteers (7.28±2.21 p.u. vs 6.35±1.88 p.u.) in the forearms, but this difference did not reach a statistically significant level.

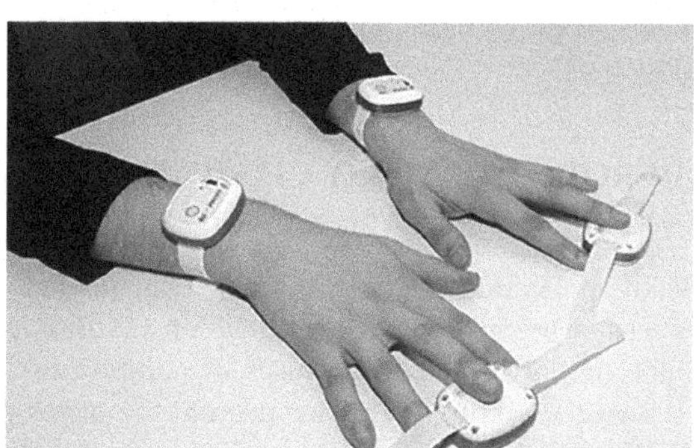

FIGURE 5.12 Placement of the LDF sensors in the measurements of the age-related changes in skin blood microcirculation. (Reprinted with permission from [65] © SPIE.)

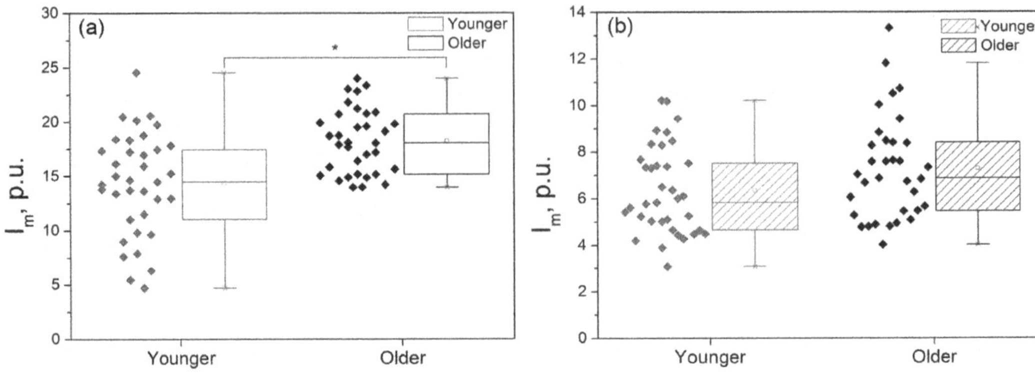

FIGURE 5.13 The microcirculation index values for the first (younger) and second (older) groups studied during measurements on the fingertip (a) and the dorsum of the forearm (b). (Reprinted with permission from [65] © SPIE.)

The absence of significant differences in the index of microcirculation in the non-glabrous skin between age groups is consistent with previous studies. At the same time, a significant increase in basal skin perfusion in glabrous skin with age is confirmed by previous studies [48,63]. The increase in the index of microcirculation in glabrous skin with age may be due to the following factors. First, these are structural age-related changes in the microcirculatory bed, in particular, an increase in the total parallel length of the vessels [48]. Second, the skin becomes thinner with age; that is, the scattering of laser radiation decreases, resulting in an increase in the diagnostic volume [61].

It is not only the absolute amplitudes of the physiological oscillations in the blood microflow that can be used for diagnostics. The mutual coherence of the oscillation measured in distant places of the skin can also be of particular interest for the quantitative characterization of microvascular disorders. One of the main approaches for such a signal comparison is the calculation of the wavelet coherence. The results obtained by this method allow one to estimate how well the fluctuations in the blood flow correspond with each other in the sense of the frequency and phase matching well, representing the coherence of the relevant biological processes going on behind the scene.

In Ref. [67], this approach was used to characterize age-related changes in the skin blood flow for the same cohort of the two age groups of healthy volunteers. The results obtained demonstrated that in myogenic oscillations, the wavelet coherence is significantly higher in young volunteers than in the second group of aged participants. Oscillations with a frequency in the region of 0.1 Hz reflect the work of smooth muscle cells [68]. It was proposed that with age, there is a change in the work of the mechanisms that synchronize fluctuations in this frequency range throughout the cardiovascular system, causing differences in the vascular tone of the right and left parts of the body, which is confirmed by other studies [69].

5.6 ANALYSIS OF MICROCIRCULATORY CHANGES DURING INTRAVENOUS INFUSION IN PATIENTS WITH DIABETES MELLITUS

One of the promising areas of application of novel wearable LDF devices in DM clinical care is the evaluation of the effectiveness of therapy for its complications. Antioxidant therapy is quite often used in the DM clinic, since one of the leading causes of complications is oxidative stress [70]. Antioxidant therapy has shown the ability to slow or prevent the development of diabetic complications in some cases [71]. The most commonly used substances include alpha-lipoic acid (ALA), which belongs to the series of strong antioxidants and has an insulin-like effect. ALA has been reported to have a number of potentially beneficial effects for both the prevention and treatment of diseases related to peroxidation.

In studies investigating the use of ALA in the treatment of DM complications, it has been shown that it promotes weight loss in obesity [72] and increases insulin sensitivity in diabetic subjects [73]. The greatest use of ALA treatment is in the field of diabetic neuropathy. Numerous studies in this direction have shown that such therapy can significantly improve the symptoms of neuropathy [74–77]. In spite of all the aforementioned positive effects of ALA, there are still insufficient research data on the effect of course therapy with this substance on changes in vascular function and microcirculation state. In our work, we have tried to conduct preliminary studies to assess the possibility of using wearable LDF devices to record changes in the microcirculation in response to ALA therapy [78].

The study was conducted using four wearable LDF devices placed on the pads of the patients' middle fingers and big toes. Each patient was examined for 5 days, starting from the second day of their hospital stay. Each study involved the recording of five LDF recordings lasting from 10 to 20 minutes. The first three recordings were made on the first day of the study (i.e., the second day of the patient's hospital stay). Recordings were made immediately before (10 minutes), during (20 minutes), and immediately after (10 minutes) the intravenous ALA infusion. The next recording (10 minutes) was done the next day, immediately before the second intravenous ALA infusion, and the last recording (10 minutes) was done on day 6 of the patient's hospital stay, the day after the last ALA infusion.

The study evaluated both changes in perfusion itself and the effect of therapy on changes in the oscillatory components of the LDF signal. Analysis of the data showed a downward trend in the patients' index of microcirculation by the end of therapy, a parameter that is known to be usually

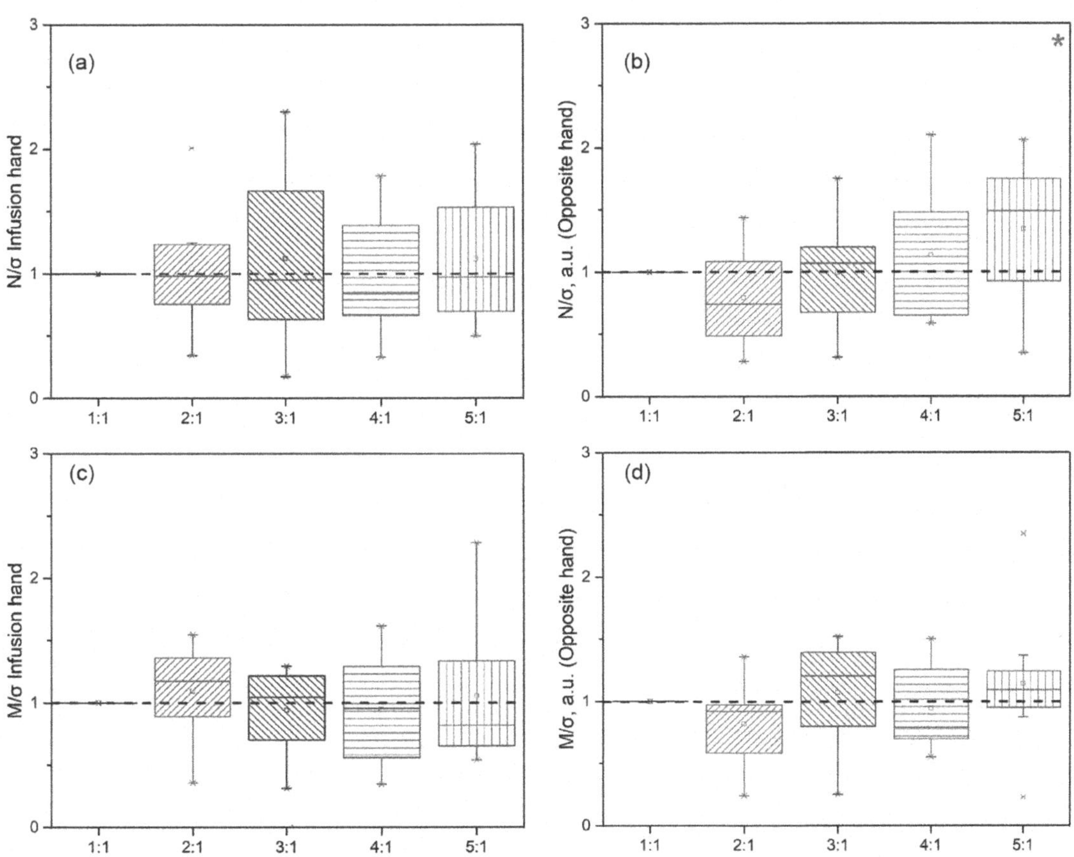

FIGURE 5.14 Relative increment of neurogenic (upper panel) and myogenic (lower panel) oscillations measured in the fingers. *The significance of the difference between the values was confirmed with $p<0.05$ according to the Friedman-ANOVA test. (Adapted with permission from [78] © SPIE.)

elevated in diabetic neuropathy [56,79]. From the analysis of the oscillatory components of the LDF signal, it is interesting to note changes in the neurogenic and myogenic ranges. Although there was an increase in these parameters by the end of therapy, a statistically significant increase was achieved only when measuring signals from the fingers and only in the neurogenic range. At the same time in the lower extremities, there is an opposite tendency, that is, a decrease in these parameters. The results of the analysis of changes in blood flow fluctuations are shown in Figure 5.14.

In general, the possibility of using LDF in the form of wearable devices to analyze the effectiveness of DM complications therapy using ALA was shown. Further study with a longer follow-up period of patients and a larger sample is needed to obtain more specific results. Previous works show that significant improvement in patients with ALA is achieved with more than an 8-week course and a combination of intravenous infusions and oral administration of ALA [80].

5.7 SUMMARY

The novel approach of wearable sensors for multipoint monitoring and recording of blood perfusion allows the acquisition of good-quality blood perfusion signals from areas of the skin with different levels of microvascular bed density. Recent studies have demonstrated that the implementation of blood perfusion sensors as fiber-free wireless wearable devices is a promising solution for application in point-of-care testing [81]. This technology makes it possible to register age-specific changes in blood perfusion as well as changes that may be associated with the development of diabetes.

ACKNOWLEDGMENTS

This work was supported by fund from the Russian Foundation for Basic Research (RFBR) Grant Number 20–08–01153. E.Z. acknowledges the support of the Academy of Finland (Grant No. 318281).

REFERENCES

1. S.M. Daly and M.J. Leahy, "'Go with the flow': A review of methods and advancements in blood flow imaging," *J. Biophotonics* **6**(3), 217–255 (2013).
2. M.V. Volkov, D.A. Kostrova, N.B. Margaryants, et al., "Evaluation of blood microcirculation parameters by combined use of laser Doppler flowmetry and videocapillaroscopy methods," *Proc. SPIE* **10336**, 1033607 (2017).
3. V.V. Dremin, E.A. Zherebtsov, I.N. Makovik, et al., "Laser Doppler flowmetry in blood and lymph monitoring, technical aspects and analysis," *Proc. SPIE* **10063**, 1006303 (2017).
4. E.V. Potapova, V.V. Dremin, E.A. Zherebtsov, et al., "A complex approach to noninvasive estimation of microcirculatory tissue impairments in feet of patients with diabetes mellitus using spectroscopy," *Opt. Spectrosc.* **123**(6), 955–964 (2017).
5. E. Zherebtsov, S. Sokolovski, V. Sidorov, et al., "Novel wearable VCSEL-based blood perfusion sensor," In *2018 International Conference Laser Optics (ICLO)*, 564 (2018).
6. E.A. Zherebtsov, E.V. Zharkikh, I. Kozlov, et al., "Novel wearable VCSEL-based sensors for multipoint measurements of blood perfusion," *Proc. SPIE* **10877**, 1087708 (2019).
7. H.L. Lutgers, E.G. Gerrits, R. Graaff, et al., "Skin autofluorescence provides additional information to the UK Prospective Diabetes Study (UKPDS) risk score for the estimation of cardiovascular prognosis in type 2 diabetes mellitus," *Diabetologia* **52**(5), 789–797 (2009).
8. E.G. Gerrits, H.L. Lutgers, N. Kleefstra, et al., "Skin autofluorescence: A tool to identify type 2 diabetic patients at risk for developing microvascular complications," *Diabetes Care* **31**(3), 517–521 (2008).
9. M.S. Hsu and M. Itkin, "Lymphatic anatomy," *Tech. Vasc. Inter. Radiol.* **19**(4), 247–254 (2016).
10. A.I. Krupatkin, and V.V. Sidorov, *Functional Diagnostics of the State of Microcirculatory-Tissue Systems: Oscillations, Information, Nonlinearity: A Guide for Physicians*, LIBROKOM Book House, Moscow (2013).
11. M.S. Shah and M. Brownlee, "Molecular and cellular mechanisms of cardiovascular disorders in diabetes," *Circ. Res.* **118**(11), 1808–1829 (2016).

12. B.M. Sörensen, A. Houben, B. Tos, et al., "Prediabetes and type 2 diabetes are associated with generalized microvascular dysfunction," *Circulation* **134**(18), 1339–1352 (2016).
13. W.D. Strain and P.M. Paldánius, "Diabetes, cardiovascular disease and the microcirculation," *Cardiovasc. Diabetol.* **17**(1), 57 (2018).
14. B.I. Levy, E.L. Schiffrin, J.J. Mourad, et al., "Impaired tissue perfusion," *Circulation* **118**(9), 968–976 (2008).
15. W.D. Strain, A.D. Hughes, J. Mayet, et al., "Attenuated systemic microvascular function in men with coronary artery disease is associated with angina but not explained by atherosclerosis," *Microcirculation* **20**(7), 670–677 (2013).
16. P. Ong, P.G. Camici, J.F. Beltrame, et al., "International standardization of diagnostic criteria for microvascular angina," *Int. J. Cardiol.* **250**, 16–20 (2018).
17. D. Rizzoni, and E. Rosei, "Agabiti small artery remodeling in hypertension and diabetes," *Curr. Hypertens. Rep.* **8**(1), 90–95 (2006).
18. D. Rizzoni, E. Porteri, D. Guelfi, et al., "Structural alterations in subcutaneous small arteries of normotensive and hypertensive patients with non–insulin-dependent diabetes mellitus," *Circulation* **103**(9), 1238–1244 (2001).
19. A. Jumar, C. Ott, I. Kistner, et al., "Early signs of end-organ damage in retinal arterioles in patients with type 2 diabetes compared to hypertensive patients," *Microcirculation* **23**(6), 447–455 (2016).
20. R. Scalia, Y. Gong, B. Berzins, et al., "Hyperglycemia is a major determinant of albumin permeability in diabetic microcirculation," *Diabetes Am. Diabetes Assoc.* **56**(7), 1842–1849 (2007).
21. T. Deckert, B. Feldt-Rasmussen, K. Borch-Johnsen, et al., "Albuminuria reflects widespread vascular damage," *Diabetologia* **32**(4), 219–226 (1989).
22. M.S. Goligorsky, "Vascular endothelium in diabetes," *Am. J. Physiol. Physiol.* **312**(2), F266–F275 (2017).
23. S. Kumar, M.N. Darisipudi, S. Steiger, et al., "Cathepsin S cleavage of protease-activated receptor-2 on endothelial cells promotes microvascular diabetes complications," *J. Am. Soc. Nephrol.* **27**(6), 1635–1649 (2016).
24. M.S. Goligorsky, J. Chen, S. Brodsky, "Workshop: Endothelial cell dysfunction leading to diabetic nephropathy," *Hypertension.* **37**(2), 744–748 (2001).
25. J. Wada, A. Nakatsuka, "Mitochondrial dynamics and mitochondrial dysfunction in diabetes," *Acta Med. Okayama. Okayama Univ. Med. School* **70**(3), 151–158 (2016).
26. M. Sezer, M. Kocaaga, E. Aslanger, et al., "Bimodal pattern of coronary microvascular involvement in diabetes mellitus," *J. Am. Heart Assoc.* **5**(11), e003995 (2016).
27. H. Jonasson, S. Bergstrand, et al., "Skin microvascular endothelial dysfunction is associated with type 2 diabetes independently of microalbuminuria and arterial stiffness," *Diabetes Vasc. Dis. Res.* **14**(4), 363–371 (2017).
28. A.J. Jaap, C.A. Pym, C. Seamark, et al., "Microvascular function in type 2 (Non-insulin-dependent) diabetes: Improved vasodilation after one year of good glycaemic control," *Diabet. Med.* **12**(12), 1086–1091 (1995).
29. A.E. Caballero, S. Arora, R. Saouaf, et al., "Microvascular and macrovascular reactivity is reduced in subjects at risk for type 2 diabetes," *Diabetes* **48**, 1856–1862 (1999).
30. R.G. Jzerman, R.T. De Jongh, M.A.M. Beijk, et al., "Individuals at increased coronary heart disease risk are characterized by an impaired microvascular function in skin," *Eur. J. Clin. Invest.* **33**(7), 536–542 (2003).
31. R.T. De Jongh, E.H. Serné, R.G. IJzerman, et al., "Impaired microvascular function in obesity," *Circulation* **109**(21), 2529–2535 (2004).
32. D. Fuchs, P.P. Dupon, L.A. Schaap, et al., "The association between diabetes and dermal microvascular dysfunction non-invasively assessed by laser Doppler with local thermal hyperemia: A systematic review with meta-analysis," *Cardiovasc. Diabetol.* **16**(1), 11 (2017).
33. V.V. Tuchin, *Handbook of Optical Biomedical Diagnostics*, SPIE Press Book, Bellingham, WA (2002).
34. X. Luo, et al., "Roles of pyruvate, NADH, and mitochondrial complex I in redox balance and imbalance in β cell function and dysfunction," *J. Diabetes Res.* **2015**, 512618 (2015).
35. A.A. Heikal, "Intracellular coenzymes as natural biomarkers for metabolic activities and mitochondrial anomalies," *Biomark. Med.* **4**(2), 241–263 (2010).
36. A. Goltsov, V.V. Sidorov, S.G. Sokolovski, et al., "Editorial: Advanced non-invasive photonic methods for functional monitoring of haemodynamics and vasomotor regulation in health and diseases," *Front. Physiol. Frontiers Media S.A.* **11**, 325 (2020).
37. A.A. Andronov, S.V. Barashkov, G.N. Lobanov, et al., "The effect of aqua-trainings are in fresh water in patients with hypertension on the microcirculation," *Bull. Restor. Med.* **100**(6), 25–32 (2020).

38. A. Goltsov, A.V. Anisimova, M. Zakharkina, et al., "Bifurcation in blood oscillatory rhythms for patients with ischemic stroke: A small scale clinical trial using laser doppler flowmetry and computational modeling of vasomotion," *Front. Physiol.* **8**, 160 (2017).
39. J.M. Johnson, C.T. Minson, and D.L. Kellogg, "Cutaneous vasodilator and vasoconstrictor mechanisms in temperature regulation," *Compr. Physiol.* **4**(1), 33–89 (2014).
40. E. Zharkikh, V. Dremin, E. Zherebtsov, et al., "Biophotonics methods for functional monitoring of complications of diabetes mellitus," *J. Biophotonics* **13**(10), e202000203 (2020).
41. Y. Kimura, M. Goma, A. Onoe, et al., "Integrated laser Doppler blood flowmeter designed to enable wafer-level packaging," *IEEE Trans. Biomed. Eng.* **57**(8), 2026–2033 (2010).
42. H. Nogami, W. Iwasaki, T. Abe, et al., "Use of a simple arm-raising test with a portable laser Doppler blood flow meter to detect dehydration," *Proc. Inst. Mech. Eng. H* **225**(4), 411–419 (2011).
43. W. Iwasaki, H. Nogami, H. Ito, et al., "Useful method to monitor the physiological effects of alcohol ingestion by combination of micro-integrated laser Doppler blood flow meter and arm-raising test," *Proc. Inst. Mech. Eng. H* **226**(10), 759–765 (2012).
44. W. Iwasaki, H. Nogami, S. Takeuchi, et al., "Detection of site-specific blood flow variation in humans during running by a wearable laser Doppler flowmeter," *Sensors (Basel)* **15**(10) 25507–25519 (2015).
45. M. Kido, S. Takeuchi, S. Hayashida, et al., "Assessment of abnormal blood flow and efficacy of treatment in patients with systemic sclerosis using a newly developed microwireless laser Doppler flowmeter and arm-raising test," *Br. J. Dermatol.* **157**(4), 690–697 (2007).
46. M. Saha, V. Dremin, I. Rafailov, et al., "Wearable laser Doppler flowmetry sensor: A feasibility study with smoker and non-smoker volunteers," *Biosensors* **10**(12), 201 (2020).
47. E.A. Zherebtsov, E.V. Zharkikh, I.O. Kozlov, et al., "Wearable sensor system for multipoint measurements of blood perfusion: Pilot studies in patients with diabetes mellitus," *Proc. SPIE* **11079**, 110791O (2019).
48. L. Li, S. Mac-Mary, J.M. Sainthillier, et al., "Age-related changes of the cutaneous microcirculation in vivo," *Gerontology* **52**(3), 142–153 (2006).
49. J.C. Schramm, T. Dinh, and A. Veves, "Microvascular changes in the diabetic foot," *Int. J. Low Extrem. Wounds* **5**(3), 149–159 (2006).
50. M.A. Mahgoub and A.S. Abd-Elfattah, "Diabetes mellitus and cardiac function," *Mol. Cell. Biochem.* **180**(1–2), 59–64 (1998).
51. R. Donnelly, A.M. Emslie-Smith, I.D. Gardner, and A.D. Morris, "Vascular complications of diabetes," *BMJ* **320**(7241), 1062–1066 (2000).
52. A. Chawla, R. Chawla, and S. Jaggi, "Microvasular and macrovascular complications in diabetes mellitus: Distinct or continuum?," *Indian J. Endocrinol. Metab.* **20**(4), 546–553 (2016).
53. H.F. Hu, H. Hsiu, C.J. Sung, and C.H. Lee, "Combining laser-Doppler flowmetry measurements with spectral analysis to study different microcirculatory effects in human prediabetic and diabetic subjects," *Lasers Med. Sci.* **32**(2), 327–334 (2017).
54. E. Makrantonaki and C.C. Zouboulis, "Characteristics and pathomechanisms of endogenously aged skin," *Dermatology* **214**(4), 352–360 (2007).
55. Y.I. Loktionova, E.V. Zharkikh, E.A. Zherebtsov, et al., "Wearable laser Doppler sensors for evaluating the nutritive and shunt blood flow," *Proc. SPIE* **11457**, 114570M (2020).
56. V. Urbančič-Rovan, A. Stefanovska, A. Bernjak, et al., "Skin blood flow in the upper and lower extremities of diabetic patients with and without autonomic neuropathy," *J. Vasc. Res.* **41**(6), 535–545 (2004).
57. A.V. Dunaev, V.V. Sidorov, A.I. Krupatkin, et al., "Investigating tissue respiration and skin microhaemocirculation under adaptive changes and the synchronization of blood flow and oxygen saturation rhythms," *Physiol. Meas.* **5**(4), 607–621 (2014).
58. A.I. Krupatkin, "Evaluation of the parameters of total, nutritive, and shunt blood flows in the skin microvasculature using laser Doppler flowmetry," *Hum. Physiol.* **31**(1), 98–102 (2005).
59. T.F. Lüscher and G. Noll, "Endothelial function as an end-point in interventional trials: Concepts, methods and current data," *J. Hypertens. Suppl.* **14**(2), S111–S121 (1996).
60. H. Tanaka, F.A. Dinenno, K.D. Monahan, et al., "Aging, habitual exercise, and dynamic arterial compliance," *Circulation* **102**(11), 1270–1275 (2000).
61. G.J. Hodges, M.M. Mallette, G.A. Tew, et al., "Effect of age on cutaneous vasomotor responses during local skin heating," *Microvasc. Res.* **112**, 47–52 (2017).
62. I.V. Tikhonova, A.V. Tankanag, and N.K. Chemeris, "Age-related changes of skin blood flow during postocclusive reactive hyperemia in human," *Ski. Res. Technol.* **19**(1), e174–e181 (2013).
63. R. Ogrin, P. Darzins, and Z. Khalil, "Age-related changes in microvascular blood flow and transcutaneous oxygen tension under basal and stimulated conditions," *J. Gerontol. A Biol. Sci. Med. Sci.* **60**(2), 200–206 (2005).

64. N.V. Baboshina, "Parameters of microcirculation in both sexes at different ages," *Hum. Physiol.* **44**(4), 466–473 (2018).
65. Y.I. Loktionova, E.V. Zharkikh, I.O. Kozlov, et al., "Pilot studies of age-related changes in blood perfusion in two different types of skin," *Proc. SPIE* **11065**, 110650S (2019).
66. J.M. Johnson and D.L. Kellogg, "Local thermal control of the human cutaneous circulation," *J. Appl. Physiol.* **109**(4), 1229–1238 (2010).
67. Y.I. Loktionova, et al., "Studies of age-related changes in blood perfusion coherence using wearable blood perfusion sensor system," *Proc. SPIE* **11075**, 1107507 (2019).
68. H.D. Kvernmo, A. Stefanovska, K.A. Kirkeboen, and K. Kvernebo, "Oscillations in the human cutaneous blood perfusion signal modified by endothelium-dependent and endothelium-independent vasodilators," *Microvasc. Res.* **57**(3), 298–309 (1999).
69. A.V. Tankanag, A.A. Grinevich, T.V. Kirilina, et al., "Wavelet phase coherence analysis of the skin blood flow oscillations in human," *Microvasc. Res.* **95**, 53–59 (2014).
70. F. Giacco and M. Brownlee, "Oxidative stress and diabetic complications," *Circ. Res.* **107**(9), 1058–1070 (2010).
71. C. Li, X. Miao, F. Li, et al., "Oxidative stress-related mechanisms and antioxidant therapy in diabetic retinopathy," *Oxidative Med. Cell. Longev.* **2017**, 9702820 (2017).
72. E.H. Koh, W.J. Lee, S.A. Lee, et al., "Effects of alpha-lipoic Acid on body weight in obese subjects," *Am. J. Med.* **124**(1), 85.e1–85.e8 (2011).
73. P. Kamenova, "Improvement of insulin sensitivity in patients with type 2 diabetes mellitus after oral administration of alpha-lipoic acid," *Hormones (Athens)* **5**(4), 251–258 (2006).
74. A.S. Ametov, A. Barinov, P.J. Dyck, et al., "The sensory symptoms of diabetic polyneuropathy are improved with α-lipoic acid: The SYDNEY trial," *Diabetes Care* **26**(3), 770–776 (2003).
75. D. Ziegler, A. Ametov, A. Barinov, et al., "Oral treatment with α-lipoic acid improves symptomatic diabetic polyneuropathy," *Diabetes Care* **29**(11), 2365–2370 (2006).
76. A. Burekovic, M. Terzic, S. Alajbegovic, et al., "The role of alpha-lipoicacid in diabetic polyneuropathy treatment," *Bosn. J. Basic Med. Sci.* **8**(4), 341–345, (2008).
77. T. Han, J. Bai, W. Liu, and Y. Hu, "A systematic review and meta-analysis of α-lipoic acid in the treatment of diabetic peripheral neuropathy," *Eur. J. Endocrinol.* **167**(4), 465–471, (2012).
78. E.V. Zharkikh, Y.I. Loktionova, I.O. Kozlov, et al., "Wearable laser Doppler flowmetry for the analysis of microcirculatory changes during intravenous infusion in patients with diabetes mellitus," *Proc. SPIE* **11363**, 113631K (2020).
79. Y.K. Jan, S. Shen, R.D. Foreman, and W.J. Ennis, "Skin blood flow response to locally applied mechanical and thermal stresses in the diabetic foot," *Microvasc. Res.* **89**, 40–46, (2013).
80. S. Saboori, E. Falahi, E. Eslampour, et al., "Effects of alpha-lipoic acid supplementation on C-reactive protein level: A systematic review and meta-analysis of randomized controlled clinical trials," *Nutr. Metab. Cardiovasc. Dis.* **28**(8), 779–786 (2018).
81. V.V. Sidorov, Yu.L. Rybakov, V.M. Gukasov, and G.S. Evtushenko, "A system of local analyzers for noninvasive diagnostics of the general state of the tissue microcirculation system of human skin," *Biomed. Eng.* 55(6), 379–382 (2022).

6 Optical Angiography at Diabetes

Dan Zhu, Jingtan Zhu, Dongyu Li, Tingting Yu, Wei Feng, and Rui Shi

CONTENTS

6.1 INTRODUCTION

Diabetes mellitus (DM) is a common and serious metabolic/endocrinological disorder with an increasing global challenge in the 21st-century healthcare. Besides tumors and cardiovascular and cerebrovascular diseases, DM has become the third biggest leading cause of mortality worldwide [1]. The global prevalence of DM has dramatically increased from 108 million in 1980 to 451 million in 2017, and this number continues to rapidly increase. Estimation from the International Diabetes Federation predicts that 693 million people will suffer from diabetes worldwide by 2045 [2]. The increase in DM is not only for type 2 diabetes (T2D), but also with an equally alarming increase in the number of younger patients diagnosed with type 1 diabetes (T1D) [3].

DM is a group of metabolic diseases characterized by chronic hyperglycemia and glucose intolerance caused by impaired insulin action and/or defective insulin secretion. Long-term hyperglycemia will lead to chronic damage and dysfunction of vasculatures within various tissues and organs, causing various structural and functional changes of vasculature and the development of various complications, such as diabetic nephropathy (DN), microvascular dysfunction of cerebrum and skin, and various neurovascular diseases including stroke, vascular dementia, or Alzheimer's disease.

DOI: 10.1201/9781003112099-6

Accurate evaluation of the structural and functional changes for DM-targeted organs/tissues can provide valuable information for the early diagnosis and therapy of diabetes. The development of modern optical imaging techniques enables not only obtaining 3D structural images with high resolution for various tissues *in vitro* [4–6], but also realizing dynamic monitoring of blood flow, blood oxygen, and vascular permeability within the skin and cerebrum *in vivo* [7–10]. Up to now, various advanced optical imaging techniques have been used to study the structural and functional changes of various organs/tissues caused by DM, which greatly promotes the understanding of diabetes [11–14].

In this chapter, the structural and functional changes of several vascular complications caused by DM were briefly introduced, including structural changes of glomerulus in DN and microvascular dysfunction in the cerebrum and skin. In addition, advanced optical imaging techniques used for studying these changes induced by DM were also discussed.

6.2 VISUALIZATION OF DIABETES-INDUCED MICROVASCULAR MORPHOLOGICAL STRUCTURE OF KIDNEY

Since the first records for DM were discovered in an Egyptian papyrus in 1550 BC, it took more than three millennia for humans to the recognition of an association between diabetes and kidney disease. However, DN has become the leading cause of end-stage renal disease (ESRD) only in the past several decades [15–17]. Generally, DN is one of the most serious chronic and deeply feared microvascular complications of DM [18,19]. Strikingly, about 30% of patients with T1D and 45% with T2D are affected by this microvascular complication [20–22]. Although most persons with diabetes reaching ESRD have T2D, T1D also occupies a large proportion of new ESRD cases per year with a growing increment since 1990. The renal cellular architecture and microvasculature of diabetic patients will suffer serious damage and disruption due to the persistent exposure to high blood glucose levels [23]. As a result, unique morphological changes will occur at the level of the glomerulus in the kidney nephron. Over the past two decades, researchers have achieved significant advances in understanding of the pathophysiology of DN [24–27].

Presently, microvascular morphological structural change caused by DN has been traditionally investigated via clinical patient observations and diabetic animal models using light and electron microscopic morphometric analyses in two-dimensional (2D); several researchers also studied DN with 2D cell culture in dish or glomerulus-on-a-chip microdevice [28–30].

Recently, tissue clearing techniques have been emerged as new powerful tools for volume imaging of intact tissue [31,32]. Up to date, many clearing methods have been used for the 3D visualization of overall kidney vasculatures of health and disease at capillary level. Several latest clearing methods have already introduced tissue clearing into studying the morphological change of kidney microvascular structures in DN. Therefore, this section will briefly review these studies contributed to the understanding of the pathophysiology of DN.

6.2.1 2D HISTOPATHOLOGY OF T1D-INDUCED MICROVASCULATURE MORPHOLOGICAL STRUCTURAL CHANGES OF KIDNEYS

The major microvascular pathology of kidneys in DN is characterized by the glomerular dysfunction and structural change induced by hyperglycemia [23,27]. Typical clinical manifestations of DN include progressive albuminuria, increased blood pressure, and cardiovascular risk followed by the progression of renal damage [33]. In both types of diabetes, kidney injury is noted in many structures, including glomeruli, arterioles, tubules, and the interstitium [34,35]. The morphologic lesions in T1D mainly affect the structures of glomeruli, with the renal tubules, podocytes, arterioles, and interstitium also undergoing substantial changes, especially within several years of diabetes onset [36,37].

Glomerular lesions consist of two forms of intracapillary glomerulosclerosis, namely, diffuse and nodular forms [24]. The diffuse type is characterized by mesangial widening and thickening

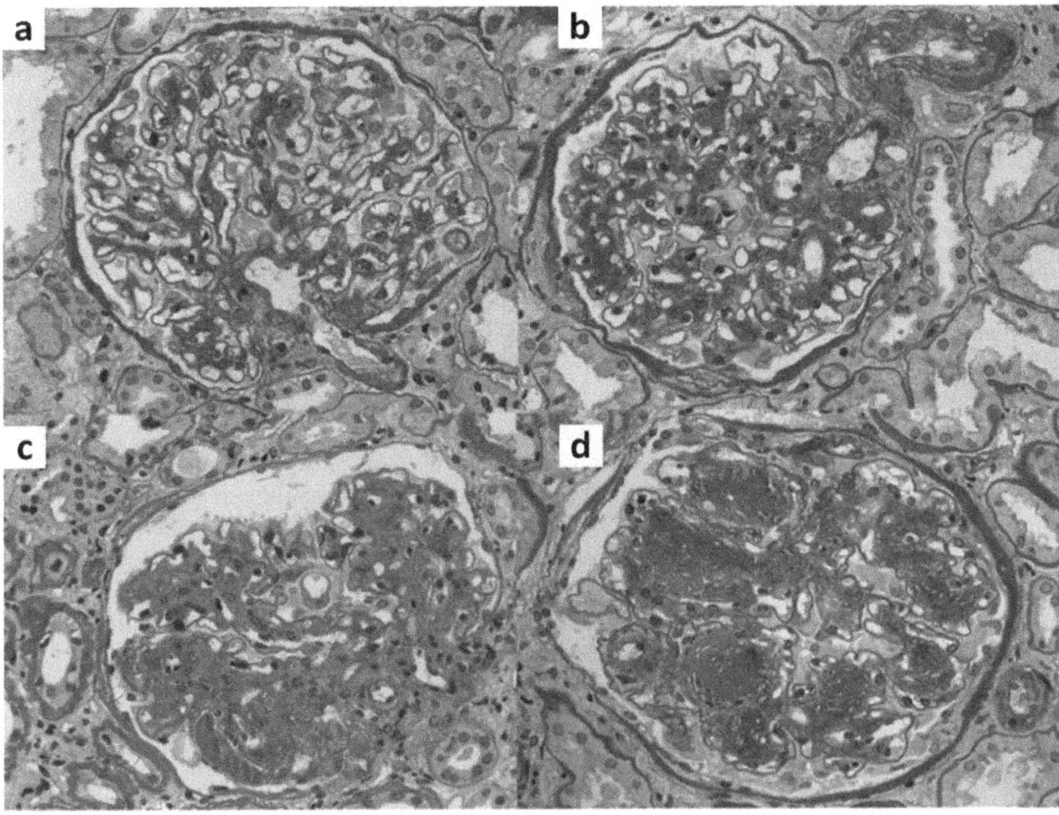

FIGURE 6.1 Representative images of different glomerular classes in DN acquired by light microscopy. (a) Mild GBM thickening. (b) Mild mesangial expansion. (c) Severe mesangial expansion. (d) Kimmelstiel–Wilson lesion. (Reprinted with permission from ref [44].)

of the capillary wall and glomerular basement membrane (GBM) (Figure 6.1a). Many studies have clinically revealed that GBM thickening might be the first measurable change shortly after the onset of T1D, which is often paralleled by capillary and tubular basement membrane thickening [38]. However, light microscopy may show only glomerular hypertrophy without any other apparent sign of injury at early stage of DN because these changes are very minor; therefore, transmission electron microscopy (TEM) is used to detect this first morphological changes of DN [39]. The width of GBM tends to increase linearly during the progress of diabetes in T1D. Other glomerular and tubular lesions can be visualized with the development of DN, including loss of endothelial fenestrations, mesangial expansion (Figure 6.1b, c), and severe tubular basement membrane thickening.

Besides GBM thickening, mesangial expansion is another typical lesion of diabetic glomerulopathy, which is progressive in duration of DN and primarily caused by an increase in mesangial matrix. Nodular lesions occur at the periphery of the glomeruli, which are well delineated by periodic acid–Schiff-positive staining that defines globular structures while exaggerating diffuse lesions. It is considered to be associated with the development of Kimmelstiel–Wilson nodules and microaneurysms, which could be seen at later stage of DN by light microscopy [40] (Figure 6.1d). Additionally, there is also an increasing recognition of lesions like glomerular endothelial injury [41], podocyte impairment [42], and glomerulotubular junction abnormalities in DN [43]. However, their value in diagnosis and classification is not as important as glomerular, tubular, and vascular lesions.

A definitive visualization and characterization of glomerular structural damage during the progress of DN must be based on clinical imaging of the renal biopsy, which is the gold standard for the diagnosis of DN [44]. However, it may be difficult to have access to large-scale clinical trials to study the developmental process of this disease and of its exact mechanism using patient renal tissues. Recently, with the development in microengineering technology, it is possible to create organ-on-a-chip microdevices lined with glomerular microtissues to mimic the microarchitectures of kidneys *in vitro* [30]. This model is successfully used to study the changes of glomerular barrier in phenotype expression, barrier integrity, and barrier permeability in the presence of high-glucose medium.

6.2.2 Tissue Optical Clearing for 3D Imaging of the Overall Kidney Structure

Though histopathological examination is the gold standard for the visualization of glomerular structural damage during the progress of DN, only a very limited number of glomeruli can be detected by this kind of 2D imaging using light microscopy or TEM. Additionally, this detection does not provide three-dimensional (3D) structural information of glomerular lesion caused by DN; thus, it is not able to provide overall changes of the vascular structures of the kidney as well.

With the advent of advanced optical imaging and fluorescent labeling techniques, 3D imaging of tissue structures with high resolution becomes popular for biomedical researches, for it can help in understanding the overall complex biological architectures and functions for whole organs and organisms [45,46]. In recent years, various tissue clearing techniques have been developed to overcome the tissue turbidity; thus, these techniques enable the interrogation of intact biological structures down to cellular and even subcellular resolution [47]. The strategy and applications of advanced tissue clearing techniques have been reviewed extensively [31,32,45,47,48].

There are now many tissue clearing methods proved to be available for 3D imaging of neural and vascular structures in kidneys. For example, 3DISCO, the typical solvent-based clearing method [49], was confirmed to be useful for imaging the vasculature of mouse kidneys labeled by lectin [50]. Qi et al. developed FDISCO method and applied it in 3D imaging and reconstruction of mouse kidneys; quantification of glomeruli was also performed [51] (Figure 6.2a). iDISCO method, recognized as a powerful whole-mount immunolabeling protocol, was also used to image the entire kidneys labeled by anti-Aquaporin2 for labeling collecting ducts and anti-Nephrin for labeling glomeruli, respectively [52]. Ethanol-ECi was developed for 3D imaging and evaluation of total glomerular number and capillary tuft size of kidneys with nephrotoxic nephritis [53]. The latest SHANEL protocol developed by Zhao et al. rendered the intact adult human kidney transparent, and 3D histology was performed with antibodies and dyes in centimeters depth [54].

As for aqueous-based methods, the first generation of CUBIC clearing protocol was used for 3D imaging of whole mouse kidneys labeled by nucleic acid dyes [55]. Matryba et al. optimized the original CUBIC protocol to image the intact rat kidneys stained by propidium iodide (PI) [56]. Recently, the upgraded CUBIC protocol was proposed and was used for volume imaging of human renal samples [57]. Additionally, CUBIC method was also specially optimized for kidney samples, named as CUBIC-kidney, realizing whole-mount immunolabeling and 3D imaging of sympathetic nerves and vessels in entire mouse kidneys for healthy and injured [58]. The very recent MACS clearing protocol developed by Zhu et al. utilized lipophilic dye DiI to label the kidney vasculature and reconstructed the glomerulus trees within mouse kidneys [11] (Figure 6.2b).

After acquisition of 3D images stacks, an image analysis pipeline can be created using commercial and open-access image analysis software [59–61] or custom-made approaches for automated segmentation and quantification of glomeruli [11, 53] in intact kidneys.

6.2.3 Applications of Tissue Optical Clearing in Studying T1D-Induced Microvascular Structural Changes in DN

Recently, a growing number of studies were conducted on 3D imaging and analysis of kidneys using tissue clearing as described above; however, most of them mainly focused on morphological

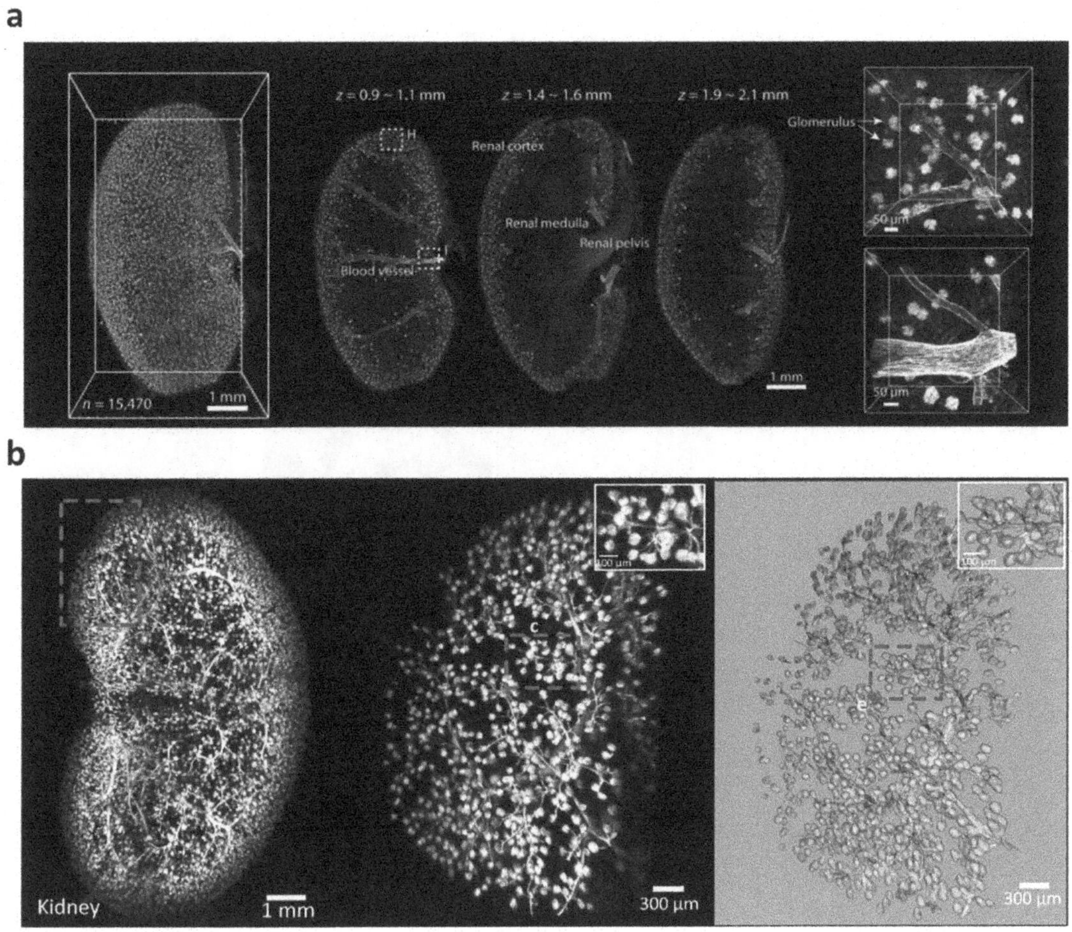

FIGURE 6.2 3D visualization of the vasculature in the mouse kidney. (a) 3D reconstruction of glomeruli labeled by the injection of CD31-A647 antibody and cleared by FDISCO. (Reprinted with permission from ref [51].) (b) 3D reconstruction of blood vessels and glomeruli labeled by DiI and cleared by MACS. (Reprinted with permission from ref [11].)

structures of normal kidneys. Up to now, there are only limited studies involving the characterization of glomerulus lesion of T1D-induced DN using tissue optical clearing techniques.

Zhu et al. constructed T1D mouse model induced by alloxan, labeled the vascular structures in healthy and diabetic mouse kidneys by DiI, and thus performed 3D pathology of glomeruli in diabetic kidneys by MACS clearing and light sheet imaging [11] (Figure 6.3). They observed the overall glomerulus trees and branches of entire normal and diabetic kidneys, as well as the fine structures of the capillary tufts of the glomerulus. Quantification of the glomerulus number and volume of entire kidneys was also performed. They found that the glomerulus number showed no obvious change in short-term of diabetes, but the average glomerulus volume increased. For long-term diabetes, both the glomerulus number and volume were reduced, which revealed that long-term T1D would lead to both individual tuft defects and glomerular loss.

Hasegawa et al. utilized T1D rat and mouse models induced by STZ or alloxan and CUBIC-kidney clearing protocol to clarify the net effects of HIF stabilization on energy metabolism in diabetic kidney [12]. They also performed comprehensive 3D analysis to visualize glomeruli in the kidney using anti-podocin antibodies. They found that renal pathological abnormalities (glomerulomegaly and GBM thickening) induced by diabetes in the early stages of DN could be mitigated by enarodustat treatment.

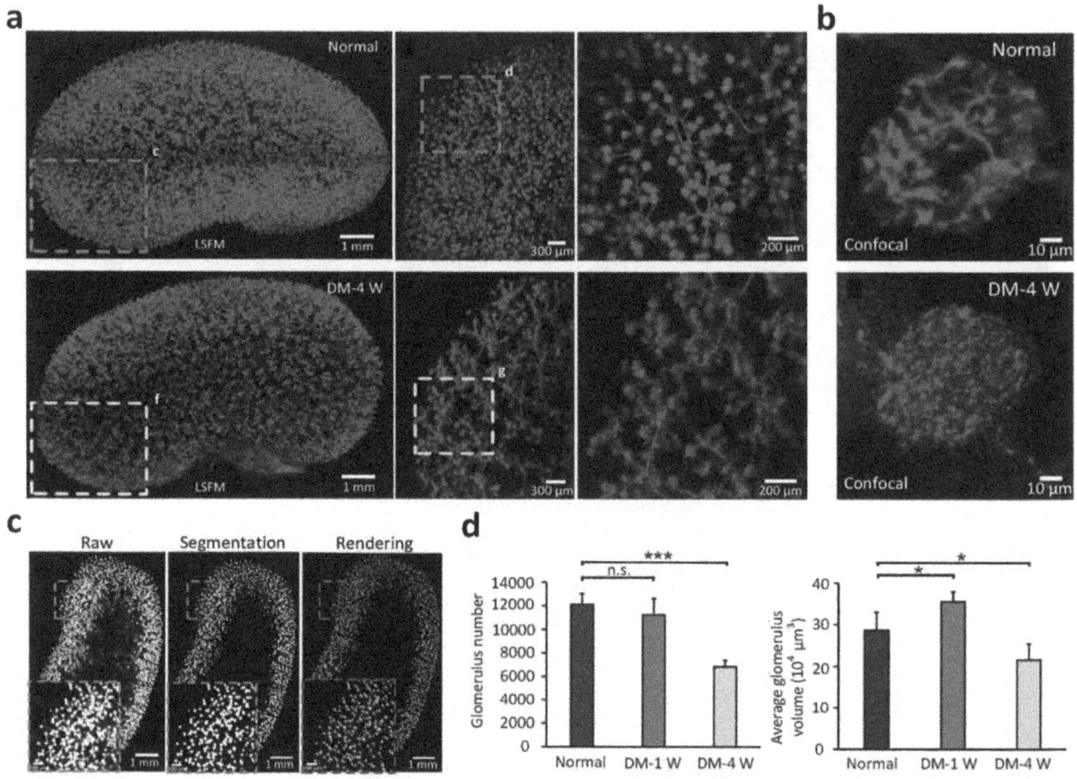

FIGURE 6.3 3D pathology of glomeruli in diabetic kidneys by MACS. (a) 3D reconstruction of glomerular tufts and vessels in normal and diabetic kidneys by LSFM imaging. (b) Confocal images of individual glomerular tufts in normal and diabetic kidneys. (c) The pipeline for image processing and counting of glomeruli. (d) Quantitative analysis of glomeruli from normal and diabetic kidneys. (Reprinted with permission from ref [11].)

These studies paved the way for tissue clearing techniques to facilitate the pathological investigations of microvascular structural changes in T1D-induced DN. However, applications of tissue clearing techniques in DN-related studies are still primitive and have limitations. For example, these studies utilized T1D animal models induced by STZ or alloxan, which still existed many pathological differences compared with clinical patients. Thus, there is still a long way to go for tissue clearing to facilitate histopathological understanding of DN. We believe that the advanced 3D optical imaging techniques are hopeful to be combined with traditional histopathology techniques, which will significantly promote the comprehensive investigation of nephrological studies.

6.3 DIABETES-INDUCED CHANGES IN SKIN MICROVASCULAR FUNCTION

The skin, as the most accessible organ, served as a model for the study of microcirculation. Studies showed that various skin vascular changes, including microvascular structure, vascular density and diameter, blood flow and blood oxygen, and vascular permeability, are associated with diabetes [7–10].

6.3.1 Diabetes-Induced Changes in Skin Microvascular Characteristics

Adamska et al. [62] showed in human samples that diabetics with diabetic kidney disease had problems with angiogenesis and vascular maturation. Compared with non-DN patients, the number of mature blood vessels in the skin of DN patients was reduced, and angiogenesis disorders occurred in the early stage of DN. Therefore, the density and maturity of skin microvessels are expected to be biomarkers for the diagnosis and treatment of DN.

Sorelli et al. [63] used a laser Doppler flowmeter (LDF) to monitor the volume of blood perfusion in patients with T1D. Studies have shown that patients with diabetes have lower baseline perfusion of skin microvessels in the big toe of the foot and respond more strongly to local thermal stimuli than the control group. In addition, through pulse decomposition analysis of LDF data, it was found that there was a significant correlation between arteriosclerosis and uncontrolled blood glucose within 1 year.

Lal et al. [64] also studied the volume of blood perfusion in foot. In their study, IDF was used to obtain the human plantar skin perfusion signal, and the time fractal analysis of the signal was carried out to study the changes of microcirculation in diabetic patients. The results showed that, compared with the control group, the Hurst coefficient of diabetic patients was significantly reduced. In addition, the authors showed that fractal analysis and wavelet analysis of Doppler signals can be used in the diagnosis of diabetes, which involved microcirculatory changes.

In study of Krumholz et al. [65], photoacoustic imaging was used to study the changes of blood flow velocity, vascular diameter, and oxygen metabolism of ear arteries and veins in STZ-modeled T1D mice during 1–6 weeks of disease. The results showed that diabetes had different effects on arteries and veins: arterial vessel diameter decreased, but no significant difference from baseline, and blood flow velocity decreased significantly. However, the diameter of venous vessels decreased and the velocity of blood flow increased significantly. In addition, the total hemoglobin concentration and oxygen saturation of the arteries and veins did not change significantly.

To obtain insights into the early microvascular deterioration resulting from prediabetes, Schaefer et al. [66] used intravital fluorescence microscopy to monitor morphological and functional microvascular parameters through a dorsal skin-fold chamber preparation in the uncoupling promotor-driven diphtheria toxin A chain (UCP1/DTA) mice. UCP1/DTA mice had increased vascular diameter and significantly reduced functional vascular density compared to normal mice, mainly due to the loss of small vessels between 3 and 12 mm in diameter. Although there was no significant change in the blood flow rate of a single vessel, the decrease in the number of vessels may still cause tissue hypoperfusion. Moreover, the skin vascular permeability of UCP1/DTA mice was significantly increased, about twice that of the normal group, and the white blood cell adhesion density was also significantly increased.

6.3.2 Diabetes-Induced Changes in Skin Microvascular Response

Diabetic foot ulcer is one of the most common complications in diabetic patients. Microvascular reactive damage will lead to the progression of diabetic foot ulcer. Jan et al. [67] used LDF to measure the amount of skin blood perfusion on the first metatarsus bone under mechanical stress (300 mmHg) and rapid thermal stress (42°C) in humans, and further evaluated metabolic, neurogenic, and myogenic responses by wavelet analysis. Results showed that metabolic neurogenic and myogenic responses to thermal stimulation were significantly reduced in diabetic patients, especially the neurogenic and myogenic control during the first vasodilation, and metabolism during the second vasodilation. In diabetic patients, myogenic control of mechanical stress stimulation was significantly reduced during reactive hyperemia. Such results suggested that diabetes would cause metabolic neurogenic and myogenic control disorders, leading to microvascular dysfunction. Therefore, the response of microvessels to locally applied mechanical and thermal stresses may be used to evaluate the risk of diabetic foot ulcers.

Argarini et al. [68] used optical coherence tomography to visualize structure and function of skin microvessels in the instep of diabetic patients. The authors measured the changes in blood vessel diameter, blood flow velocity, and density after thermal stimulation of the foot skin in healthy (CON), diabetic patients with foot ulcer (DFU), or patients without foot ulcer (DNU). They found at baseline temperature, vascular density in the DFU group was higher than in the CON group, and local heating resulted in a significant increase in blood flow velocity and density in each group, while compared with CON group and DNU group, DFU group showed less changes in diameter, lower blood flow velocity, and lower blood vessel density.

Tehrani et al. [69] studied the correlation between cutaneous microvascular reactivity and clinical microvascular lesions in patients with T1D. They used LDF to measure the skin blood flow flux response of the forearm after acetyl choline (ACh) and sodium nitroprusside (SNP) introduction. The results showed that compared with the healthy control group, the response peak of skin blood flow flux to Ach and SNP introduction in diabetic patients was reduced, while the response in patients with microvascular disease was further lower than that in patients without microvascular disease, in which the SNP-mediated response peak was significantly decreased and the Ach-mediated response showed a downward trend. In addition, the degree of vascular response in patients with microangiopathy decreased with the increase of microangiopathy score.

Feng et al. [70] used laser speckle contrast imaging and hyperspectral imaging to study the response of skin blood flow and blood oxygen to noradrenaline (NE). With the assistance of *in vivo* skin optical clearing technique [71–73], the distribution of blood flow and blood oxygen in skin vasculature could be clearly observed. The main results showed that venous and arterious blood flow decreased without recovery in diabetic mice after the injection of NE; furthermore, the decrease of arterious blood oxygen induced by NE greatly weakened, especially for 2- and 4-week diabetic mice. This change in vasoconstricting effect of NE was related to the expression of α1-adrenergic receptor. Feng et al. [74] further quantitatively analyzed the response of skin blood flow and blood oxygen to SNP and Ach. In T1D, after the injection of SNP and Ach, the results showed that there were no obvious differences in cutaneous arteriovenous blood flow response between T1D 1- and 2-week mice and non-T1D 1- and 2-week mice, respectively (Figure 6.4). However, in 3- and 4-week T1D mice, the recovery time of blood flow response obviously advanced after stimulation with SNP. For the corresponding cutaneous arteriovenous blood oxygen response, the SNP-induced increased amplitude of blood oxygen response observed in T1D was less than that observed in non-T1D from 1 to 4 weeks. Similarly, T1D would disturb the blood flow and blood oxygen response.

Such works suggest that visual monitoring of skin microvascular function response has guiding significance for early diagnosis of diabetes and clinical research.

6.3.3 Diabetes-Induced Changes in Skin Microvascular Permeability

Diabetes can also induce the increase of skin vascular permeability. Schaefer et al. [75] used skinfold chamber to observe that the vascular permeability to fluorescent dye increased due to diabetes. Besides, diabetes can facilitate inflammation of wound caused by the surgery, which may further influence the vascular permeability. Feng et al. [13] used reflective microscopy imaging through skin optical clearing window to monitor skin microvascular permeability *in vivo* and quantitatively evaluated the leakage of Evans Blue molecules from skin vasculature. As shown in Figure 6.5, for the normal mice, there was no significant difference in the vascular and extravascular Evans Blue concentration at different time points. However, for T1D mice, the vascular and extravascular Evans Blue concentration started to significantly increase at 10 minutes, indicating that T1D caused the increase in cutaneous microvascular permeability.

In summary, diabetes can lead to various skin microvascular dysfunctions, including reduced angiogenesis and vascular maturation, lower blood perfusion and stronger response to local thermal stimulation in foot, decreased diameter and increased blood flow of venous vessels in ear, increased vascular diameter and reduced functional vascular density in the back, abnormal response to drugs that constrict or dilate blood vessels, and enhanced vascular permeability. All of these may be potential diagnostic targets for diabetes.

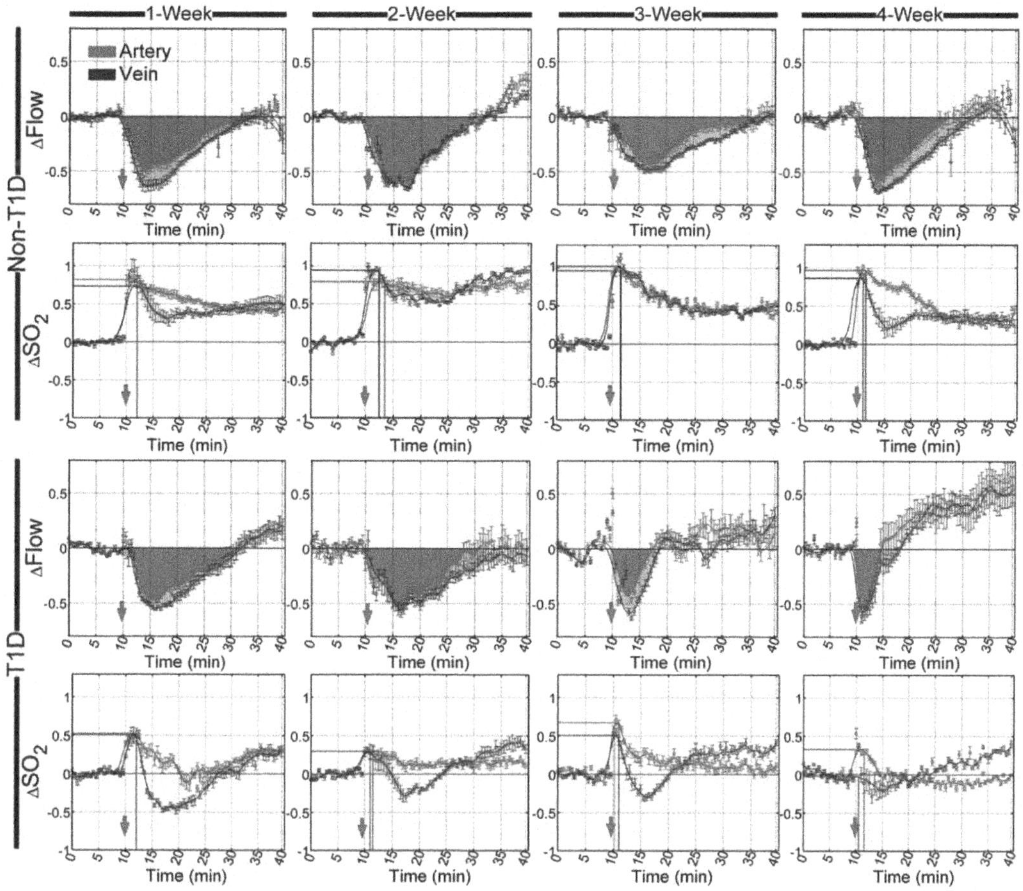

FIGURE 6.4 Time lapse of relative changes in cutaneous vascular blood flow and corresponding blood oxygen saturation in arteries and veins after the injection of SNP at different stages of T1D. The red arrows refer to the time of injection. The shadowed areas indicate the areas under curves of relative changes in blood flow. The lines perpendicular to the x- and y-axes represent the position of the maximum value of relative changes in blood oxygen saturation ($n=8$, mean ± standard error). (Reprinted with permission from Ref. [74].)

6.4 DIABETES-INDUCED DYNAMICAL CHANGES IN CEREBRAL MICROVASCULAR DYSFUNCTION

Diabetes is associated with various neurovascular diseases such as stroke, vascular dementia, or Alzheimer's disease. In diabetic stroke models, hyperglycemia can exacerbate damaging processes such as acidosis, accumulation of reactive oxides or nitrides, immune responses, and mitochondrial dysfunction. In addition, diabetes is also associated with more potential ischemic damage to the brain, primarily in small blood vessel disease and an increased risk of cognitive decline and dementia [76–81]. Therefore, it is of significance to study diabetes-induced cerebrovascular abnormality and their complications. With the assistance of modern optical techniques, it is possible to visualize changes in cerebrovascular structure and functions.

6.4.1 Diabetes-Induced Abnormal Vascular Structure

Aisha et al. [82] studied the changes of middle cerebral arteries in diabetic rats to investigate the underlying mechanism of ischemic stroke caused by DM. They used optical microscopy to visualize

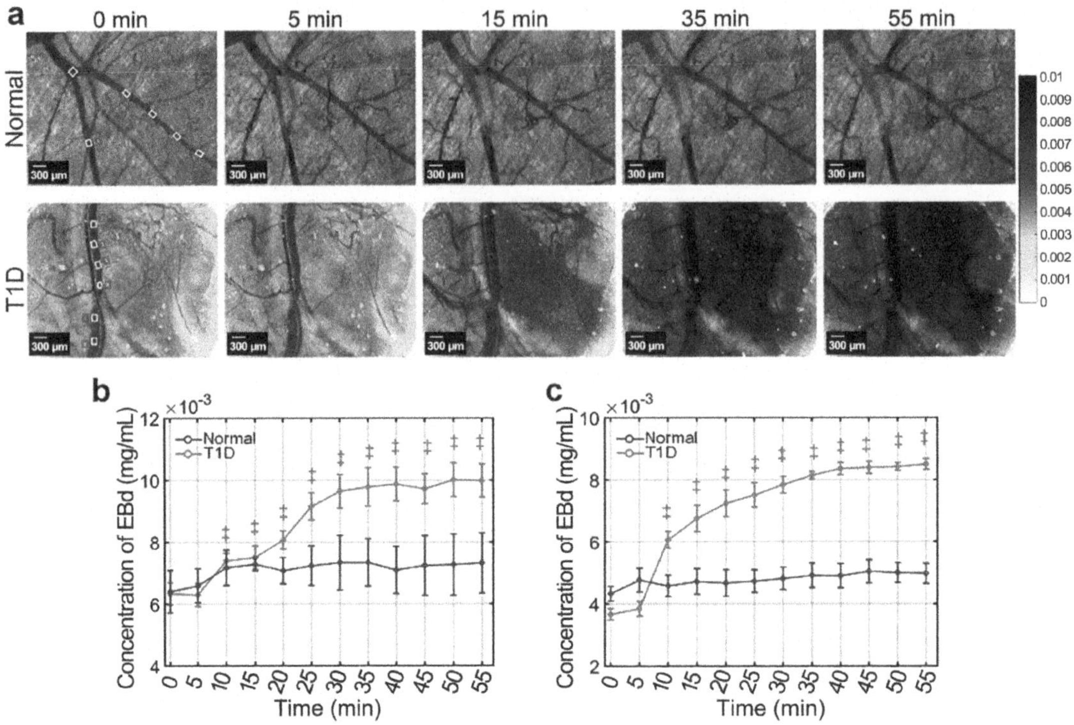

FIGURE 6.5 The permeability of cutaneous microvessels. (a) The skin microvascular EBd concentration maps obtained by spectral imaging with the help of *in vivo* skin optical clearing method (The color bar represents the EBd concentration). The time-dependent changes of EBd concentration at vascular positions (b) and extravascular positions (c) (The vascular positions and extravascular positions were white and red rectangular areas as indicated in (a), respectively. Mean ± standard deviations). ++ p <0.01 compared with EBd concentration at 5 minutes. (Reprinted with permission from [13].)

that diabetic rats exhibited significantly increased wall thickness and wall/lumen ratio, and further results showed that it was associated with increased MMP-2 activity and collagen deposition but reduced MMP-13 activity.

Roshini et al. [83] studied cerebral neovascularization and remodeling patterns in T2D with confocal fluorescence imaging for tissue slices, where vascular volume, surface area, and structural parameters including microvessel/macrovessel ratio, non-FITC (fluorescein) perfusing vessel abundance, vessel tortuosity, and branch density were measured by 3D reconstruction of FITC-stained vasculature in GK rats or Lepr$^{db/db}$ mice.

For the research, FITC-Fluorescein IsoThioCyanate-dextran was perfused before sectioning, and the brain sections were then labeled with isolectin B4 to differentiate vessels not perfused with FITC. Vascular volume refers to the ratio of the volume of the FITC-stained vasculature to the total volume. Surface area represents the area available for the diffusion/exchange of vascular nutrients to the surrounding brain tissue. Macrovasculature represents the penetrating arterioles (PA) and immediate first-order branches (PA1). Microvascular measurements were obtained by subtracting the macrovascular measures (PA and PA1) from the total vascular parameters (Figure 6.6).

Vascular measurements were made in the cortical and striatal regions. The GK model displayed increased total vascular density, volume, and surface area in both cortex and striatum. The results showed that there was a linear correlation between the surface area and volume in control rats but there was a disproportional increase in these parameters in the GK group. The results also suggested that in GK rats there were some vessels that are getting larger without a parallel increase in

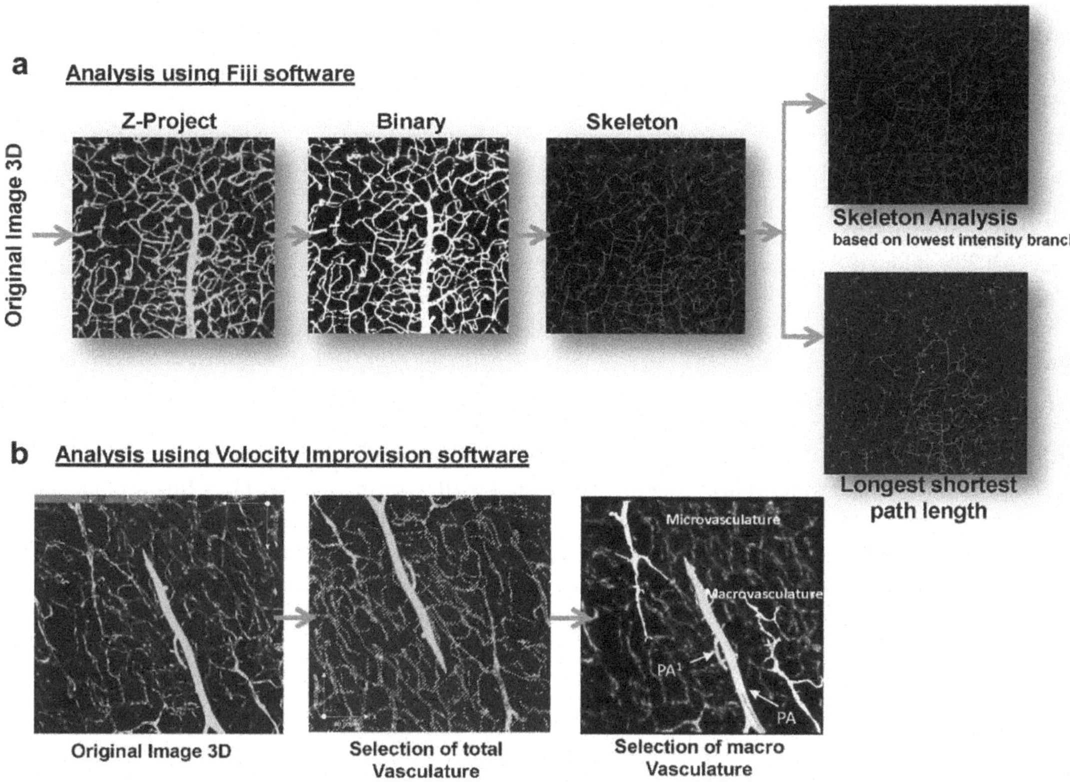

FIGURE 6.6 Schematics showing tissue procession performed using Fiji and Volocity software explained in the methodology. (Reprinted with permission from [83].)

area indicative of larger vessel remodeling. Besides, Roshini et al. found that GK rats displayed an increase in both micro- and macrovessel volume and surface area. In addition, glucose control with metformin initiated right after the onset of diabetes could prevent the increase in vascular density and correct the relationship between vascular volume and surface area. Furthermore, the nonperfused/perfused vessel ratio was observed greater in GK rats normalized with glycemic control. In the research of $Lepr^{db/db}$ mice, it was found no difference in vascular density and nonperfused/perfused ratio, but an increase in microvascular volume and area compared to control.

In their study, GK rats exhibited significantly enhanced branch density of the PAs with an increase of the diameter as well as the tortuosity, while glycemic control could only prevent the increase of the diameter and tortuosity but did not impact branch density. $Lepr^{db/db}$ mice exhibited enhanced branch density and tortuosity but there was a trend for decreased lumen diameter of the PAs.

6.4.2 Diabetes-Induced Blood-Brain Barrier Dysfunction

Blood-brain barrier (BBB) acts as a gate keeper allowing nutrient transportation while preventing harmful substances from entering the brain [84]. However, reports have shown that diabetic conditions could impair BBB function [85], where BBB permeability was increased in hyperglycemia or experimental diabetic animal models [86]. Optical microscopy provides a powerful tool to observe leakage from cerebral vasculature or abnormity BBB-related proteins caused by diabetes.

Alexis et al. [87] studied protein kinase Cβ (PKCβ)-mediated cerebrovascular breakdown in the retina in diabetic mice, and used a PKCβ inhibitor enzastaurin to re-seal the BBB. In their study, the effects of PKCβ antagonism on BBB integrity were examined. Fluorescein penetrance into the neuropil was evident on hippocampal sections from db/Veh mice, but in sections from db/Enz mice, NaFl fluorescence was confined to the vasculature in a pattern similar to Wt/Veh mice. Qualitative observations on hippocampal sections were upheld with direct quantification of NaFl in homogenates, as enzastaurin blocked increases in hippocampal NaFl penetrance. There was no effect of genotype or PKCβ antagonism on cortical NaFl concentrations, suggesting that there may be regional heterogeneity in susceptibility to BBB breakdown. Such results suggested that db/db mice exhibit PKCβ-mediated BBB breakdown.

Zhanyang et al. [88] tested the hypothesis that recombinant FGF21 (rFGF21) administration may reduce T2D-induced BBB disruption via NF-E2-related factor-2 (Nrf2) upregulation. They studied a series of BBB-related tight junction proteins, including ZO1, VE-cadherin, and occludin with immunofluorescence and immunocytochemistry in diabetic mice, and found that in db/db mice, the staining is fragmented, losing continuity. rFGF21 treatment reversed this process, making occludin and VE-cadherin staining more integral and continuous compared to untreated db/db group. rFGF21 administration may decrease T2D-induced BBB permeability.

Using both T1D and T2D mice models, Slava et al. [89] found enhanced BBB permeability and memory loss (Y maze, water maze) that are associated with hyperglycemia. They further made neuropathologic analyses of brain tissue from T1D and T2D mice and observed microglial activation, expression of ICAM-1 and attenuated coverage of pericytes compared to controls. Their findings indicate that BBB compromises in DM *in vivo* models and its association with memory deficits, gene alterations in brain endothelium, and neuroinflammation.

6.4.3 Diabetes-Induced Changes in Cortical Microvascular Blood Flow/Oxygen Response to Drugs

Diabetes is a chronic metabolic disease [90], and it is therefore necessary to monitor changes in cerebrovascular dysfunction *in vivo* during different developmental stages of diabetes. Feng et al. performed *in vivo* monitoring of SNP- and ACh-induced cerebral microvascular responses (including blood flow and blood oxygen saturation) during the development of T1D with the assistance of *in vivo* skull optical clearing techniques.

Unlike open-skull window or thinned-skull window, *in vivo* skull optical clearing technique provides an easy and surgery-free skull window for optical visualizing cortical neurovascular structure and function [91,92]. Based on the optical clearing skull window, Feng et al. [74] used a home-built laser speckle/hyperspectral two-mode imaging system to visualize vascular response to the drugs (Figure 6.7).

In their research, the relative changes in blood flow (ΔFlow) and in blood oxygen saturation (ΔSO_2) recorded before and after the injection of SNP and ACh were quantitatively analyzed. In the non-T1D mice, cerebral microvascular ΔFlow quickly increased after an initial SNP-induced decrease, and the shapes of the time-lapse curves of ΔSO_2 were slightly similar to those observed for ΔFlow. Similarly, ACh caused ΔFlow and ΔSO_2 to decrease. Additionally, the cerebral arteriovenous range of ΔSO_2 was smaller than that of ΔFlow after SNP and ACh injection.

In the T1D mice, the cerebral arteriovenous ΔFlow increased directly, and the initial decrease almost disappeared after the injection of SNP in the T1D mice at 1, 2, 3 and 4 weeks. In the 3- and 4-week T1D mice, the range of blood flow response induced by SNP became dramatically weak. Similarly, the diminished blood flow response after the injection of ACh was also observed in the T1D mice. Compared to the non-T1D mice, during the development of T1D, the corresponding cerebral arteriovenous ΔSO_2 also increased directly, and the initial decrease disappeared after the injection of SNP. Furthermore, in the T1D 1-, 2-, 3- and 4-week mice, the increasing SNP-induced cerebral arteriovenous range of ΔSO_2 was larger than that observed in the non-T1D mice. In addition, the ACh-induced decrease in ΔSO_2 occurred only in veins, while the changes of ΔSO_2 in artery almost disappeared in the T1D 1-, 2-, 3- and 4-week mice. These data indicate that T1D can cause

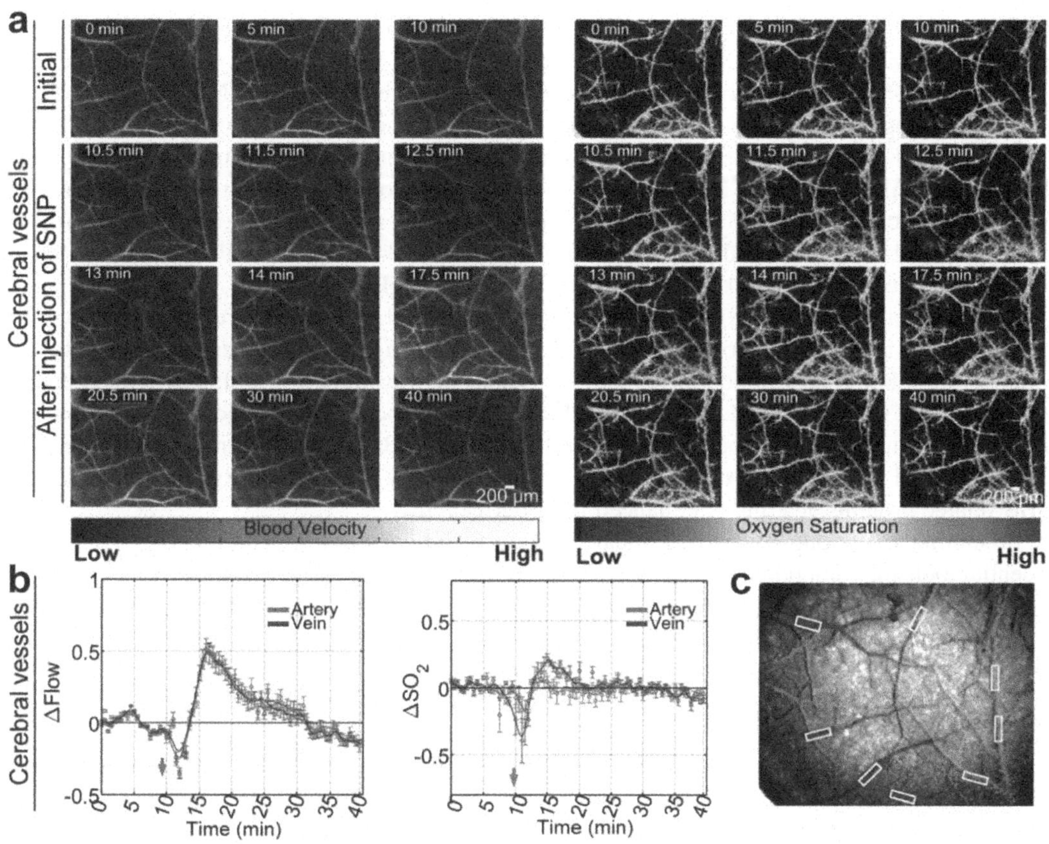

FIGURE 6.7 Maps of dynamic responses of blood flow and blood oxygen saturation in non-T1D mice. (a) Typical blood flow and blood oxygen saturation maps of cerebral vessels viewed through an optical clearing skull window before and after the injection of SNP. (b) Statistical analyses based on the yellow rectangular boxes containing vessel positions were performed to show the time courses of relative changes in blood flow and blood oxygen saturation in cerebral microvessels (mean ± standard error). (c) Arteriovenous segmentation maps of cerebral microvessels (rectangular boxes were used for statistical analysis). (Reprinted with permission from [74].)

abnormal changes in cerebrovascular blood flow and blood oxygen responses at the early stage of T1D. In addition, insulin treatment could alleviate this abnormal performance.

The results suggested that abnormal changes occurred in cerebral vascular blood flow and blood oxygen responses in the early stage of diabetes (1 week), and became more and more serious as diabetes progressed.

In conclusion, diabetes can influence brain function by causing cerebral vascular dysfunction from aspects of structural changes, BBB breakdown, and abnormal response. It is associated with increase of vessel wall, vessel tortuosity, branch density, and decrease of the BBB-related proteins, leading to enhanced BBB permeability. In addition, diabetes also influences the cerebral vascular response to vasodilators.

6.5 SUMMARY

In this chapter, we have shortly reviewed several kinds of chronic damage and dysfunction of vasculatures induced by DM, including DN and microvascular dysfunction of cerebrum/skin.

Additionally, the structural and functional changes of these complications can be comprehensively studied via advanced optical imaging techniques.

DN is one of the most serious chronic microvascular complications of DM. The morphological structural changes caused by DN include GBM thickening, mesangial expansion, and development of Kimmelstiel–Wilson nodules, which have been clinically investigated by traditional histopathology via light microscopy or TEM. Recently, the burgeoning optical clearing techniques have been used for 3D reconstruction of neural and vascular structures within entire kidneys combined with modern optical imaging techniques. Several optical clearing methods have already been developed for studying the glomerular structural changes induced by T1D in 3D.

Except for *in vitro* studies for DN, structural and functional changes of cerebrum and skin vasculature induced by DM are also reviewed. Modern optical imaging techniques can realize dynamic monitoring of blood flow, blood oxygen, and vascular permeability within the skin and cerebrum *in vivo*. Recent studies have revealed that diabetes can influence brain function by causing cerebral vascular dysfunction from aspects of structural changes, BBB breakdown, and abnormal response. It is associated with increase of vessel wall, vessel tortuosity, branch density, and decrease of the BBB-related proteins, leading to enhanced BBB permeability.

In summary, various advanced optical imaging techniques developed recently have provided essential tools for comprehensively studying the pathological investigations of pathological changes for many kinds of tissues/organs in T1D-induced DN, as well as greatly facilitating the understanding of diabetes.

ACKNOWLEDGMENT

This chapter was supported by the National Science Foundation of China (Grant Nos. 61860206009, 81171376, 82001877, 62105113), Innovation Fund of WNLO, and Innovation Project of Optics Valley Laboratory (Grant No. OVL2021BG011).

REFERENCES

1. NCD Risk Factor Collaboration, "Worldwide trends in diabetes since 1980: A pooled analysis of 751 population-based studies with 4.4 million participants," *Lancet (London, England)* **387**(10027), 1513–1530 (2016).
2. T. Tuomi, N. Santoro, S. Caprio et al., "The many faces of diabetes: A disease with increasing heterogeneity," *The Lancet* **383**(9922), 1084–1094 (2014).
3. T. H. Lipman, L. E. Levitt Katz, S. J. Ratcliffe et al., "Increasing incidence of type 1 diabetes in youth: Twenty years of the Philadelphia Pediatric Diabetes Registry," *Diabetes Care* **36**(6), 1597–1603 (2013).
4. F. Helmchen, and W. Denk, "Deep tissue two-photon microscopy," *Nature Methods* **2**(12), 932–940 (2005).
5. H.-U. Dodt, U. Leischner, A. Schierloh et al., "Ultramicroscopy: Three-dimensional visualization of neuronal networks in the whole mouse brain," *Nature Methods* **4**(4), 331–336 (2007).
6. Y.-C. Liu, and A.-S. Chiang, "High-resolution confocal imaging and three-dimensional rendering," *Methods* **30**(1), 86–93 (2003).
7. S. P. M. Hosking, R. Bhatia, P. A. Crock et al., "Non-invasive detection of microvascular changes in a paediatric and adolescent population with type 1 diabetes: A pilot cross-sectional study," *Bmc Endocr Disord* **13**(1), 41 (2013).
8. C. Schaefer, T. Biermann, M. Schroeder et al., "Early microvascular complications of prediabetes in mice with impaired glucose tolerance and dyslipidemia," *Acta Diabetol* **47**, S19–S27 (2010).
9. A. Krumholz, L. D. Wang, J. J. Yao, and L. H. V. Wang, "Functional photoacoustic microscopy of diabetic vasculature," *J Biomed Opt* **17**(6), 060502 (2012).
10. L. Khaodhiar, T. Dinh, K. T. Schomacker et al., "The use of medical hyperspectral technology to evaluate microcirculatory chaves in diabetic foot ulcers and to predict clinical outcomes," *Diabetes Care* **30**(4), 903–910 (2007).
11. J. Zhu, T. Yu, Y. Li et al., "MACS: Rapid aqueous clearing system for 3D mapping of intact organs," *Adv Sci (Weinh)* **7**(8), 1903185 (2020).

12. S. Hasegawa, T. Tanaka, T. Saito et al., "The oral hypoxia-inducible factor prolyl hydroxylase inhibitor enarodustat counteracts alterations in renal energy metabolism in the early stages of diabetic kidney disease," *Kidney Int* **97**(5), 934–950 (2020).
13. W. Feng, C. Zhang, T. Yu, and D. Zhu, "Quantitative evaluation of skin disorders in type 1 diabetic mice by in vivo optical imaging," *Biomed Opt Express* **10**(6), 2996–3008 (2019).
14. W. Feng, S. Liu, C. Zhang et al., "Comparison of cerebral and cutaneous microvascular dysfunction with the development of type 1 diabetes," *Theranostics* **9**(20), 5854–5868 (2019).
15. L. Stapley, "The history of diabetes mellitus," *Trends Endocrinol Metab* **12**(6), 277 (2001).
16. J. S. Cameron, "The discovery of diabetic nephropathy: From small print to centre stage," *J Nephrol* **19**(Suppl 10), S75–87 (2006).
17. M. Akhtar, N. M. Taha, A. Nauman, I. B. Mujeeb, and A. D. M. H. Al-Nabet, "Diabetic kidney disease: Past and present," *Adv Anat Pathol* **27**(2), 87–97 (2020).
18. H. L. Minuk, "Diseases of the kidneys and urinary tract," in *Brackenridge's Medical Selection of Life Risks*, R. D. C. Brackenridge, R. S. Croxson, and R. MacKenzie, Eds., pp. 611–654, Palgrave Macmillan, London (2006).
19. Y. An, F. Xu, W. Le et al., "Renal histologic changes and the outcome in patients with diabetic nephropathy," *Nephrol Dial Transplant* **30**(2), 257–266 (2015).
20. R. Z. Alicic, M. T. Rooney, and K. R. Tuttle, "Diabetic kidney disease: Challenges, progress, and possibilities," *Clin J Am Soc Nephrol* **12**(12), 2032–2045 (2017).
21. A. T. Reutens, "Epidemiology of diabetic kidney disease," *Med Clin North Am* **97**(1), 1–18 (2013).
22. R. Saran, B. Robinson, K. C. Abbott et al., "US renal data system 2018 annual data report: Epidemiology of kidney disease in the United States," *Am J Kidney Dis* **73**(3, Supplement 1), A7–A8 (2019).
23. W. K. C. Leung, L. Gao, P. M. Siu, and C. W. K. Lai, "Diabetic nephropathy and endothelial dysfunction: Current and future therapies, and emerging of vascular imaging for preclinical renal-kinetic study," *Life Sci* **166**, 121–130 (2016).
24. A. J. Joseph, and E. A. Friedman, "Diabetic nephropathy in the elderly," *Clin Geriatr Med* **25**(3), 373–389 (2009).
25. Y. S. Kanwar, J. Wada, L. Sun et al., "Diabetic nephropathy: Mechanisms of renal disease progression," *Exp Biol Med (Maywood)* **233**(1), 4–11 (2008).
26. N. Torkamani, G. Jerums, P. Crammer et al., "Three dimensional glomerular reconstruction: A novel approach to evaluate renal microanatomy in diabetic kidney disease," *Sci Rep* **9**(1), 1829 (2019).
27. F. C. Brosius 3rd, "New insights into the mechanisms of fibrosis and sclerosis in diabetic nephropathy," *Rev Endocr Metab Disord* **9**(4), 245–254 (2008).
28. M. Mauer, M. L. Caramori, P. Fioretto, and B. Najafian, "Glomerular structural-functional relationship models of diabetic nephropathy are robust in type 1 diabetic patients," *Nephrol Dial Transplant* **30**(6), 918–923 (2015).
29. H. Peng, P. Luo, Y. Li et al., "Simvastatin alleviates hyperpermeability of glomerular endothelial cells in early-stage diabetic nephropathy by inhibition of RhoA/ROCK1," *PloS One* **8**(11), e80009 (2013).
30. L. Wang, T. Tao, W. Su et al., "A disease model of diabetic nephropathy in a glomerulus-on-a-chip microdevice," *Lab on a Chip* **17**(10), 1749–1760 (2017).
31. T. Yu, Y. Qi, H. Gong, Q. Luo, and D. Zhu, "Optical clearing for multiscale biological tissues," *J Biophoton* **11**(2), e201700187 (2018).
32. H. R. Ueda, A. Erturk, K. Chung et al., "Tissue clearing and its applications in neuroscience," *Nat Rev Neurosci* **21**(2), 61–79 (2020).
33. Y. Chen, K. Lee, Z. Ni, and J. C. He, "Diabetic kidney disease: Challenges, advances, and opportunities," *Kidney Dis (Basel)* **6**(4), 215–225 (2020).
34. K. E. White, and R. W. Bilous, "Type 2 diabetic patients with nephropathy show structural-functional relationships that are similar to type 1 disease," *J Am Soc Nephrol (JASN)* **11**(9), 1667–1673 (2000).
35. R. Osterby, M. A. Gall, A. Schmitz et al., "Glomerular structure and function in proteinuric type 2 (non-insulin-dependent) diabetic patients," *Diabetologia* **36**(10), 1064–1070 (1993).
36. P. L. Brito, P. Fioretto, K. Drummond et al., "Proximal tubular basement membrane width in insulin-dependent diabetes mellitus," *Kidney Int* **53**(3), 754–761 (1998).
37. R. D. Harris, M. W. Steffes, R. W. Bilous, D. E. R. Sutherland, and S. M. Mauer, "Global glomerular sclerosis and glomerular arteriolar hyalinosis in insulin dependent diabetes," *Kidney Int* **40**(1), 107–114 (1991).
38. P. Fioretto, and M. Mauer, "Histopathology of diabetic nephropathy," *Semin Nephrol* **27**(2), 195–207 (2007).
39. C. Carrara, M. Abbate, S. Conti et al., "Histological examination of the diabetic kidney," in *Diabetic Nephropathy: Methods and Protocols*, L. Gnudi, and D. A. Long, Eds., pp. 63–87, Springer US, New York (2020).

40. S. Olsen, "Light microscopy of diabetic glomerulopathy: The classic lesions," in *The Kidney and Hypertension in Diabetes Mellitus*, C. E. Mogensen, Ed., pp. 217–226, Springer US, Boston, MA (1998).
41. Y. Maezawa, M. Takemoto, and K. Yokote, "Cell biology of diabetic nephropathy: Roles of endothelial cells, tubulointerstitial cells and podocytes," *J Diabetes Investig* **6**(1), 3–15 (2015).
42. K. E. White, and R. W. Bilous, "Structural alterations to the podocyte are related to proteinuria in type 2 diabetic patients," *Nephrol Dial Transplant* **19**(6), 1437–1440 (2004).
43. B. Najafian, J. T. Crosson, Y. Kim, and M. Mauer, "Glomerulotubular junction abnormalities are associated with proteinuria in type 1 diabetes," *J Am Soc Nephrol (JASN)* **17**(4 Suppl 2), S53–60 (2006).
44. Y. An, F. Xu, W. Le et al., "Renal histologic changes and the outcome in patients with diabetic nephropathy," *Nephrol Dial Transplant* **30**(2), 257–266 (2014).
45. H. R. Ueda, H. U. Dodt, P. Osten et al., "Whole-brain profiling of cells and circuits in mammals by tissue clearing and light-sheet microscopy," *Neuron* **106**(3), 369–387 (2020).
46. M. V. Gomez-Gaviro, D. Sanderson, J. Ripoll, and M. Desco, "Biomedical applications of tissue clearing and three-dimensional imaging in health and disease," *iScience* **23**(8), 101432 (2020).
47. D. Zhu, K. V. Larin, Q. Luo, and V. V. Tuchin, "Recent progress in tissue optical clearing," *Laser Photon Rev* **7**(5), 732–757 (2013).
48. K. Tainaka, A. Kuno, S. I. Kubota, T. Murakami, and H. R. Ueda, "Chemical principles in tissue clearing and staining protocols for whole-body cell profiling," *Annu Rev Cell Dev Biol* **32**, 713–741 (2016).
49. A. Ertürk, K. Becker, N. Jährling et al., "Three-dimensional imaging of solvent-cleared organs using 3DISCO," *Nat Protoc* **7**(11), 1983–1995 (2012).
50. J. Xu, Y. Ma, T. Yu, and D. Zhu, "Quantitative assessment of optical clearing methods in various intact mouse organs," *J Biophoton* **12**(2), e201800134 (2019).
51. Y. Qi, T. Yu, J. Xu et al., "FDISCO: Advanced solvent-based clearing method for imaging whole organs," *Sci Adv* **5**(1), eaau8355 (2019).
52. N. Renier, Z. Wu, D. J. Simon et al., "iDISCO: A simple, rapid method to immunolabel large tissue samples for volume imaging," *Cell* **159**(4), 896–910 (2014).
53. A. Klingberg, A. Hasenberg, I. Ludwig-Portugall et al., "Fully automated evaluation of total glomerular number and capillary tuft size in nephritic kidneys using lightsheet microscopy," *J Am Soc Nephrol (JASN)* **28**(2), 452–459 (2017).
54. S. Zhao, M. I. Todorov, R. Cai et al., "Cellular and molecular probing of intact human organs," *Cell* **180**(4), 796–812 (2020).
55. K. Tainaka, S. I. Kubota, T. Q. Suyama et al., "Whole-body imaging with single-cell resolution by tissue decolorization," *Cell* **159**(4), 911–924 (2014).
56. P. Matryba, L. Bozycki, M. Pawlowska, L. Kaczmarek, and M. Stefaniuk, "Optimized perfusion-based CUBIC protocol for the efficient whole-body clearing and imaging of rat organs," *J Biophoton* **11**(5), e201700248 (2018).
57. K. Tainaka, T. C. Murakami, E. A. Susaki et al., "Chemical landscape for tissue clearing based on hydrophilic reagents," *Cell Rep* **24**(8), 2196–2210 (2018).
58. S. Hasegawa, E. A. Susaki, T. Tanaka et al., "Comprehensive three-dimensional analysis (CUBIC-kidney) visualizes abnormal renal sympathetic nerves after ischemia/reperfusion injury," *Kidney Int* **96**(1), 129–138 (2019).
59. N. Renier, E. L. Adams, C. Kirst et al., "Mapping of brain activity by automated volume analysis of immediate early genes," *Cell* **165**(7), 1789–1802 (2016).
60. F. de Chaumont, S. Dallongeville, N. Chenouard et al., "Icy: An open bioimage informatics platform for extended reproducible research," *Nat Methods* **9**(7), 690–696 (2012).
61. H. Peng, A. Bria, Z. Zhou, G. Iannello, and F. Long, "Extensible visualization and analysis for multidimensional images using Vaa3D," *Nat Protoc* **9**(1), 193–208 (2014).
62. A. Adamska, S. Pilacinski, D. Zozulinska-Ziolkiewicz et al., "Disturbances in angiogenesis and vascular maturation in the skin are associated with diabetic kidney disease in type 1 diabetes," *Clin Diabetol* **8**(5), 231–237 (2019).
63. M. Sorelli, P. Francia, L. Bocchi, A. De Bellis, and R. Anichini, "Assessment of cutaneous microcirculation by laser Doppler flowmetry in type 1 diabetes," *Microvasc Res* **124**, 91–96 (2019).
64. C. Lal, and S. N. Unni, "Correlation analysis of laser Doppler flowmetry signals: A potential non-invasive tool to assess microcirculatory changes in diabetes mellitus," *Med Biol Eng Comput* **53**(6), 557–566 (2015).
65. A. Krumholz, L. D. Wang, J. J. Yao, and L. H. V. Wang, "Functional photoacoustic microscopy of diabetic vasculature," *J Biomed Opt* **17**(6), 060502 (2012).

66. C. Schaefer, T. Biermann, M. Schroeder et al., "Early microvascular complications of prediabetes in mice with impaired glucose tolerance and dyslipidemia," *Acta Diabetol* **47**(s1), 19–27 (2010).
67. Y. K. Jan, S. Shen, R. D. Foreman, and W. J. Ennis, "Skin blood flow response to locally applied mechanical and thermal stresses in the diabetic foot," *Microvasc Res* **89**(9), 40–46 (2013).
68. R. Argarini, R. A. Mclaughlin, S. Z. Joseph, L. H. Naylor, and D. J. Green, "Optical coherence tomography: A novel imaging approach to visualize and quantify cutaneous microvascular structure and function in patients with diabetes," *BMJ Open Diabetes Res Care* **8**(1), e001479 (2020).
69. S. Tehrani, K. Bergen, L. Azizi, and G. Jorneskog, "Skin microvascular reactivity correlates to clinical microangiopathy in type 1 diabetes: A pilot study," *Diabetes Vasc Dis Res* **17**(3), (2020). DOI: 10.1177/1479164120928303.
70. W. Feng, R. Shi, C. Zhang et al., "Visualization of skin microvascular dysfunction of type 1 diabetic mice using in vivo skin optical clearing method," *J Biomed Opt* **24**(3), 1–9 (2018).
71. D. Zhu, J. Wang, Z. W. Zhi, X. Wen, and Q. M. Luo, "Imaging dermal blood flow through the intact rat skin with an optical clearing method," *J Biomed Opt* **15**(2), 026008 (2010).
72. J. Wang, R. Shi, and D. Zhu, "Switchable skin window induced by optical clearing method for dermal blood flow imaging," *J Biomed Opt* **18**(6), 061209 (2013).
73. R. Shi, M. Chen, V. V. Tuchin, and D. Zhu, "Accessing to arteriovenous blood flow dynamics response using combined laser speckle contrast imaging and skin optical clearing," *Biomed Opt Express* **6**(6), 1977–1989 (2015).
74. W. Feng, S. Liu, C. Zhang et al., "Comparison of cerebral and cutaneous microvascular dysfunction with the development of type 1 diabetes," *Theranostics* **9**(20), 5854–5868 (2019).
75. F. Liu, L. Zhang, R. M. Hoffman, and M. Zhao, "Vessel destruction by tumor-targeting Salmonella typhimurium A1-R is enhanced by high tumor vascularity," *Cell Cycle* **9**(22), 4518–4524 (2010).
76. T. Hardigan, R. Ward, and A. Ergul, "Cerebrovascular complications of diabetes: Focus on cognitive dysfunction," *Clin Sci (Lond)* **130**(20), 1807–1822 (2016).
77. S. Ma, J. Wang, Y. Wang et al., "Diabetes mellitus impairs white matter repair and long-term functional deficits after cerebral ischemia," *Stroke* **49**(10), 2453–2463 (2018).
78. S. P. Rensma, T. T. van Sloten, J. Ding et al., "Type 2 diabetes, change in depressive symptoms over time, and cerebral small vessel disease: Longitudinal data of the AGES-reykjavik study," *Diabetes Care* **43**(8), 1781–1787 (2020).
79. T. T. van Sloten, S. Sedaghat, M. R. Carnethon, L. J. Launer, and C. D. A. Stehouwer, "Cerebral microvascular complications of type 2 diabetes: Stroke, cognitive dysfunction, and depression," *Lancet Diabetes Endocrinol* **8**(4), 325–336 (2020).
80. A. Ergul, A. Kelly-Cobbs, M. Abdalla, and S. C. Fagan, "Cerebrovascular complications of diabetes: Focus on stroke," *Endocr Metab Immune Disord Drug Targets* **12**(2), 148 (2012).
81. N. M. Bornstein, M. Brainin, A. Guekht, I. Skoog, and A. D. Korczyn, "Diabetes and the brain: Issues and unmet needs," *Neurol Sci* **35**(7), 995–1001 (2014).
82. A. I. Kelly-Cobbs, A. K. Harris, M. M. Elgebaly et al., "Endothelial endothelin B receptor-mediated prevention of cerebrovascular remodeling is attenuated in diabetes because of up-regulation of smooth muscle endothelin receptors," *J Pharmacol Exp Ther* **337**(1), 9–15 (2011).
83. R. Prakash, M. Johnson, S. C. Fagan, and A. Ergul, "Cerebral neovascularization and remodeling patterns in two different models of type 2 diabetes," *PLoS One* **8**(2), e56264 (2013).
84. R. Haddad-Tovolli, N. R. V. Dragano, A. F. S. Ramalho, and L. A. Velloso, "Development and function of the blood-brain barrier in the context of metabolic control," *Front Neurosci-Switz* **11**, 224 (2017).
85. S. Prasad, R. K. Sajja, P. Naik, and L. Cucullo, "Diabetes mellitus and blood-brain barrier dysfunction: An overview," *J Pharmacovigil* **2**(2), 125 (2014).
86. B. T. Hawkins, T. F. Lundeen, K. M. Norwood, H. L. Brooks, and R. D. Egleton, "Increased blood-brain barrier permeability and altered tight junctions in experimental diabetes in the rat: Contribution of hyperglycaemia and matrix metalloproteinases," *Diabetologia* **50**(1), 202–211 (2007).
87. A. M. Stranahan, S. Hao, A. Dey, X. Yu, and B. Baban, "Blood-brain barrier breakdown promotes macrophage infiltration and cognitive impairment in leptin receptor-deficient mice," *J Cereb Blood Flow Metab* **36**(12), 2108–2121 (2016).
88. Z. Y. Yu, L. Lin, Y. H. Jiang et al., "Recombinant FGF21 protects against blood-brain barrier leakage through Nrf2 upregulation in type 2 diabetes mice," *Mol Neurobiol* **56**(4), 2314–2327 (2019).
89. S. Rom, V. Zuluaga-Ramirez, S. Gajghate et al., "Hyperglycemia-driven neuroinflammation compromises BBB leading to memory loss in both diabetes mellitus (DM) type 1 and type 2 mouse models," *Mol Neurobiol* **56**(3), 1883–1896 (2019).

90. L. P. Reagan, "Diabetes as a chronic metabolic stressor: Causes, consequences and clinical complications," *Exp Neurol* **233**(1), 68–78 (2012).
91. Y. J. Zhao, T. T. Yu, C. Zhang et al., "Skull optical clearing window for in vivo imaging of the mouse cortex at synaptic resolution," *Light-Sci Appl* **7**, 17153 (2018).
92. C. Zhang, W. Feng, Y. J. Zhao et al., "A large, switchable optical clearing skull window for cerebrovascular imaging," *Theranostics* **8**(10), 2696–2708 (2018).

7 Noninvasive Sensing of Serum sRAGE and Glycated Hemoglobin by Skin UV-Induced Fluorescence

Vladimir V. Salmin, Tatyana E. Taranushenko, Natalya G. Kiseleva, and Alla B. Salmina

CONTENTS

7.1 INTRODUCTION

Type 1 diabetes mellitus is a disease that develops in genetically susceptible individuals due to chronic inflammation and destruction of pancreatic beta-cells. At present, type 1 diabetes is one of the most common chronic diseases, and vascular complications of diabetes are among the most important medical and social problems of medicine. International epidemiological studies done in recent years indicated a progressive increase in the incidence of type 1 diabetes, primarily in childhood. The average annual increase in the incidence of type 1 diabetes in Russia is close to the level of European countries and is 4%. It is important that the highest rates of increase in the incidence of diabetes are observed in young children. According to the EURODIAB research group, the incidence of type 1 diabetes under the age of 5 will double by 2020. The manifestation of diabetes in early childhood is accompanied by a lengthening of the duration of the illness, often a labile course, and the risk of early disability. The mortality rate in children exceeds that in the population without diabetes by 2 - 9 times (0.1 - 0.2 per 100,000). The main causes of mortality are acute conditions such as ketoacidosis, hypoglycemia, and progression of chronic vascular complications. Acute complications of diabetes are the result of untimely diagnosis or inadequate therapy, and late complications are a consequence of unsatisfactory compensation of carbohydrate metabolism. Most microangiopathies in children are reversible when detected at the preclinical stage; therefore, screening and subsequent dynamic monitoring of metabolic compensation should be carried out from the moment of debut of the disease. However, mainly invasive and expensive techniques are currently used to assess the compensation of carbohydrate metabolism in children. These circumstances justify the

DOI: 10.1201/9781003112099-7

need to search for and study new markers of carbohydrate metabolism compensation, as well as to improve diagnostics with noninvasive or minimally invasive examination methods applicable in everyday pediatric practice. Glycated proteins should be considered as such markers.

7.2 GLYCATED PRODUCTS AS INDUCERS AND MARKERS OF INFLAMMATION IN DYSMETABOLIC CONDITIONS

It is known that in type 1 diabetes mellitus, there is an increased content of advanced glycation end products (AGEs), which are the result of non-enzymatic glycation of cell proteins. The accumulation of AGE in the vascular wall leads to the disruption of its structure and functioning and plays a key role in the occurrence of vascular complications. According to the literature, increased accumulation of AGEs in the vessel wall in diabetes is associated with changes in the functional activity of endothelial cells, macrophages, and smooth muscle cells, and leads to the development of endothelial dysfunction, thickening of the basement membrane, and impaired vasoconstriction and vasodilation. Specific AGE receptors are RAGE (Receptors for Advanced Glycation End Products) that belong to the immunoglobulin superfamily. When AGE binds to RAGE, "metabolic memory" is formed as the leading mechanism of the pathogenesis of specific vascular complications. It is important that AGE and RAGE levels are independent of current blood glucose levels and are markers of chronic hyperglycemia over the previous several months [1].

It has been established that glycated proteins being accumulating in the skin can cause fluorescence. One of the methods for the quantitative determination of glycated proteins in the skin is fluorescence spectroscopy. This type of examination is noninvasive, allows regular dynamic monitoring of carbohydrate metabolism on an outpatient basis, and reveals the degree of metabolic changes at the preclinical stage. Accumulation of AGEs in tissues is a result of chronic hyperglycemia – some general alterations lead to aberrant metabolism and clearance of AGEs and impaired expression of RAGE receptors. Then, deposition of AGEs in the skin causes an increase in skin autofluorescence, which correlates with microcirculatory alterations, increase in RAGE expression, and shifts in key biochemical parameters [2,3]. The level of skin autofluorescence is considered as an integral indicator of dysmetabolic changes in kidneys; brain; and endocrine, vascular, and respiratory systems. Previously, a significant correlation was shown in mortality rate and skin autofluorescence in patients with terminal renal failure, diabetes, and severe chronic pathology of the cardiovascular system.

At present, it is well established that AGEs are endogenous toxins accumulating in chronic diseases associated with the development of oxidative and carbonyl stress, ischemia and reperfusion, and inflammation. Among AGEs, there are fluorescent compounds that accumulate in degenerating brain tissue, in sputum in bronchial asthma, in peripheral blood in diabetes mellitus, uremia and acute coronary syndrome or other pathologies as well as in various tissues with aging, and their high concentration corresponds to the severity of disease or the development of complications [4–6].

The main biological effects of glycated products are mediated through the cellular receptor RAGE, which can be found in biological fluids in a soluble form – sRAGE [7,8]. Appearance of soluble RAGE is a result of alternative splicing occurred in RAGE gene transcription, or proteolysis of membrane-bound RAGE with sheddases (i.e., of ADAM family) leading to the release of extracellular part of the RAGE molecule to the extracellular space [9–11]. It should be noted that alternative splicing-derived (esRAGE) or proteolysis-originated (sRAGE) soluble RAGE receptors have quite different antigen determinants; therefore, they could not be identified with the same antibodies (for diagnostic purposes) [10].

At present, it is commonly accepted that the activation of RAGE on cell plasma membrane (triggered by AGEs or other ligands released from damaged cells) leads to the development of inflammation. It is associated with elevated levels of sRAGE in biological fluids, and since sRAGE cannot induce signal transduction in target cells, they are considered as decoy receptors that can trap some endogenous and exogenous RAGE ligands to prevent RAGE-driven inflammation [12–14].

However, it is rather possible that dramatic elevation of sRAGE levels might be caused by massive expression of RAGE in inflammatory conditions associated with excessive activation of numerous metalloproteinases/sheddases [15,16]. Thus, it is not surprising that some new therapeutic strategies in inflammation, particularly, in systemic inflammation, are based on the prevention of RAGE-mediated mechanisms [17]. At the same time, it should be taken into consideration that RAGEs possess multiple physiological functions in various cells and tissues, i.e., they mediate the transport of neuropeptide oxytocin through the intestine wall and blood-brain barrier [18,19], and regulate immune cells' activity and brain development [20,21].

An increase in RAGE expression is recorded during inflammation, and cell damage is a result of the release of RAGE ligands like AGEs, high-mobility group box 1 proteins (HMGB1), amyloid, and other molecules. Thus, it is not surprising that elevated levels of RAGE and sRAGE are observed not only in conditions associated with abnormal glycation, but also in situations when cell injury leads to a dramatic increase in RAGE ligand levels in tissues or in biological fluids. Thereby, accumulating evidence confirms the role of AGEs and (s)RAGE in the pathogenesis of chronic kidney disease [22,23], diabetes mellitus [24–26], atherosclerosis [27], cardiac diseases [28], ischemia-reperfusion injury of the heart and blood vessels [29], dementia and Alzheimer's type neurodegeneration [30,31], respiratory pathology, including COVID-19 lung injury [32–35]. Usually, deposition of AGEs in tissues corresponds to an increase in UV-induced autofluorescence of the skin, whereas levels of soluble RAGE in the peripheral blood have an inverse correlation with the levels of RAGE expressed in tissues.

In recent years, attempts have been made to assess the level of sRAGE in various pathological conditions as a marker of systemic inflammatory and vascular reactions. In diabetes mellitus, glycated proteins accumulate in tissues even with satisfactory compensation, but during decompensation stage, their levels increase significantly. In addition to tissue expression, sRAGEs circulate in the peripheral blood being originated from proteolysis-mediated release of receptor molecules from the cell surface. Thus, RAGE and sRAGE have been proposed as potential biomarkers of inappropriate control of glycemia.

The most susceptible to glycation are "long-lived" proteins such as skin collagen or crystalline in lens. The end products of non-enzymatic stable glycation and glycooxidation are pentosidine and carboxymethyl lysine. Being accumulated in the collagen of the skin, these compounds change their structure, disrupt the morphology of the dermis, and lead to the development of diabetic dermopathy. It was found that changes in skin collagen correlate with the content of glycated hemoglobin (HbA1c) in the blood, and the degree of hyperglycemia-induced alterations in microvessel endothelial cells, and can be considered in patients with diabetes as a predictor of microangiopathies and the risk of developing specific complications, or as a marker for early diagnostics even at the preclinical stage. It is important that glycated proteins are stable compounds and characterize glycation processes over several months and/or years. Therefore, the assessment of these molecules is informative for the early diagnosis of carbohydrate metabolism disorders before the onset of clinical symptoms of the disease, as well as for identifying the risk of developing microvascular complications in patients with diabetes.

7.3 METHODS FOR NONINVASIVE ASSESSMENT OF PROTEIN GLYCATION END PRODUCTS IN DIABETES MELLITUS

The main advantages of noninvasive techniques are the absence of pain and other unpleasant sensations; a decrease in the risk of infection with viruses, bacteria, and foreign substances; cost-effectiveness; simplicity; and availability of use.

According to the literature, glycated proteins cause fluorescence (collagen-associated fluorescence of the skin). One of the promising methods for the quantitative determination of glycated proteins in the skin is real-time fluorescence spectrometry. UV-induced fluorescence of a number of end products of protein glycation provides noninvasiveness; therefore, it could be successfully applied in the pediatric practice [36]. The accumulation of protein glycation end products in the skin causes an increase in skin autofluorescence in the ultraviolet region. For example, AGE readers are increasingly used in some countries being able to register the skin autofluorescence signal [37]. Currently, these devices demonstrate rather nice prospects not only for inpatient but also for outpatient use.

It is known that glycated collagen has specific spectral characteristics. The intensity of the fluorescence is determined by the degree of accumulation of glycated proteins in the skin and does not depend on environmental factors, microcirculation and blood supply to tissues, metabolic rate, comorbidity, drug intake, and the sensor interface. In addition, in children and adolescents, there are no collagen changes that are noted in adulthood and can affect AGE-associated skin fluorescence; this aspect increases the reliability of the technique used in pediatric practice.

There are some data on the development of novel noninvasive methods for the assessment of glycated proteins, i.e., pulsed terahertz spectroscopy, infrared and ultraviolet spectroscopy, bioimpedancemetry, ultrasonic, and electromagnetic methods. These data have been obtained in experimental studies only and have not yet been introduced into clinical pediatric practice. At the same time, reliable correlations in skin autofluorescence and the levels of glycated hemoglobin, glucose concentrations, development of microangiopathies and diabetic dermopathy were found. These data confirm that noninvasive methods might be applied for assessing the state of carbohydrate metabolism (for screening and early diagnosis of hyperglycemia in risk groups, assessment of compensation in diabetes mellitus). Moreover, painlessness and absence of the risk of infection make it possible to carry out regular dynamic monitoring in hospitals and on an outpatient basis (for improving the risk identification and preclinical diagnosis of specific vascular complications of diabetes).

At present, the main focus of research is the analysis of gender-specific effects; the influence of puberty, skin type, and patient phototype on skin autofluorescence; and suitability of other body regions for measurements.

The literature contains a lot of information on analyzers that determine the end products of collagen glycation. As an example, US patent [38] shows the characteristics of such device. The method is based on the registration of diffuse fluorescence of protein glycation products in tissues caused by broadband light. The device detects end products of collagen glycation and measures their concentrations. The device consists of a flexible optical probe with a light source and a radiation receiver. Oral mucosa, sclera, and skin are recommended for measurements.

Another patent [39] describes the principle of the method based on the registration of the intensity of skin fluorescence excited by UV or IR light. The analyzer consists of an excitation light source, a detector that receives the fluorescence signal, and a processor for measuring the attenuation of the fluorescence signal intensity. The excitation pulse is directed to the skin surface of the patient's forearm, leg, or palm. The fluorescence signal reflects the qualitative presence of protein glycation products.

The AGE-Reader medical analyzer of end products of glycation [37] is a diagnostic device that noninvasively measures the content of AGEs in tissues as a parameter of UV-induced autofluorescence in the real-time mode. It includes a spectrometer, fiber probe, and UV fluorescent lamp with the emission range of 300–420 nm. The measurement of radiation is carried out in the area of the patient's forearm; the result reflects the ratio of the intensity of autofluorescence (420–600 nm) to the intensity of the reflected exciting radiation (300–420 nm). In the clinical practice, this device is used to assess the risk of cardiovascular complications in patients with diabetes mellitus.

Another example is the medical spectrometer analyzing both fluorescence and reflection spectra known as "Skin-AGE" [40]. The device is based on local spectroscopy assessing with the fiber multichannel spectrometer. To excite fluorescence, radiation with a wavelength of 365 nm is used. Fluorescence spectra are measured in radiometric mode and recorded in units of the spectral power density of the radiation. Examination can be taken on various areas of the skin (fingertips, palms, wrist, shoulder, forearm, elbows, knee, lower leg, and ankle). The analyzer has been used in patients with diabetes mellitus. Experimental models of "Skin-AGE" spectrometers with external and built-in computers have been created [41].

Assessment of the level of RAGE expression in tissues may have some methodological difficulties (i.e., the need to obtain a histological sample and to perform immunohistochemistry, western blotting, or molecular genetic studies), while in the peripheral blood, sRAGE levels can be determined with the enzyme-linked immunoassay. A significant number of studies are currently focused on determining the diagnostic potential of data on actual sRAGE levels and their correlations with the expression of RAGE in tissues in normal and pathological conditions [42–44]. However, despite the well-established methods for the noninvasive assessment of collagen glycation products in the skin, there is a lack of noninvasive methods for the evaluation of sRAGE levels in the blood.

7.4 METHODS FOR NONINVASIVE ASSESSMENT OF GLYCATED HEMOGLOBIN

Noninvasive assessment of blood glucose levels is now possible with several protocols and devices; however, there are no analogous methods for noninvasive evaluation of HbA1c in the peripheral blood. Thus, some groups reported on the progress in the development of novel approaches aimed to solve this problem.

As an example, results of clinical trials to test the efficacy and reliability of the original technique for a comprehensive assessment of microcirculatory alterations in patients with diabetes mellitus have been reported. The diffuse reflectance coefficient of skin in the spectral range from 500 to 600 nm was determined based on diffuse reflectance spectroscopy. An experimental setup includes a spectrometer (FLAME, Ocean Optics, USA), an illumination device (broadband tungsten halogen radiation source HL-2000-HP-232R, Ocean Optics, USA), a fiber-optic probe (R400-7, Ocean Optics, USA) with seven fibers (1 – readout, 6 – lighting), personal computer, and Ocean View software (Ocean Optics). To assess the fluorescence of biological tissue, the method of fluorescence spectroscopy was used, and analysis of blood flow and tissue perfusion was done with the laser Doppler flowmeter. Registration of the amplitude of fluorescence signal of biological tissue and perfusion parameters was carried out simultaneously in one tissue area using the LAZMA MC diagnostic complex (NPP LAZMA LLC, Moscow, Russia). The device includes a laser module and an optical fiber probe. It was found that the diagnostic volume and depth of probing in the spectral region of the Q absorption bands of total hemoglobin (540–580 nm) were 1.0–1.3 mm^3 and 0.4–0.5 mm, respectively. Additionally, according to the diffuse reflectance spectra obtained at a specific point of the skin, the hemoglobin index was calculated to assess its quantitative content in the tissue, and the degree of oxygenation of hemoglobin was determined. The authors found that hemoglobin was involved in the formation of diffuse reflectance spectra of the skin (in the visible and near-infrared ranges), proved that the calculation of the total hemoglobin index reflects the amount of blood in the skin, and suggested that high values of HbA1c have specific optical properties and can change diffuse reflectance spectra [45].

The US patent [46] describes that glycosylated hemoglobin is determined by the difference in absorption spectra of reflected radiation of Hb and HbA1c.

There is a combined photosensitive linear charge-coupled device that allows noninvasive simultaneous determination of levels of hemoglobin (λ=0.55 μm), glucose (λ=0.94 μm), and bilirubin (λ=0.46 μm) in relation to the skin reflectance (λ=0.7 μm). The device includes an optoelectronic unit (fiber focal point, interference filters, linear CCD, light filter, flash lamp with a reflector), an information processing unit (microcontroller, analog-to-digital converter), an indication unit on an LCD indicator, and a power supply with a stabilizer of five batteries. The device registers scattered and reflected optical signals excited by broadband radiation; the information is presented in the form of glucose, hemoglobin, and bilirubin levels in absolute values (hemoglobin 0 - 300 g/l, glucose 0 - 40 mmol/l, bilirubin 0 - 400 μmol/l) [47].

A device for noninvasive analysis of blood parameters by means of optical control and spectroscopy has been announced [48]. The device consists of an optical examination system with a dichroic mirror (for spectral decomposition of the image of the object of study), lighting devices (standard illuminator, at least one laser radiation source), image receivers connected to an electronic signal processing unit, and a lodgment for placing the subject's finger. The principle of the technique is visualization of the microvessels using optical microscopy, and analysis of diffuse scattering spectra obtained from vascular structures. The beam of light radiation is reflected from the capillaries under the nail plate and, after passing through the optical system, is recorded by a CCD matrix with subsequent conversion into an electrical signal. This device allows a long-term noninvasive monitoring of the parameters of microcirculation, spectral characteristics of blood (dynamic characteristics of cells within the bloodstream, internal dynamics of erythrocytes) and adjacent tissues in the real-time mode, but is not capable of assessing the quantitative content of the blood (levels of glycation and oxygenation of hemoglobin). The possibility of determining the actual glucose levels using this device is under the discussion.

Similar devices for the noninvasive determination of blood parameters were developed by Sysmex Corporation known as "Noninvasive blood analyzer" [49] and Abbott Laboratories [50].

7.5 UV-INDUCED FLUORESCENCE SPECTROSCOPY FOR THE DIAGNOSTICS OF SKIN AUTOFLUORESCENCE IN DIABETES MELLITUS

We developed the original compact spectrofluorometer with UV LED excitation for the assessment of skin autofluorescence. The device (Figure 7.1) includes the following elements: Information about the spectra can be obtained using the spectrometer STS-VIS USA (1). Radiation from the object under study enters the spectrometer through a condenser lens (2). To excite fluorescence, frontal illumination of UV LEDs with different wavelengths, crossing beams at an acute angle (3–5), is used. A white LED (6) is used to record the reflection spectra. For synchronous with the start of the spectrometer on and off the LEDs, a hardware-software device for controlling LEDs is used (7). The measurement signal is triggered by a command from the computer. For the convenience of storing information (measurement data, software), a flash memory module is used (8). A USB hub (9) is used to connect USB devices. Data transfer to a computer and power supply of the device is carried out via a USB cable (10). Fixation of the device on the skin area is carried out using a lens hood (11).

The study involved 46 patients aged 5 - 18 years that were divided into two groups: (1) group 1 – a control group, without diabetes mellitus and skin pathology, $n=15$ and 2) group 2 – a group of patients with type 1 diabetes mellitus without skin pathology, $n=31$. All studies were conducted with the permission of the local ethics committee, as well as on the basis of informed voluntary parental consent to participate in the study for a scientific purpose, in compliance with all ethical standards related to the organization and implementation of such research. For each group, the fluorescence spectra were recorded on the inner surface of the patient's shoulder.

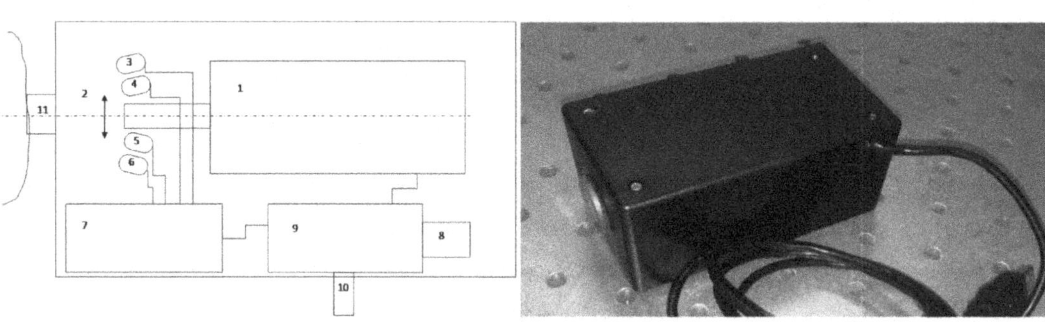

FIGURE 7.1 Functional diagram and photo of the spectrofluorometer with UV-LED excitation.

The spectra were processed in the following order:

The data were presented as a two-dimensional array (7.1):

$$I_{ij} = \begin{matrix} I_{11} & \dots & I_{i1} & \dots & I_{N1} \\ \dots & & & & \dots \\ I_{1j} & \dots & I_{ji} & \dots & I_{Nj} \\ \dots & & & & \dots \\ I_{1K} & \dots & I_{iK} & \dots & I_{NK} \end{matrix} \tag{7.1}$$

The spectra were normalized to the maximum of the reflected radiation signal of the UV LED according to the formula (7.2):

$$I_{ij} = \frac{I_i - I_{\min}}{I_r} \tag{7.2}$$

As a result, the normalized fluorescence spectra were averaged over groups according to the formula (7.3):

$$I_{\text{average}} = \frac{\sum_{k=1}^{n} I_k}{n} \tag{7.3}$$

Evaluation of sRAGE levels in the samples of peripheral blood was carried out by the method of enzyme-linked immunoassay. Assessment of the HbA1c level has been made with the standard diagnostic protocol. Besides general clinical data, we registered the gender, age, sexual maturity, and skin phototype of the patients.

The STATISTICA 10 package was used for statistical data processing. Descriptive statistics methods, correlation, and discriminant and multivariate regression analyses were used.

Average-normalized spectra for the control group and the group with diabetes mellitus are presented in Figure 7.2. The difference spectrum was found from the average-normalized spectra (Figure 7.3). As it can be seen from the difference spectrum, the greatest differences are achieved at wavelengths of 430 and 613 nm. According to the experimental data, peak at 430 nm corresponds to the luminescence peak of pentosidine.

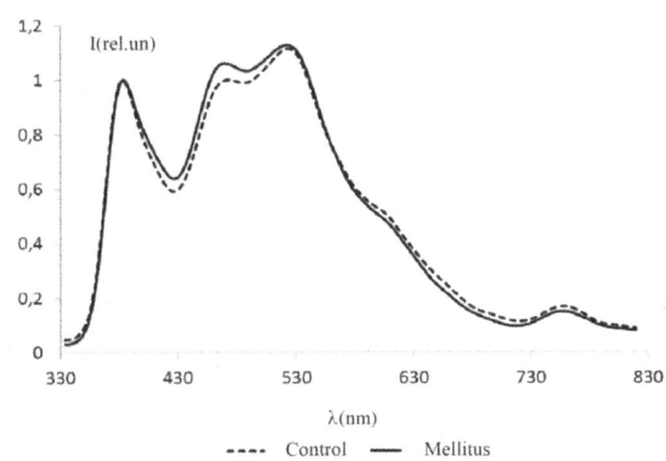

FIGURE 7.2 Normalized averaged skin fluorescence spectra in the control group and in the group with diabetes mellitus.

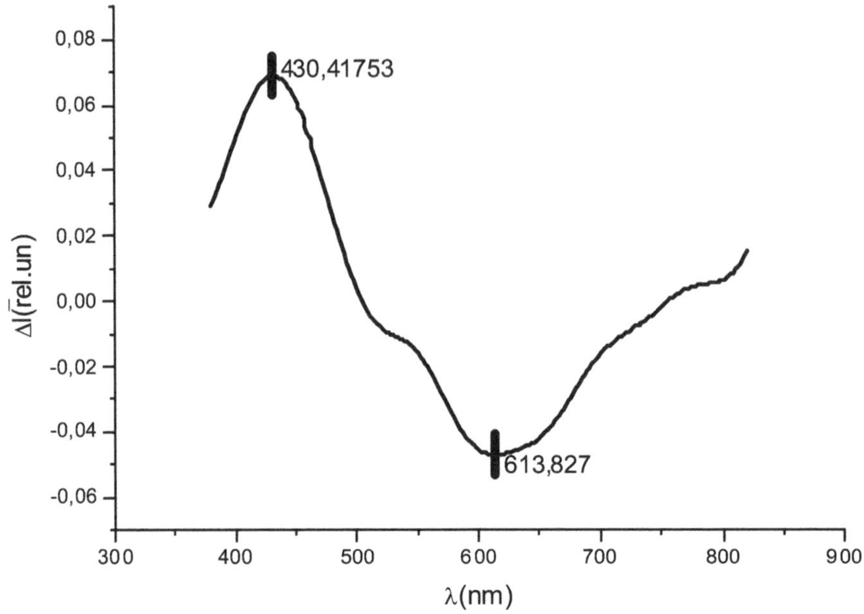

FIGURE 7.3 Difference spectrum of skin fluorescence in the group with diabetes mellitus in relation to the control group.

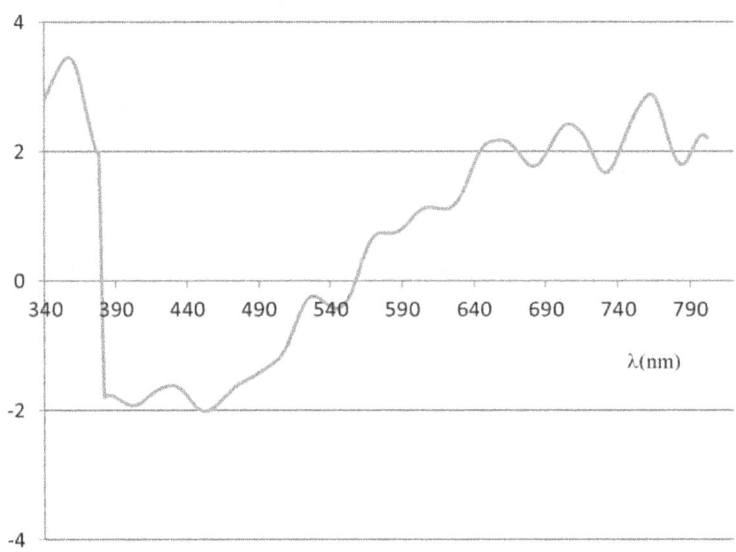

FIGURE 7.4 *t*-value spectra difference of skin fluorescence in the group with diabetes mellitus in relation to the control group.

Optimization of the choice of reference wavelengths for differential diagnosis of diabetes mellitus based on the spectrum of reflection and fluorescence was performed by calculating the dependence of the t - value on the wavelength (Figure 7.4).

As can be seen from the obtained diagram, significant differences between the normalized fluorescence spectra of the skin of patients with diabetes mellitus and the control group are observed not only at a wavelength close to the maximum fluorescence of pentosidine (450 nm), but also at a wavelength of 356 nm, which corresponds to the absorption peak of pentosidine as well as in the red area. However, taking into account that fluorescence intensity in this region is rather low, applying the data obtained in this wavelength region seems to be irrational for the reasons of signal-to-noise ratio.

The method of discriminant analysis (forward stepwise) made it possible to reveal very close wavelengths that were useful for differentiating the skin fluorescence in diabetes mellitus patients and in healthy persons. Significance criteria are shown in Table 7.1.

At the same time, the classification matrix shows that using the intensities at the indicated wavelengths helps in verifying the diabetes mellitus without taking into account gender and age characteristics, or skin phototype (Table 7.2).

However, discriminant analysis used for determining the best subset of combined effects revealed the need to take into account the skin phototype (Table 7.3).

The next problem to be solved was to find spectral and other signs that allow noninvasive determination of sRAGEs levels and HbA1c levels in patients with diabetes mellitus. We compared the

TABLE 7.1
Significance Criteria of Discriminant Analysis (Forward Stepwise) at Different Wavelengths

	Test	Value	F	Effect	Error	*p*
Intercept	Wilks	0.917056	3.88917	1	43	0.055052
I(363)	Wilks	0.767564	13.02140	1	43	0.000798
I(450)	Wilks	0.835898	8.44171	1	43	0.005773

TABLE 7.2
Classification Matrix of Discriminant Analysis (Forward Stepwise)

	%	Control	Mellitus
Control	68.42	13	6
Mellitus	92.59	2	25
Total	82.60	15	31

TABLE 7.3
Significant Combined Effects of Discriminant Analysis

	Test	Value	F	Effect	Error	*p*
Intercept	Wilks	0.920695	3.273168	1	38	0.078338
Phototype^2	Wilks	0.950934	1.960720	1	38	0.169549
I363	Wilks	0.960201	1.575040	1	38	0.217140
I363	Wilks	0.943006	2.296648	1	38	0.137928
I450	Wilks	0.825518	8.031729	1	38	0.007320
I450^2	Wilks	0.848292	6.795910	1	38	0.012982
Phototype*I363	Wilks	0.810078	8.909055	1	38	0.004941
Phototype*I450	Wilks	0.813431	8.715707	1	38	0.005383

sRAGE concentration values in the control and experimental groups (Figure 7.5). The p value <0.02 was obtained with t - test (Table 7.4).

It is clear that the development of diabetes mellitus leads to reduced levels of sRAGEs in the peripheral blood. By comparison of the difference spectrum data, we found that progression of diabetes mellitus was associated with an increase in skin fluorescence at the pentosidine peak.

We revealed significant correlations of fluorescence intensities, sRAGE levels, and HBA1c concentrations. The diagram (Figure 7.6) shows the dependence of the correlation coefficients on the wavelength. These data allow us to propose the values of the reference wavelengths for solving the problem of evaluation of sRAGE and HbA1c levels. Thus, the intensity of fluorescence at a wavelength of 408 nm demonstrates the most significant negative correlation with sRAGE levels, whereas the intensity of fluorescence at a wavelength of 440 nm demonstrates the positive correlation with HbA1c concentrations in the blood. The correlations found support the idea on sRAGE level alterations along the progression of diabetes mellitus in children: as it was demonstrated before [51], sRAGE concentrations decline with the appearance of diabetes-predictive autoantibodies in

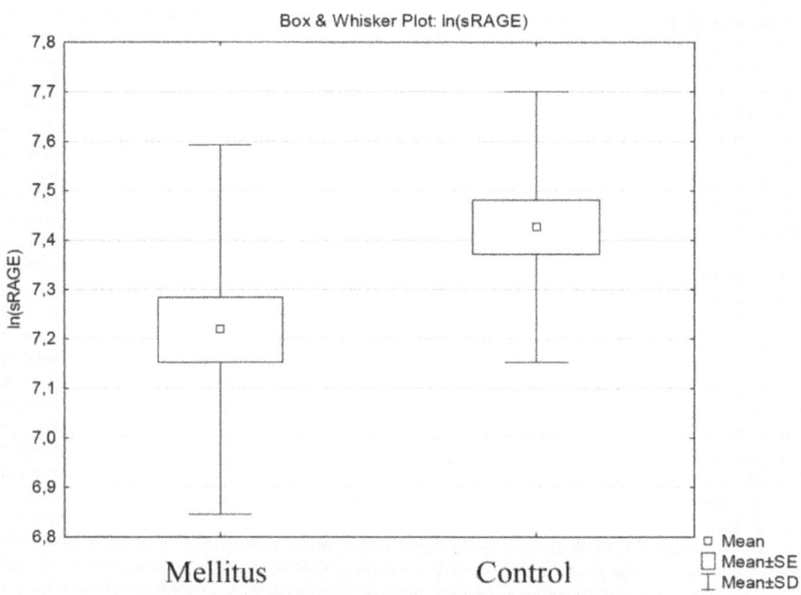

FIGURE 7.5 sRAGE levels in the peripheral blood in the control and experimental groups.

TABLE 7.4
Classification Matrix Reconstructed with Best Combined Effects When Phototype was Taken into the Consideration

	%	Control	Mellitus
Control	73.68	14	5
Mellitus	96.29	1	26
Total	86.95	15	31

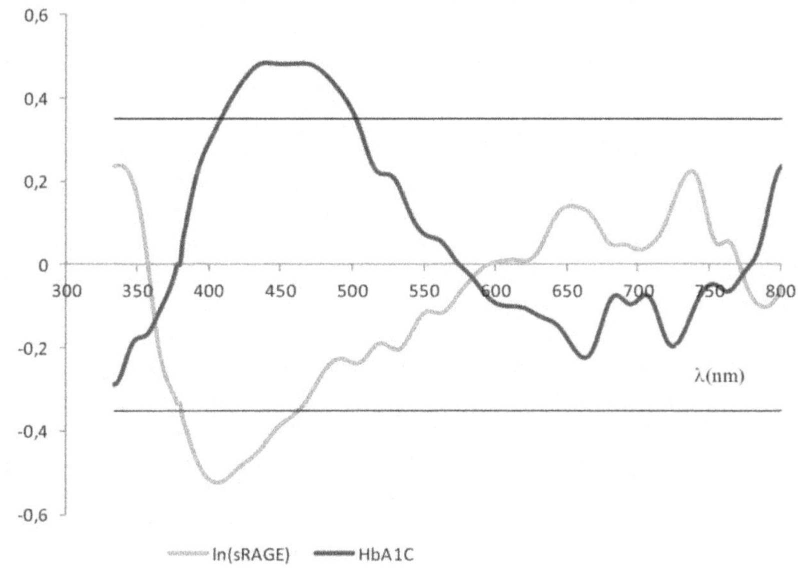

FIGURE 7.6 Dependences of Pearson's correlation coefficients on wavelength obtained from the values of fluorescence intensities, sRAGE and HbA1c levels.

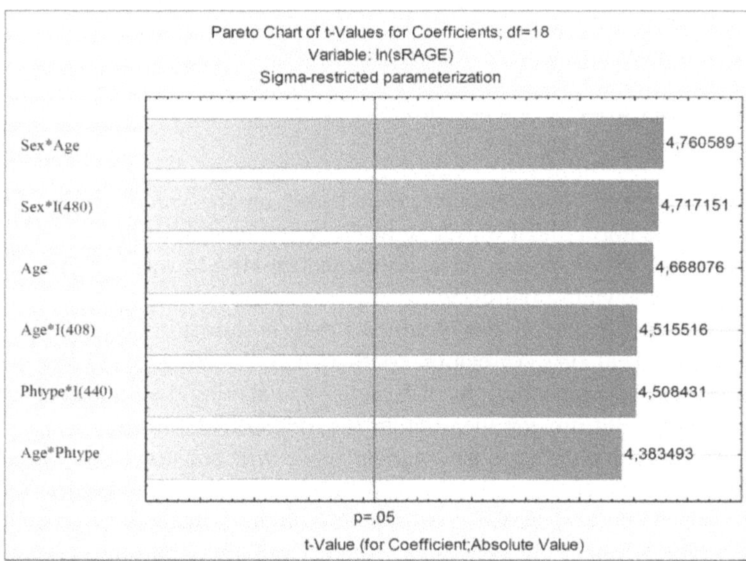

FIGURE 7.7 Best predictors of sRAGE concentrations based on the analysis of fluorescence intensity at reference wavelengths (408 and 440 nm).

children progressing to type 1 diabetes. So, the higher levels of sRAGE correspond to their better AGEs scavenging ability, whereas lower levels of circulating sRAGE result in excessive AGEs deposition in peripheral tissues and promotion of inflammation.

The analysis of the best predictors of the sRAGE concentrations value done with factorial combinations of the second order gives the set presented in Figure 7.7.

A plot of experimental and model sRAGE concentrations is shown in Figure 7.8. In this case, the Pearson's correlation coefficient was not less than $r=0.87$. Thus, we proposed a method for

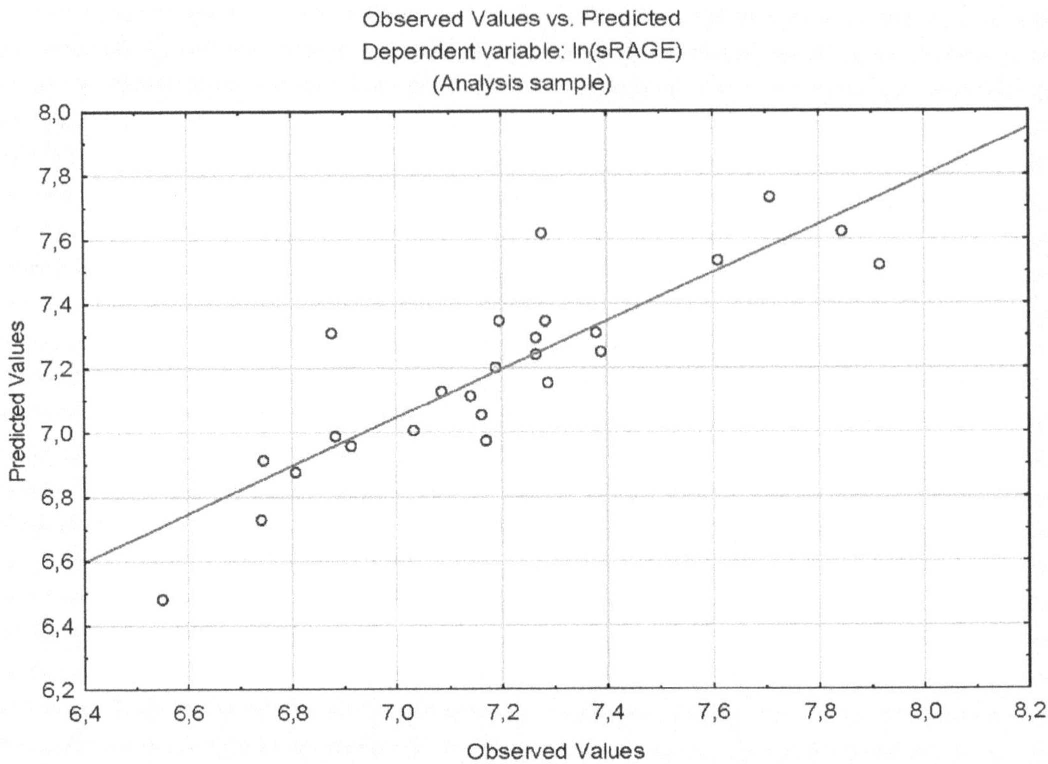

FIGURE 7.8 Experimental and model-obtained values of sRAGE concentration.

noninvasive assessment of sRAGE concentrations based on the values of the skin fluorescence intensity at reference and normalized wavelengths (380, 408, and 440 nm).

The regression model for the noninvasive assessment of HbA1c was analyzed in a similar way. The best predictors are shown in Figure 7.9.

A diagram of experimental and model-obtained values is shown in Figure 7.10.

The high value of the Pearson's correlation coefficient $r=0.9$ allows us to propose a noninvasive method for determining the concentrations of HbA1c by analyzing the fluorescence intensity at the normalized wavelength at 380 nm and the reference wavelength at 440 nm.

To take into account the phototype automatically, we will construct the Pearson's correlation spectrum of the normalized fluorescence intensity and the phototype (Figure 7.11). Our studies included patients of 1–4 phototypes. We used spectra from both the control group and the diabetic group, and the correlation spectrum demonstrated clear maxima and minima at 659 and 452 nm, respectively.

Let use the ratio of intensities at wavelengths I_{659}/I_{452} as well as the logarithm of the ratio for the calibration on phototype. Pearson's correlation coefficients for the indicated ratios are presented in Table 7.5.

Figure 7.12 shows a diagram of the dependence of the ratio ln(I659/I452) logarithm on the phototype.

Using an additional set of variables by the forward stepwise method, a multiple regression model was built that allows determining the concentrations of HbA1c within 5%–12% of the total hemoglobin content from the fluorescence intensities at wavelengths of 408, 452, 659 nm and data on the gender, age, and sex maturity of patients. T\he quality parameters of the multivariate regression model are presented in Table 7.6. Based on the value of the F-parameter, it is possible to enter at least 36-rank scale.

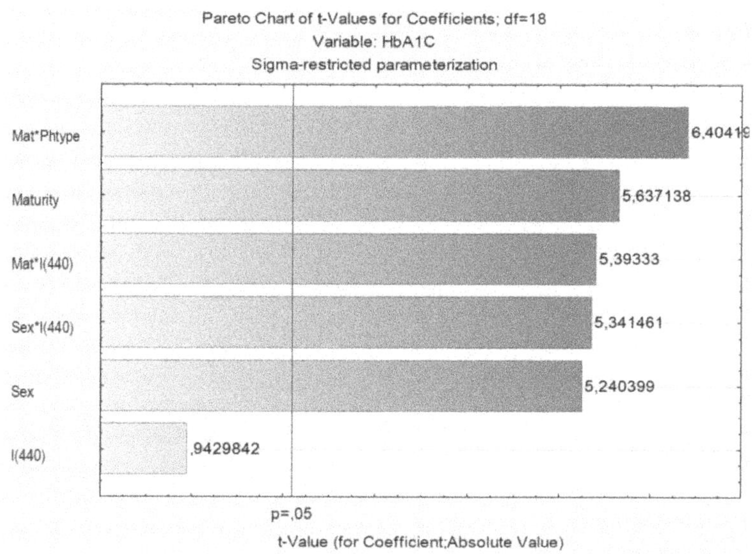

FIGURE 7.9 Predictors for noninvasive assessment of glycated hemoglobin based on the analysis of fluorescence intensity at 440 nm.

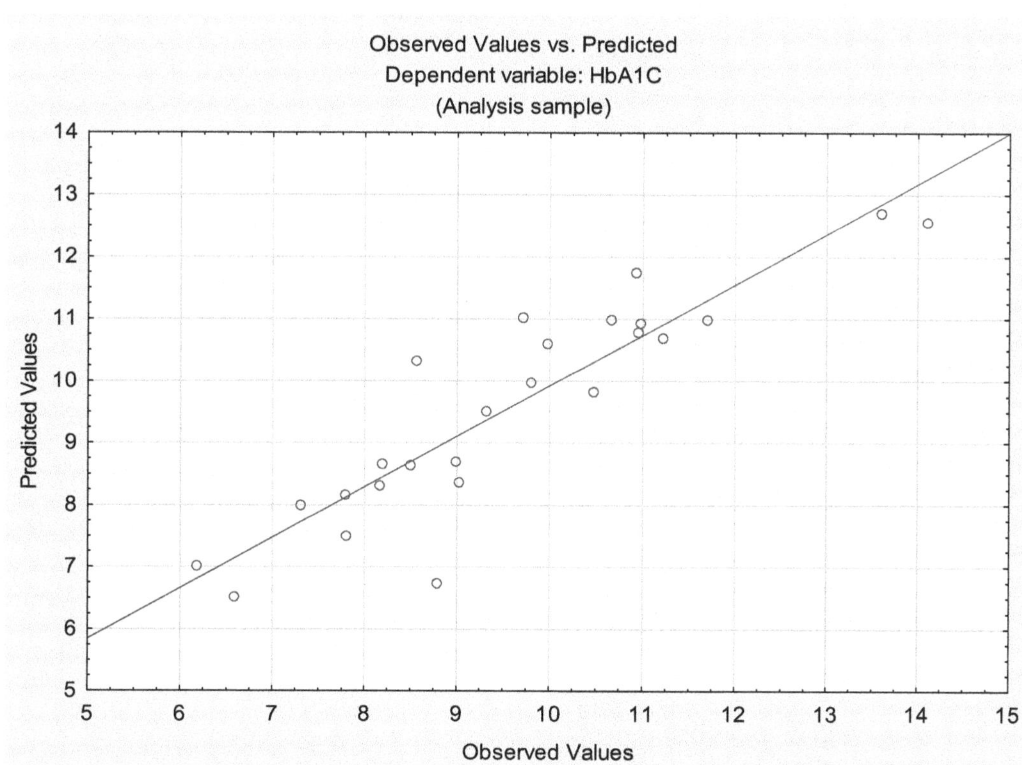

FIGURE 7.10 Experimental and model-obtained values of glycated hemoglobin concentrations.

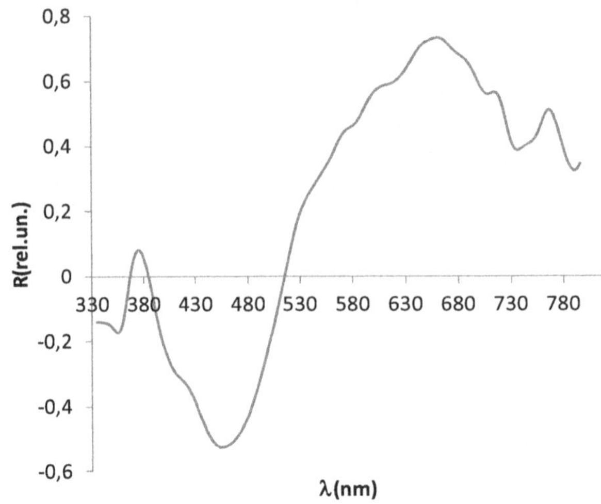

FIGURE 7.11 Pearson's correlation spectrum of normalized fluorescence intensity and phototype.

TABLE 7.5
Pearson's Correlation Coefficients for Various Predictors of Phototype

Variable	r
I_{452}	−0.525
I_{659}	0.732
I_{659}/I_{452}	0.814
$\ln(I_{659}/I_{452})$	0.834

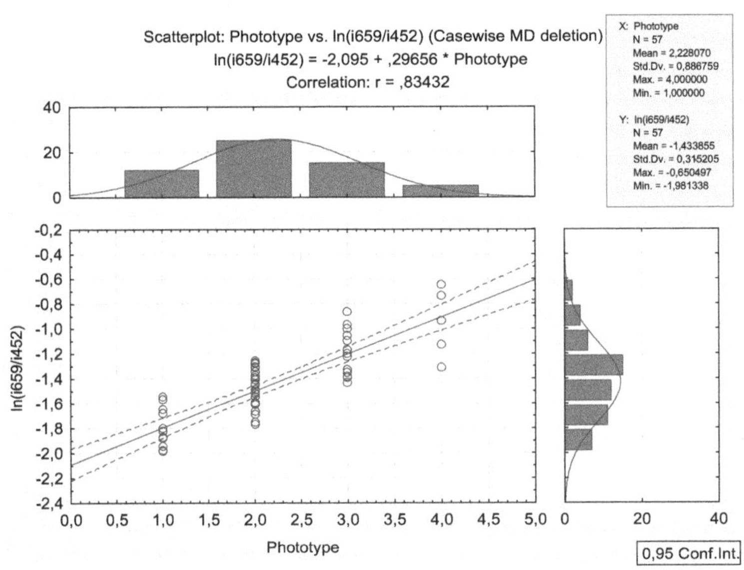

FIGURE 7.12 Dependence of the logarithm of the ratio of intensities at the reference wavelengths $\ln(I_{659}/I_{452})$ on the phototype.

TABLE 7.6
The Quality Parameters of the Multiple Regression Model C (HbA1c)

Dependent Variable	Multiple R	Multiple R2	Adjusted R2	SS Model	df Model	MS Model	SS Residual	df Residual	MS Residual	F	p
	Test of SS Whole Model vs. SS Residual										
C(HbA1C)	0.962462	0.926333	0.900549	59.97478	7	8.567826	4.769540	20	0.238477	35.92726	0.000000

The multiple regression coefficients are shown in Table 7.7, and the significance of the t-parameters of the model is presented in the diagram (Figure 7.13). The quality of the model can also be assessed at the diagram (Figure 7.14).

The multivariate regression model constructed by the method of best subsets for noninvasive determination of sRAGE concentrations uses the values of fluorescence intensities at wavelengths of 363, 408, 452, and 659 nm, as well as the values of age and gender. The significance of the combined predictors of the model is shown in Figure 7.15.

The parameters of the regression model quality are presented in Table 7.8. Based on the data obtained, it was possible to introduce a 45-rank scale to determine the sRAGE concentrations. Conformity of model-obtained and experimental values is shown in Figure 7.16.

The values of the coefficients of the regression model are presented in Table 7.9.

7.6 SUMMARY

Using original spectrofluorometer, we have developed new approach to accurate noninvasive determination of HbA1c concentration in children with type 1 diabetes mellitus. The method of UV-induced fluorescence spectroscopy provides a detection rate of 5%12% of HbA1c that is quite enough for routine diagnostic purposes. To measure HbA1c levels, the fluorescence intensity should be assessed at three wavelengths (408, 452, and 659 nm) and analyzed in relation to age and gender of a patient. The model allows using the 35-rank scale.

For noninvasive assessment of sRAGE levels in children with diabetes mellitus, it is necessary to collect data on fluorescence intensities at four wavelengths (363, 408, 452, and 659 nm) and analyzed in relation to age and gender of a patient. The range of the obtained sRAGE concentrations ln (C [pg/ml]) = 6.5–7.7, which corresponds to the actual concentrations within the 650–2200 pg/ml. In this case, the 45-rank logarithmic scale can be used.

An increase in the convergence of the methods was demonstrated due to the automatic measurement of the skin phototype by determining the ratio of fluorescence intensities at wavelengths of 452 and 659 nm. This increase has been clearly shown for all four phototypes.

The method of multivariate regression analysis is effective in developing protocols of noninvasive detection of carbohydrate metabolism-related parameters with UV-induced fluorescence spectroscopy.

ACKNOWLEDGMENT

The authors express their thanks to Dr. Diana P. Skomorokha and Dr. Olga L. Lopatina for their assistance in getting the experimental data.

TABLE 7.7
Multiple Regression Coefficients of the Model C (HbA1c)

	Parameter Estimates Sigma-Restricted Parameterization									
Effect	**C(HbA1C) Param.**	**C(HbA1C) Std. Err**	**C(HbA1C) t**	**C(HbA1C) p**	**−95,00% Cnf. Lmt**	**+95,00% Cnf.Lmt**	**C(HbA1C) Beta (?)**	**C(HbA1C) St.Err.?**	**−95,00% Cnf. Lmt**	**+95,00% Cnf.Lmt**
Intercept	9.7570	0.126384	77.20097	0.000000	9.4933	10.0206				
Gender*Matur	−1.2897	0.209952	−6.14289	0.000005	−1.7277	−0.8518	−0.84598	0.137717	−1.13325	−0.55871
gender*Age	0.5621	0.068696	8.18287	0.000000	0.4188	0.7054	4.79053	0.585434	3.56933	6.01172
Matur*I659/I452	8.1734	1.775822	4.60260	0.000172	4.4691	11.8777	1.31779	0.286314	0.72055	1.91503
Gender*I408	−9.2668	1.181811	−7.84118	0.000000	−11.7320	−6.8016	−4.58587	0.584845	−5.80584	−3.36591
Matur*ln(408)	5.4121	1.482661	3.65024	0.001591	2.3193	8.5048	0.97814	0.267965	0.41917	1.53710
Gender*ln(452)	−2.7535	1.067144	−2.58026	0.017873	−4.9795	−0.5275	−0.19777	0.076648	−0.35766	−0.03789
ln(408)*ln(452)	−40.3845	4.238403	−9.52823	0.000000	−49.2256	−31.5433	−0.73227	0.076853	−0.89258	−0.57196

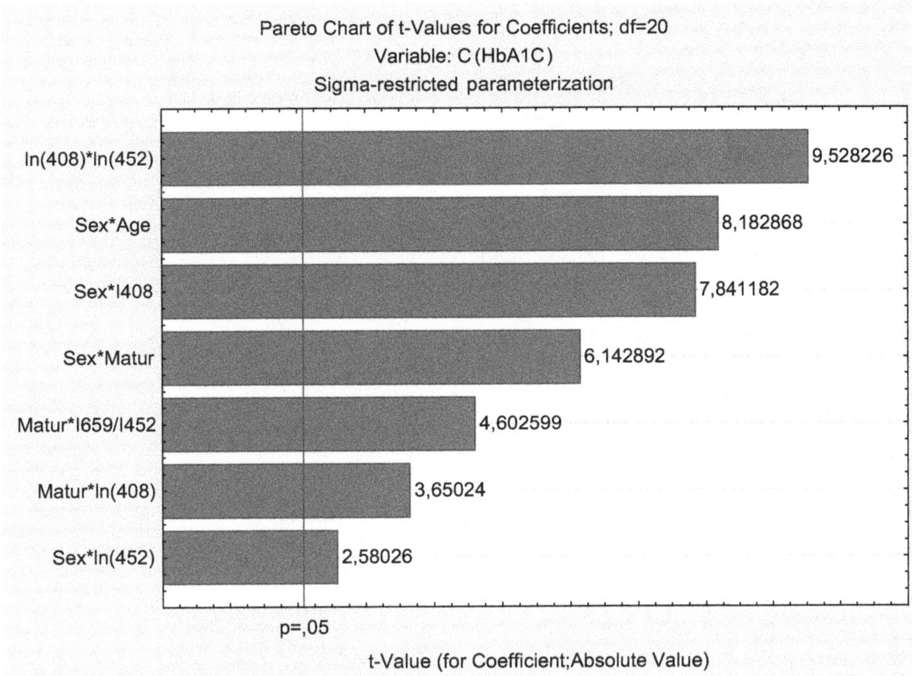

FIGURE 7.13 The best predictors of multiple regression model for evaluating the concentrations of glycated hemoglobin with automatic determination of skin phototype.

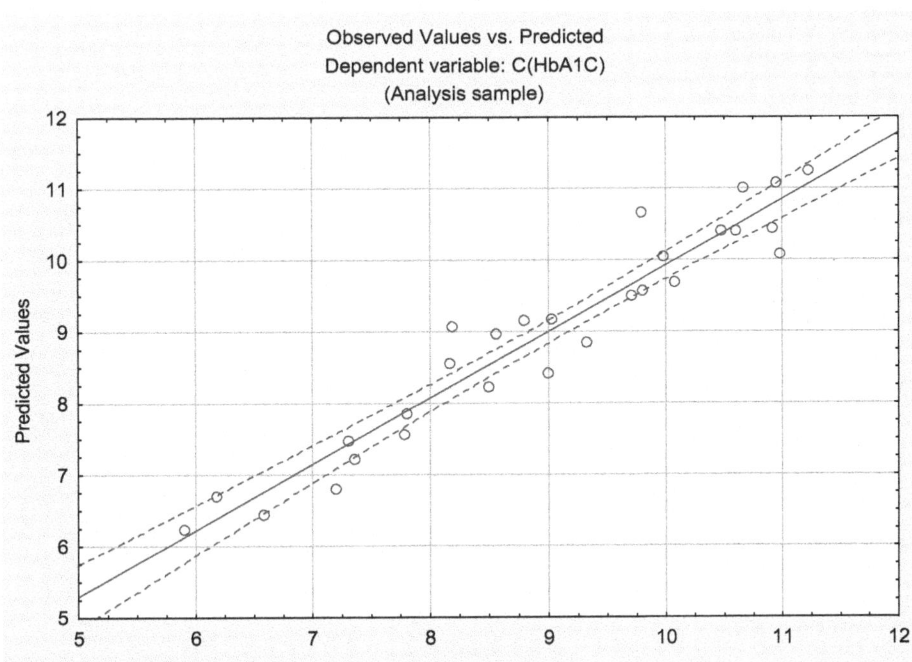

FIGURE 7.14 Experimental and model-obtained values of glycated hemoglobin concentrations with automatic recognition of skin phototype.

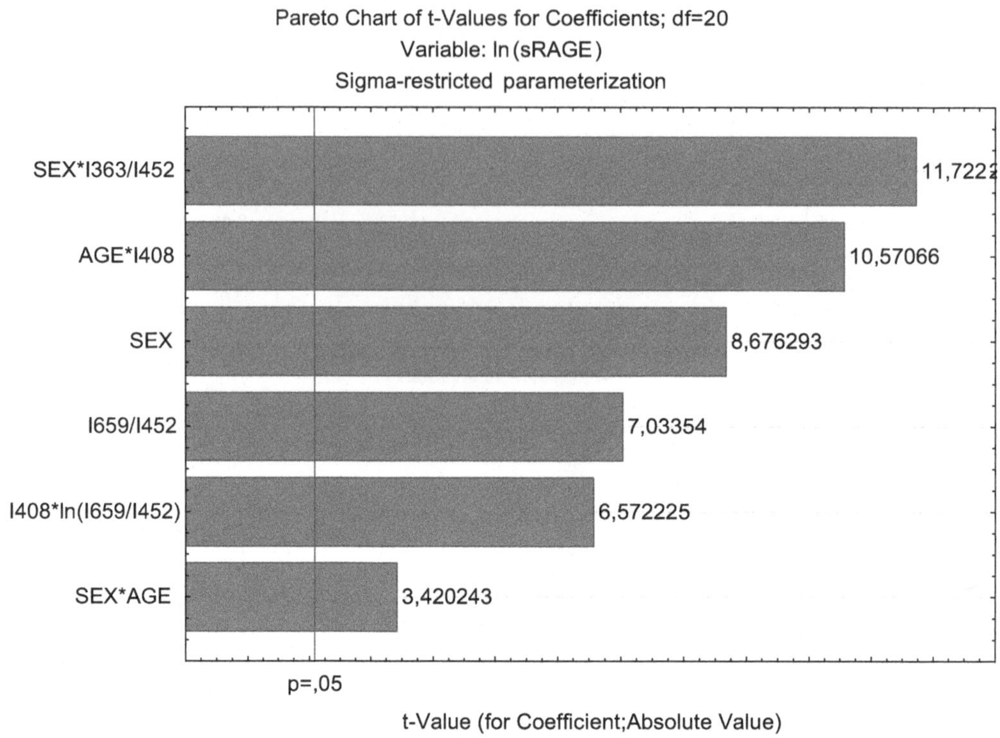

FIGURE 7.15 Best multiple regression model predictors for evaluating sRAGE concentrations with automatic determination of skin phototype.

TABLE 7.8
The Parameters of the Quality of the Regression Model ln(sRAGE)

	Test of SS Whole Model vs. SS Residual										
Dependent Variable	**Multiple R**	**Multiple R2**	**Adjusted R2**	**SS Model**	**df Model**	**MS Model**	**SS Residual**	**df Residual**	**MS Residual**	**F**	**p**
ln(sRAGE)	0.965603	0.932389	0.912106	1.959977	6	0.326663	0.142125	20	0.007106	45.96848	0.000000

TABLE 7.9
The Values of the Regression Model Coefficients ln(sRAGE)

Effect	Parameter Estimates Sigma-Restricted Parameterization									
	ln(sRAGE) Param.	ln(sRAGE) Std.Err	ln(sRAGE) t	ln(sRAGE) p	−95,00% Cnf. Lmt	+95,00% Cnf. Lmt	ln(sRAGE) Beta (?)	ln(sRAGE) St.Err.?	−95,00% Cnf. Lmt	+95,00% Cnf. Lmt
Intercept	10.96830	0.451634	24.2858	0.000000	10.02621	11.91039				
GENDER	1.63816	0.188809	8.6763	0.000000	1.24431	2.03201	5.86696	0.676206	4.45642	7.27750
I659/I452	−6.03784	0.858435	−7.0335	0.000001	−7.82850	−4.24717	−1.51923	0.215998	−1.96980	−1.06867
GENDER*AGE	0.02469	0.007218	3.4202	0.002712	0.00963	0.03974	1.17604	0.343847	0.45879	1.89329
GENDER*I363/I452	−0.75331	0.064264	−11.7222	0.000000	−0.88737	−0.61926	−7.09716	0.605446	−8.36010	−5.83422
AGE*I408	−0.07668	0.007254	−10.5707	0.000000	−0.09181	−0.06155	−0.71989	0.068103	−0.86195	−0.57783
I408*ln(I659/I452)	1.45300	0.221082	6.5722	0.000002	0.99183	1.91417	1.56627	0.238316	1.06915	2.06339

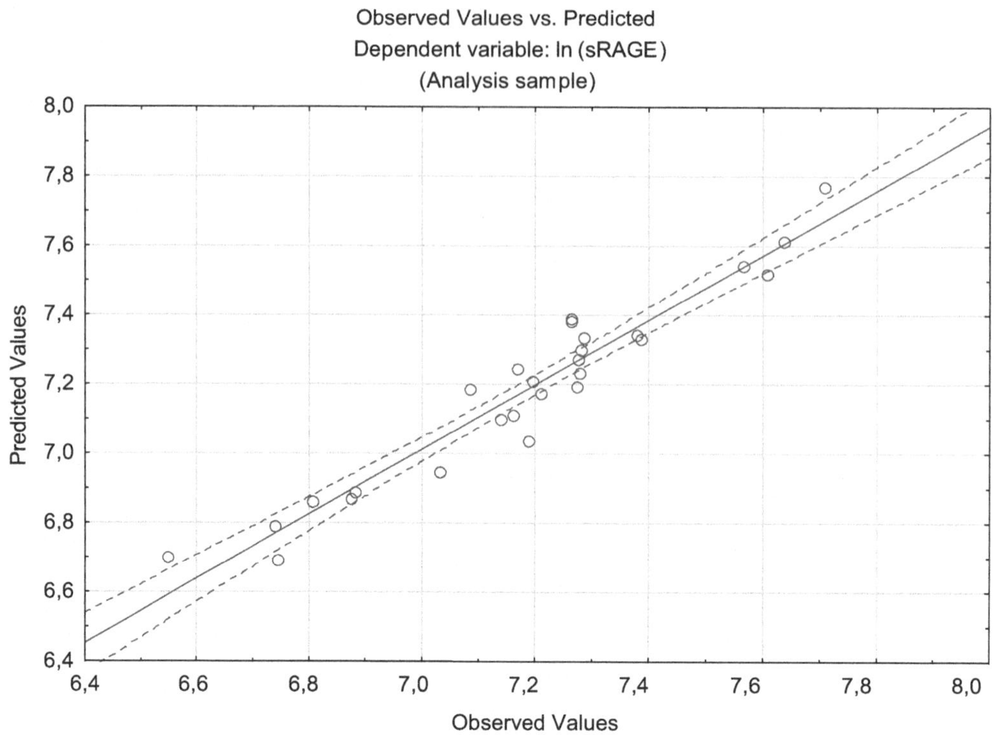

FIGURE 7.16 Experimental and model-obtained values of sRAGE concentrations with automatic recognition of skin phototype.

REFERENCES

1. D.L. Felipe, J.M. Hempe, S. Liu, et al., "Skin intrinsic fluorescence is associated with hemoglobin A1c and hemoglobin glycation index but not mean blood glucose in children with type 1 diabetes," *Diabetes Care* **34**(8), 1816–1820 (2011).
2. H. Nienhuis, K. de Leeuw, J. Bijzet, et al., "Skin autofluorescence is increased in systemic lupus erythematosus but is not reflected by elevated plasma levels of advanced glycation end products," *Rheumatology* **47**(10), 1554–1558 (2008).
3. J. Skrha, Jr., J. Soupal, G. Loni Ekali, et al., "Skin autofluorescence relates to soluble receptor for advanced glycation end-products and albuminuria in diabetes mellitus," *J. Diabetes Res.* **2013**, 650694–650694 (2013).
4. F. Kipfmueller, K. Heindel, A. Geipel, et al., "Expression of soluble receptor for advanced glycation end products is associated with disease severity in congenital diaphragmatic hernia," *Am J Physiol. Lung Cell. Mol. Physiol.* **316**(6), L1061–L1069 (2019).
5. M. Kopytek, M. Ząbczyk, P. Mazur, A. Undas, and J. Natorska, "Accumulation of advanced glycation end products (AGEs) is associated with the severity of aortic stenosis in patients with concomitant type 2 diabetes," *Cardiovasc. Diabetol.* **19**(1), 1–12 (2020).
6. T. Hirano, Y. Toriyama, and T. Murata, "Association of diabetic retinopathy severity with lens fluorescence ratio reflecting advanced glycation end product level," *Investig. Ophthalmol. Visual Sci.* **57**(12), 5441–5441 (2016).
7. Y. Yamamoto and H. Yamamoto, "RAGE-mediated inflammation, type 2 diabetes, and diabetic vascular complication," *Front. Endocrinol.* **4**, 105 (2013).
8. A.M. Maresca, L. Guasti, S. Bozzini, et al., "sRAGE and early signs of cardiac target organ damage in mild hypertensives," *Cardiovasc. Diabetol.* **18**(1), 17 (2019).
9. A.Z. Kalea, N. Reiniger, H. Yang, et al., "Alternative splicing of the murine receptor for advanced glycation end-products (RAGE) gene," *FASEB J.* **23**(6), 1766–1774 (2009).

10. A. Raucci, S. Cugusi, A. Antonelli, et al., "A soluble form of the receptor for advanced glycation end-products (RAGE) is produced by proteolytic cleavage of the membrane-bound form by the sheddase a disintegrin and metalloprotease 10 (ADAM10),!" *FASEB J.* **22**(10), 3716–3727 (2008).
11. N. Yamakawa, T. Uchida, M.A. Matthay, and K. Makita, "Proteolytic release of the receptor for advanced glycation end products from in vitro and in situ alveolar epithelial cells," *Am. J. Physiol. Lung Cell. Mol. Physiol.* **300**(4), L516–L525 (2011).
12. M. Kerkeni and J. Gharbi, "RAGE receptor: May be a potential inflammatory mediator for SARS-COV-2 infection?" *Med.Hypotheses* **144**, 109950–109950 (2020).
13. J. Jeong, J. Lee, J. Lim, et al., "Soluble RAGE attenuates AngII-induced endothelial hyperpermeability by disrupting HMGB1-mediated crosstalk between AT1R and RAGE," *Exp. Mol. Med.* **51**(9), 1–15 (2019).
14. K. Kierdorf and G. Fritz, "RAGE regulation and signaling in inflammation and beyond," *J Leukoc. Biol.* **94**(1), 55–68 (2013).
15. B.N. Lambrecht, M. Vanderkerken, and H. Hammad, "The emerging role of ADAM metalloproteinases in immunity," *Nat. Rev. Immunol.* **18**(12), 745–758 (2018).
16. S.F. Lichtenthaler, M.K. Lemberg, and R. Fluhrer, "Proteolytic ectodomain shedding of membrane proteins in mammals—hardware, concepts, and recent developments," *EMBO J.* **37**(15), e99456 (2018).
17. B.I. Hudson and M.E. Lippman, "Targeting RAGE signaling in inflammatory disease," *Ann. Rev. Med.* **69**, 349–364 (2018).
18. H. Higashida, K. Furuhara, A.-M. Yamauchi, et al., "Intestinal transepithelial permeability of oxytocin into the blood is dependent on the receptor for advanced glycation end products in mice," *Sci. Rep.* **7**(1), 1–15 (2017).
19. Y. Yamamoto, M. Liang, S. Munesue, et al., "Vascular RAGE transports oxytocin into the brain to elicit its maternal bonding behaviour in mice," *Commun. Biol.* **2**(1), 1–13 (2019).
20. A. Bierhaus, P.M. Humpert, M. Morcos, et al., "Understanding RAGE, the receptor for advanced glycation end products," *J. Mol. Med.* **83**(11), 876–886 (2005).
21. J. Kim, C.K. Wan, S. J. O'Carroll, S.B. Shaikh, and L.F. Nicholson, "The role of receptor for advanced glycation end products (RAGE) in neuronal differentiation," *J Neurosci. Res.* **90**(6), 1136–1147 (2012).
22. E.J. Lee and J.H. Park, "Receptor for advanced glycation endproducts (RAGE), its ligands, and soluble RAGE: Potential biomarkers for diagnosis and therapeutic targets for human renal diseases," *Genom. Inform.* **11**(4), 224 (2013).
23. S.K. Malin, S.D. Navaneethan, C.E. Fealy, et al., "Exercise plus caloric restriction lowers soluble RAGE in adults with chronic kidney disease," *Obes. Sci. Pract.* **1**, 1–6 (2020).
24. A. Stirban, "Measurement of lens autofluorescence for diabetes screening," *J. Diabetes Sci. Technol.* **8**(1), 50–53 (2014).
25. J.N. Tsoporis, E. Hatziagelaki, S. Gupta, et al., "Circulating ligands of the receptor for advanced glycation end products and the soluble form of the receptor modulate cardiovascular cell apoptosis in diabetes," *Molecules* **25**(22), 5235 (2020).
26. E.O. Melin, J. Dereke, and M. Hillman, "Higher levels of the soluble receptor for advanced glycation end products and lower levels of the extracellular newly identified receptor for advanced glycation end products were associated with lipid-lowering drugs in patients with type 1 diabetes: A comparative cross-sectional study," *Lipids Health Dis.* **19**(1), 1–10 (2020).
27. S. Moriya, M. Yamazaki, TH. Murakami, K. Maruyama, and S. Uchiyama, "Two soluble isoforms of receptors for advanced glycation end products (RAGE) in carotid atherosclerosis: The difference of soluble and endogenous secretory RAGE," *J. Stroke Cerebrovasc. Dis.* **23**(10), 2540–2546 (2014).
28. S. Raposeiras-Roubín, B.K. Rodiño-Janeiro, L. Grigorian-Shamagian, et al., "Soluble receptor of advanced glycation end products levels are related to ischaemic aetiology and extent of coronary disease in chronic heart failure patients, independent of advanced glycation end products levels: New Roles for Soluble RAGE," *Eur. J. Heart Fail.* **12**(10), 1092–1100 (2010).
29. L. Jensen, K. Munk, A. Flyvbjerg, H.E. Bøtker, and M. Bjerre, "Soluble receptor of advanced glycation end-products in patients with acute myocardial infarction treated with remote ischaemic conditioning," *Clin. Lab.* **61**(3–4), 323–328 (2015).
30. J. Chen, S.S. Mooldijk, S. Licher, et al., "Assessment of advanced glycation end products and receptors and the risk of dementia," *JAMA Netw. Open* **4**(1), e2033012 (2021).
31. L. Gao, J. Wang, Y. Jiang, et al., "Relationship between peripheral transport proteins and plasma amyloid-β in patients with Alzheimer's disease were different from cognitively normal controls: A propensity score matching analysis," *J. Alzheimer's Dis.* **78**(2), 699–709 (2020).

32. D. Yalcin Kehribar, M. Cihangiroglu, E. Sehmen, et al., "The receptor for advanced glycation end product (RAGE) pathway in COVID-19," *Biomarkers* **1**, 1–5 (2021).
33. M. Sierra-Colomina, A. García-Salido, I. Leoz-Gordillo, et al., "sRAGE as severe acute bronchiolitis biomarker, prospective observational study," *Pediatr. Pulmonol.* **55**(12), 3429–3436 (2020).
34. J.T. Patregnani, B.A. Brooks, E. Chorvinsky, and D.K. Pillai, "High BAL sRAGE is associated with low serum eosinophils and IgE in children with asthma," *Children* **7**(9), 110 (2020).
35. E. Dozio, C. Sitzia, L. Pistelli, et al., "Soluble receptor for advanced glycation end products and its forms in COVID-19 patients with and without diabetes mellitus: A pilot study on their role as disease biomarkers," *J. Clin. Med.* **9**(11), 3785 (2020).
36. S. Shah, E.A. Baez, D.L. Felipe, et al., "Advanced glycation endproducts in children with diabetes," *J. Pediatr.* **163**(5), 1427–1431 (2013).
37. Diagnoptics_Technologies, "AGE Reader," 2018, available from: https://www.diagnoptics.com/age-reader/.
38. J.D. Maynard, M.N. Ediger, R.D. Johnson, and M.R. Robinson, "Determination of a measure of a glycation end-product or disease state using a flexible probe to determine tissue fluorescence of various sites," USA, patent US 8140147 (2012).
39. L. Pilon and K. Katika, "Time-resolved non-invasive optometric device for detecting diabetes," USA, patent US20070156036A1 (2007).
40. K. Uk, V. Berezin, G. Papayan, N. Petrishchev, and M. Galagudza, "Spectrometer for fluorescence–reflection biomedical research," *J. Opt. Technol.* **80**(1), 40–48 (2013).
41. G.V. Papayan, V.B. Berezin, N.N. Petrishchev, and M.M. Galagudza, "Spectrometer for fluorescent reflective biomedical research," *Opticheskiy Zhurnal* **80**(1), 56–67 (2013).
42. S. Ahmad, H. Khan, Z. Siddiqui, et al., "AGEs, RAGEs and s-RAGE; friend or foe for cancer," *Semin. Cancer Biol.* **49**, 44–55 (2018).
43. R. Klein, K. Horak, K.E. Lee, et al., "The relationship of serum soluble receptor for advanced glycation end products (sRAGE) and carboxymethyl lysine (CML) to the incidence of diabetic nephropathy in persons with type 1 diabetes," *Diabetes Care* **40**(9), e117–e119 (2017).
44. E.R. Miranda, V.S. Somal, J.T. Mey, et al., "Circulating soluble RAGE isoforms are attenuated in obese, impaired-glucose-tolerant individuals and are associated with the development of type 2 diabetes," *Am. J. Physiol. Endocrinol. Metabol.* **313**(6), E631–E640 (2017).
45. E.V. Potapova, V.V. Dremin, E.A. Zherebtsov, et al., "A complex approach to noninvasive estimation of microcirculatory tissue impairments in feet of patients with diabetes mellitus using spectroscopy," *Opt. Spectrosc.* **123**(6), 955–964 (2017).
46. J. Samsoondar, R. Pawluczyk, P. Borge, et al., "Method and apparatus for rapid measurement of HbA1c," USA, patent US6582964B1 (2003).
47. M.A. Mezentseva and T.A. Bukrina, "Non-invasive methods for measuring blood sugar," in *VI Scientific and Practical Conference Information and Measuring Equipment and Technologies*, Tomsk Polytechnic University, Tomsk, Russia (2015).
48. V.G. Pevgov and Yu.I. Gurfinkel', "Device for non-invasive determination of blood parameters," RU, patent 2373846 (2008).
49. K. Ishihara, K. Asano, Y. Kouchi, and H. Kusuzawa, "Noninvasive blood analyzer," USA, patent US 6061583 (2000).
50. E.M. Lee, D.A. Westerberg, H.H. Yao, J. Adamczyk, and M.A. Christensen, "Determination of % glycated hemoglobin," USA, patent US 6316265 (2001).
51. K.M. Salonen, S.J. Ryhänen, J.M. Forbes, et al., "Decrease in circulating concentrations of soluble receptors for advanced glycation end products at the time of seroconversion to autoantibody positivity in children with prediabetes," *Diabetes Care* **38**(4), 665–670 (2015).

8 Hyperspectral Imaging of Diabetes Mellitus Skin Complications

Viktor V. Dremin, Evgenii A. Zherebtsov, Alexey P. Popov, Igor V. Meglinski, and Alexander V. Bykov

CONTENTS

8.1 INTRODUCTION

Over the past decades, diabetes mellitus (DM) became one of the most widespread and serious diseases [1]. DM leads to glycation of main blood proteins, namely, hemoglobin and albumin. Reacting with glucose, these proteins form glycated hemoglobin and glycated albumin. The accompanying structural and functional changes of collagen significantly contribute to the development of various pathological malformations affecting the collagen-containing tissues, especially in skin tissues and blood vessels, causing serious micro- and macrovascular complications [2], increasing disability risks and threat to life, and as a result morbidity and mortality of patients. The main goal of diabetes care is the prevention of these complications and/or early-stage diagnostics of the disease.

Quantitatively, the measure of diabetic complications, including retinopathy, neuropathy, and cardiovascular disease, is represented by the level of glycated (glycosylated) hemoglobin and albumin. It has been shown that changes caused by glycation of proteins cannot be observed clearly due to high scattering and absorption of human skin tissues [3], whereas optical techniques showed a high potential of evaluating the skin complications of DM at an early stage [4].

Doppler optical coherence tomography (DOCT), as a functional extension of optical coherence tomography (OCT) [5], is an optic, non-contact, noninvasive technique able to achieve a detailed analysis of the blood flow and subcutaneous blood vessels for various applications [6,7]. This technique has been adopted and used extensively for the visualization and quantitative evaluation of cutaneous microvascular structure and function in patients with diabetes [8]. However, these techniques are suffering with some serious drawbacks, including inaccurate assessment of ischemia and low blood supply at ulcer area [9,10] and healing probabilities [11].

DOI: 10.1201/9781003112099-8

Hyperspectral imaging (HSI) is a modern implementation of the diffuse reflectance spectroscopy (DRS) method, providing information about the oxygenation state of tissue. This approach is natural extension of the spectrophotometry [12] that was translated to the measurements of the diffuse reflectance spectra of human skin in optical-to-near-infrared (NIR) spectral range [13], whereas multiregression analysis was used for the quantitative analysis of the measured spectra [13,14]. While due to recent developments in the photonic-based technologies, the optical/NIR spectroscopy is progressed to the HSI approach, the recent developments in the computational technologies are also evolved, i.e., the concept of multiregression analysis to the machine learning regression. Combined with the artificial neural network (ANN), the machine learning algorithms are well suited for stand-alone classification of spectral data that opens new horizons for the optical/NIR spectroscopy in tissue diagnosis and clinical applications [15,16].

This approach is able to detect systemic and local microcirculatory changes associated with DM [2,4]. The technology is most commonly used to analyze the development of wounds, including diabetic ulcers and the results of burns [17]. Using this technology, both the spatial and spectral information of an object can be acquired. The obtained 3D image (two spatial and one spectral dimension), or the so-called hypercube, consists of about a hundred or more spectral bands for every measured pixel of an object. The analysis of the data contained in the hypercube allows us to assess the content of chromophores in the biological tissue studied, the thickness of the epidermis, blood filling, and oxygen saturation.

Studies conducted using this technology have shown the ability of the method to evaluate the effectiveness of treatment of diabetic ulcers, and predict their development even before they appear. The study carried out by Yudovski et al. showed that, by using HSI, it was possible to register the changes in thickness of the epithelium that precede the formation of an ulcer [18].

The study conducted by Khaodhiar et al. investigated the healing process of diabetic wounds over 6 months, using hyperspectral medical technology [19]. The study was conducted on patients with type 1 diabetes and registered values corresponding to concentrations of oxy- and deoxyhemoglobin, with subsequent calculation of the healing index. It was shown that HSI successfully distinguished between healing ulcers and ones that failed to heal.

Another study that aimed at studying the healing of diabetic ulcers through HSI included patients with both type 1 and type 2 diabetes [20]. Tissue oxygenation maps were constructed from oxy- and deoxyhemoglobin values determined for each pixel in the image. Linear discriminant analysis was used to calculate the healing index, by separating healed and unhealed ulcers. It was concluded that the technique could accurately predict wound healing in advance, providing information necessary for the treatment and monitoring of diabetic complications.

HSI can accurately predict ulcer healing in a few months and identify ulcers that are at risk of not healing, and therefore, it could be used to screen for lower-extremity complications due to diabetes. Novel technologies for the collection and analysis of hyperspectral data, including artificial intelligence systems, suggest an even wider introduction of this technology in clinical diagnosis.

8.2 HYPERSPECTRAL IMAGING SYSTEM AIDED BY ARTIFICIAL NEURAL NETWORKS

A compact hand-held imaging system was constructed on the basis of the hyperspectral snapshot camera (Senop Optronics, Finland) utilizing a micro Fabry-Perot tunable filter providing a spectral resolution of 6–10 nm within the total range of 500–900 nm. The schematic layout of the constructed device is presented in Figure 8.1.

A broadband illumination unit utilizing 50W halogen lamp was constructed, based on a fiber-optic ring illuminator (Edmund Optics, USA), providing a uniform distribution of light intensity in the camera focal plane with the average irradiance of $4.3 \pm 0.5\,mW/cm^2$ in the camera field of view (FOV) on the skin surface. The use of the ring illuminator made it possible to combine the axes of illumination and detection. The illumination ring and the camera were equipped with rotatable

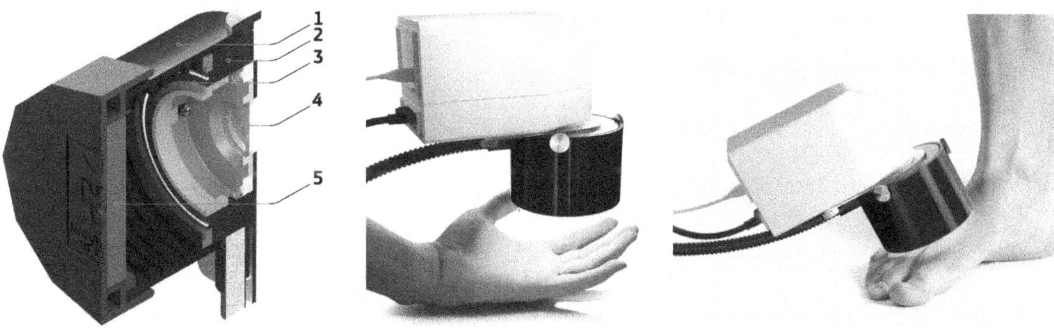

FIGURE 8.1 Schematic design of the developed prototype of hyperspectral imaging system: 1 – lens hood (spacer), 2 – fiber ring illuminator delivering broadband radiation from a 50W halogen lamp, 3 – ring-shaped polarizer of illumination unit, 4 – camera polarizer; 5 – built-in diffuse reflectance standard (gray Spectralon – 50% reflection). (Adapted with permission from [15] © The Optical Society.)

broadband linear polarizers fixed at the crossed position to reduce specular reflection from the measured object. A 3D model of the unit was implemented using CAD software and printed out with a 3D material printer. The measurement approach considered allows the capture of the reflected signal of the entire FOV at a certain waveband. The scanning is performed in the spectral domain. Generally, the constructed device is capable of recording spatially and spectrally resolved reflectance used for further ANN analysis. Thus, diffuse reflectance is recorded with a spectral step of 5 nm from an area of $8\times8\,cm^2$ at 1010×1010 pixels resolution by the CMOS sensor. Normalized spatially resolved tissue reflectance is calculated as a pointwise ratio of dark-noise-corrected light intensity reflected from the object to that reflected from the diffuse reflectance standard. A hypercube from the reflectance standard is recorded every time before the measurement of an object. Gray Spectralon (50% reflection, Labsphere, Inc., USA) is used as a calibration standard. Use of the gray standard allows us to avoid the possible oversaturation of the CMOS sensor matrix for white reflectance standard (100% reflection) at a fixed integration time.

8.2.1 Human Skin Model and Monte Carlo Simulations of Diffuse Reflectance Spectra

For ANN training, the diffuse reflectance spectra were imitated by using GPU-accelerated Monte Carlo (MC) technique [21,22]. A seven-layer model of skin (Figure 8.2) was used for the calculations. The model utilized accounts for the in-depth distribution of blood in the dermis that is missing from the basic two-layer models, which consider dermal blood distribution as uniform. A record of the dermal blood distribution is essential due to the dependence of the light penetration depth on its wavelength (see, e.g., [23]). Thus, nonuniform blood distribution in the dermis will have an effect on the shape of the reflectance spectrum.

The basics of the considered model have been described in Refs. [15,21,24]. The absorption coefficient of each layer takes into account the blood volume fraction (BVF) in various vascular beds, oxygen saturation, water content, and melanin fraction. Twenty-five non-zero parameters are required by the model to describe the absorption of the skin. To make the model more practical, some groups of parameters were reduced by the introduction of the coefficients that link the parameters inside the groups. The main idea of this simplification was to consider a minimal set of parameters influencing the reflectance spectra shape and changing independently. This approach assumes that the parameters of each group can vary coherently. Thus, parameter K_b links BVFs of dermal and subdermal layers. The average value of blood oxygen saturation S for all layers was taken into account. The parameter K_{epi} describes the thickness of the epidermis

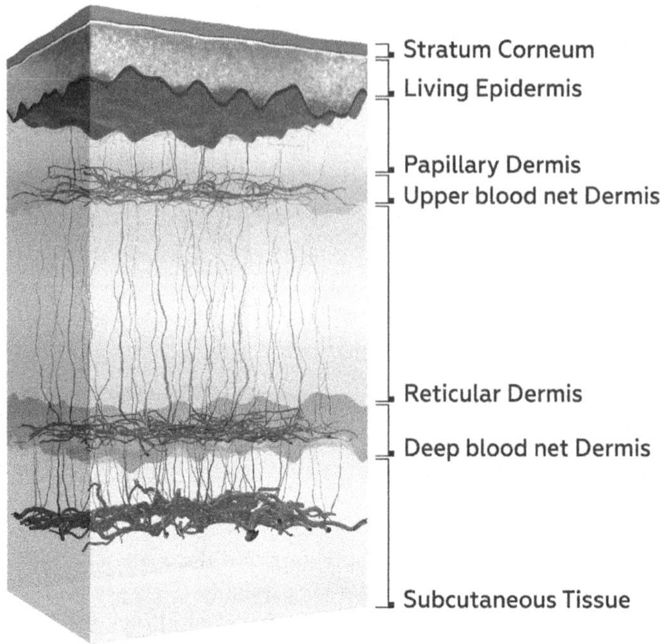

FIGURE 8.2 Skin tissue layers taken into account in the seven-layer model used. (Reprinted with permission from [15] © Optica.)

and stratum corneum. Finally, four parameters, K_b, S, K_{mel}, and K_{epi}, were considered as variables to model the diffuse reflectance of the skin. The remaining parameters were considered to be fixed and equal to their typical values. The differences in the spatial distribution of blood, melanin, blood oxygen saturation, hematocrit, the water content of the skin, as well as the numerical aperture of the detector, were taken into account. All possible combinations of four variable parameters were obtained for the simulated skin reflectance spectra accounting for blood volume coefficient K_b = [0–2] with a step of 0.1, blood oxygen saturation S = [30–100]% with a step of 1%, melanin concentration K_{mel} = [0–0.1] with a step of 0.01, and epidermis thickness coefficient K_{epi} = [1–5] with a step of 1. The complete training data set contained 82,005 spectra in the range of 500–900 nm with a step of 5 nm.

8.2.2 Neural Networks Processing

Matlab Deep Learning Toolbox was chosen to build and train an interconnected set of neural networks for the retrieval of skin parameters K_b, S, K_{mel}, and K_{epi}, as well as biotissue phantom parameters d and S (see the phantom parameters below). The toolbox allows the multilayer perceptron (MLP) network to be used for curve fitting and regression. The MLP network has an input layer, one hidden layer, and a linear output. It can be regarded as a nonlinear, parameterized mapping of the input feature vector to the output space of data-fitting results. In the problem statement under consideration, the inverse solution for the characterization of tissue parameters can be classified as a construction of a multiple regression model. The favorable number of neurons in the hidden layer was evaluated in a series of tests. In the tests, the number of neurons was sequentially increased from 1 to N, where N is the number of data points in the spectrum range considered for use in the estimation of skin/phantom parameters. The number of hidden neurons providing the best-fitting performance was selected for further use.

A general flowchart of data processing is shown in Figure 8.3.

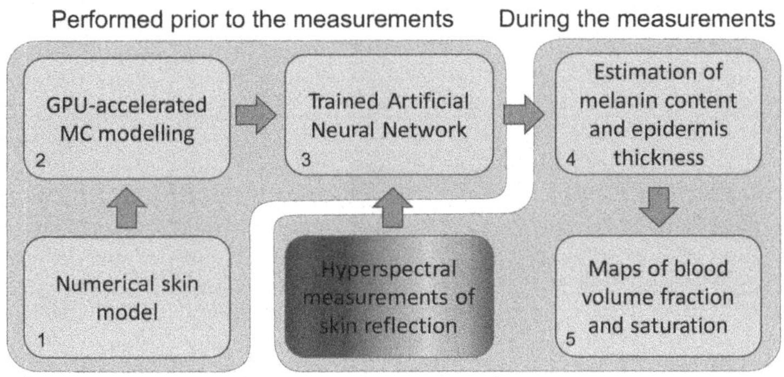

FIGURE 8.3 Flowchart of hyperspectral data processing. Steps 1–3 are performed prior to the measurements. Steps 4 and 5 – in line with the data acquisition. Note: step 4 could be omitted for repeated measurements of the same area of interest. (Reprinted with permission from [15] © Optica.)

All the steps in Figure 8.3 can be divided into two major groups. The first group, consisting of numerical tissue model adjustment, simulation of the training dataset, and training the ANN, is the most time-consuming but performed prior to the measurements. The second group is the application of the trained ANN to the results of the hyperspectral measurements and obtaining the required parameters. This group does not demand significant computational resources and can be performed in line with the experimental data acquisition. At first, the epidermal thickness and melanin content are estimated (step 4 in Figure 8.3) using the corresponding trained ANN. The best fit for these parameters was obtained in the spectral range of 680–725 nm and 725–770 nm, respectively. Then, for the estimated epidermal thickness and melanin content, the blood BVF and skin blood oxygenation (SBO) were retrieved for the spectral range of 535–705 nm and 675–825 nm, respectively (step 5 in Figure 8.3). Preprocessing of the measurement data, aimed at improving the signal-to-noise ratio, consists of the spatial averaging of the input hyperspectral cube in a 5×5-pixel window. All the input measurement data vectors for the considered spectral ranges, as well as the training dataset, were normalized prior to the network training procedure. The normalization procedure was implemented in the data processing algorithm and dedicated to the mitigation of possible height variations of the skin surface in the camera FOV. Vector normalization (by L^2-norm) for both the training dataset and measured data was used to obtain the corresponding unit vectors.

8.3 VALIDATION OF TISSUE BLOOD OXYGEN ASSESSMENT APPROACH WITH THE PHANTOM STUDIES

First, a solid biotissue phantom with predefined optical properties equal to those of bloodless human dermis was used to validate the ability of the proposed approach to measure blood oxygen saturation in the embedded vessels. The layout and the dimensions of the developed phantom are shown in Figure 8.4a and b. The phantom contains two tilted plain hollow channels (0.25×1 mm^2 cross section) located at different angles. The embedding depth increases linearly from 0.3 to 2 mm for the superficial channel and from 0.3 to 4 mm for the deep one. A detailed description of the manufacturing and characterization of biotissue phantoms is given in Refs. [25–27]. In short, a polyvinyl chloride (PVC)-based matrix was used as a transparent host for ZnO nanoparticles introducing scattering. The proper amount of added scattering particles was estimated on the basis of Mie theory, taking into account their size distribution. The average diameter of the particles was 0.34 μm. To control the absorption coefficient of the phantom, a black plastic color composed of CI Pigment Black 7 was added. Two glass capillaries were installed at different angles, as described above, inside the phantom mold prior to solidification of the phantom. After solidification, the capillaries

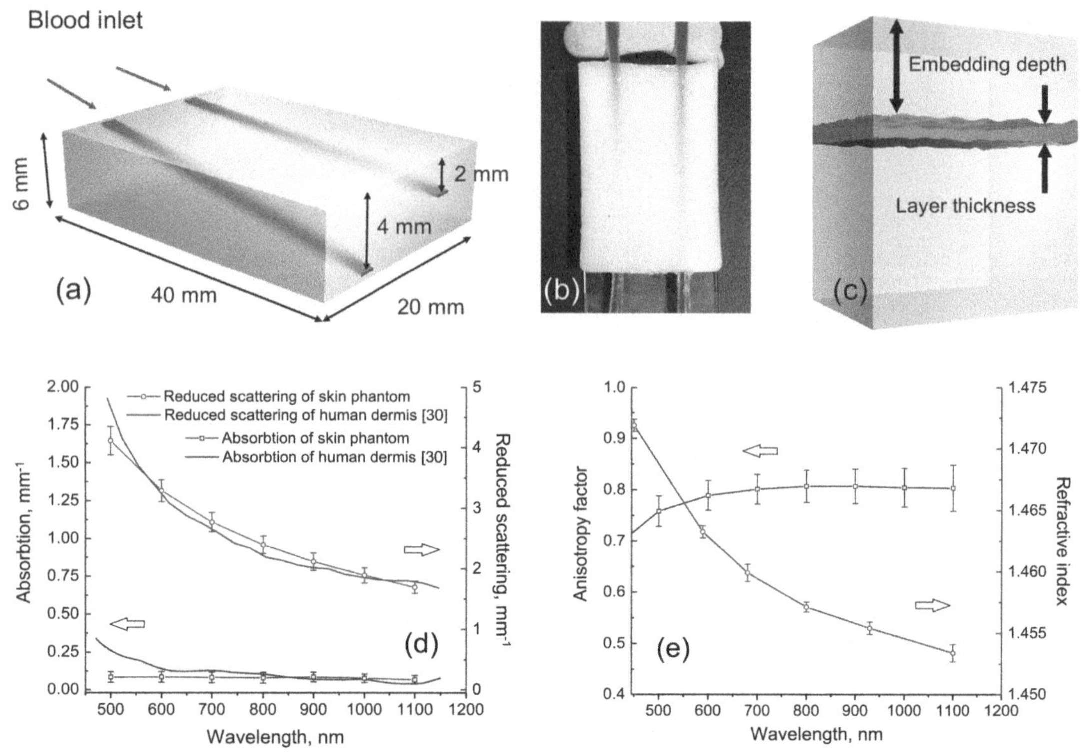

FIGURE 8.4 3D schematics of the biotissue phantom with the embedded blood vessels (a); photograph (top view) of the manufactured phantom with the channels filled with fully oxygenated blood (b); three-layer tissue phantom model used for calculations of diffuse reflectance spectra (c); measured optical properties of the biotissue phantom (d, e). (Reprinted with permission from [15] © Optica.)

were gently removed, thus forming hollow channels. The channels were further connected via micropipette tips and plastic tubing to the syringe pump.

A suspension of washed human red blood cells (RBC) (45% v/v), obtained from a healthy donor, was pumped through the channels at a rate of 10 ml/h to avoid sedimentation. The used RBC suspension was prepared according to the following procedure. First, a whole blood (hematocrit 45%) from a healthy donor was drawn by venipuncture and put into ethylenediaminetetraacetic acid (EDTA)-covered tubes to prevent coagulation. Written consent was obtained from the volunteer. The studies were conducted in accordance with ethical permission obtained from the Finnish Red Cross. Second, to prevent uncontrolled aggregation, the RBC were washed in phosphate-buffered saline (PBS, pH 7.4). For that purpose, the whole blood was centrifuged at 3000 g for 10 minutes, and plasma supernatant was replaced with PBS. This procedure was repeated three times to remove residual plasma containing the proteins inducing uncontrolled aggregation. Finally, the washed RBC mass was suspended in PBS to the initial value of hematocrit. All the measurements were taken at a room temperature of 20°C.

For the training of the ANN and estimation of blood oxygen saturation, the diffuse reflectance spectra of the biotissue phantom were also simulated by MC technique using a three-layer tissue model (see Figure 8.4c) with the blood layer embedding depth d and its oxygen saturation S as the variable parameters. The selected thickness of the blood layer was equal to the channel thickness of the biotissue phantom. A GPU-accelerated MC model of photon migration in scattering tissue-like media [22] was used for routine simulation of the spectra in the training dataset for all possible combinations of the considered parameters and cross-validated for biotissue phantoms with work [28]. In total, 2,091 spectra were simulated at $d = [0–4]$ mm with a step of 0.1 mm and $S = [50–100]\%$ with a step of 1% for the spectral range 500–900 nm with a step of 5 nm. The measured optical properties of the biotissue phantom presented above were utilized for the simulations. The optical properties of blood (hematocrit 45%) for different oxygen levels were adopted from Ref. [29].

The optical properties of the fabricated phantom, including absorption and scattering coefficients, as well as the anisotropy factor in the visible and NIR spectral region, were derived on the basis of the inverse adding-doubling method from the diffuse reflectance, diffuse transmittance, and collimated transmittance measured with the spectrophotometer (Gooch & Housego, USA), equipped with integrating spheres. Additionally, the refractive index of the phantom was measured using the multiwavelength Abbe refractometer (Atago, Japan).

The results of the characterization of the phantom optical properties are presented in Figure 8.4d, e. Figure 8.4d also shows the comparison of the retrieved absorption and reduced scattering coefficients of the created phantom with those measured from the human *ex vivo* dermis. The data for the human dermis are taken from Ref. [30]. The measurements for the reduced scattering (see Figure 8.4d) are in good agreement for the whole spectral range 500–1100 nm. For the absorption coefficient, good agreement is observed for the spectral range 650–1100 nm. The measured scattering anisotropy factor (Figure 8.4e) is on the level of 0.8 for the considered range, which is also typical for biotissues, including skin [31,32]. Thus, the developed phantom can be considered a relevant model of bloodless human dermis/skin for red and infrared spectral range.

Before the measurements, the prepared RBC suspension was carefully stirred in air to increase its oxygen saturation up to 100%. To confirm the oxygen level, 0.5 mL of the suspension was placed into a sealed 1-mm-thick plain glass cuvette, and the optical properties of the suspension for the spectral range of 600–1000 nm were estimated on the basis of the spectrophotometric measurements and the inverse adding-doubling procedure, as described above. Good agreement of the obtained absorption spectrum with the absorption of oxygenated blood (hematocrit 45%) [29] confirmed the predetermined oxygen saturation level.

The simulated diffuse reflectance spectra for the biotissue phantom at two values of blood layer embedding depth *d* and different blood oxygen level *S* are presented in Figure 8.5.

Strong blood absorption in the spectral region $\lambda < 600$ nm is the cause of low reflection in this range. It is clearly seen that the increase of the embedding depth also increases the reflection in the range of 500–600 nm by making the absorbing layer deeper, so that fewer photons are capable of reaching it and getting absorbed. Reflection in the range of 600–800 nm is less depth sensitive and depends mostly on the blood oxygen level. With the decrease in the blood oxygen level, the reflection in the range of 600–800 nm drops and the characteristic deoxyhemoglobin absorption peak at 760 nm also becomes visible as a pit on the phantom reflection spectra.

The simulated reflectance spectra were used to train the ANN for the recovery of the oxygen level in the blood channels inside the biotissue phantom. The trained ANN was applied to process

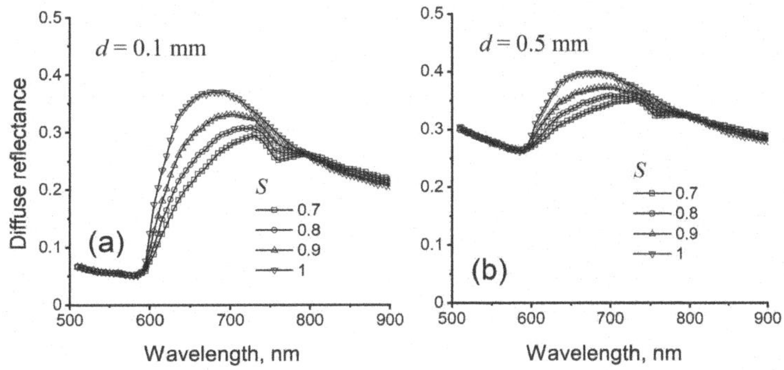

FIGURE 8.5 Simulated spectra of diffuse reflectance from biotissue phantom (a, b) at different embedding depths *d* of blood layer and oxygen saturation *S*. (Adapted with permission from [15] © Optica.)

the hyperspectral measurements of phantom diffuse reflection. The spectral region of 550–800 nm was selected for the blood oxygen estimation, as the most prominent changes of reflectance are observed in this region, as can be seen from Figure 8.5a and b. Figure 8.6a shows the measured diffuse reflectance along the superficial channel. The distance along the channel linearly corresponds to the channel embedding depth. Thus, the increase in the channel embedding depth causes a significant increase of the reflection in the range 500–600 nm, similar to observations made during the simulations. Figure 8.6b shows the comparison of the measured reflectance spectra (dots) with the simulated observations (lines), selected as a best fit by the trained ANN for three values of the capillary embedding depth. In all cases, the recovered blood oxygen saturation differs less than 4% from the actual value. The map of the recovered blood oxygen saturation for the whole biotissue phantom is presented in Figure 8.6c. The corresponding values for the superficial and deep channels are shown in Figure 8.6d. It is seen that the correct value of blood oxygenation, in this case, can be recovered up to a depth of about 2 mm, which corresponds approximately to the middle of channel 2. Further increase of the embedding depth causes immediate signal loss.

The results described above were obtained using a two-parameter fit accounting for the dependence of the simulated spectra on the channel embedding depth d and blood oxygen saturation S.

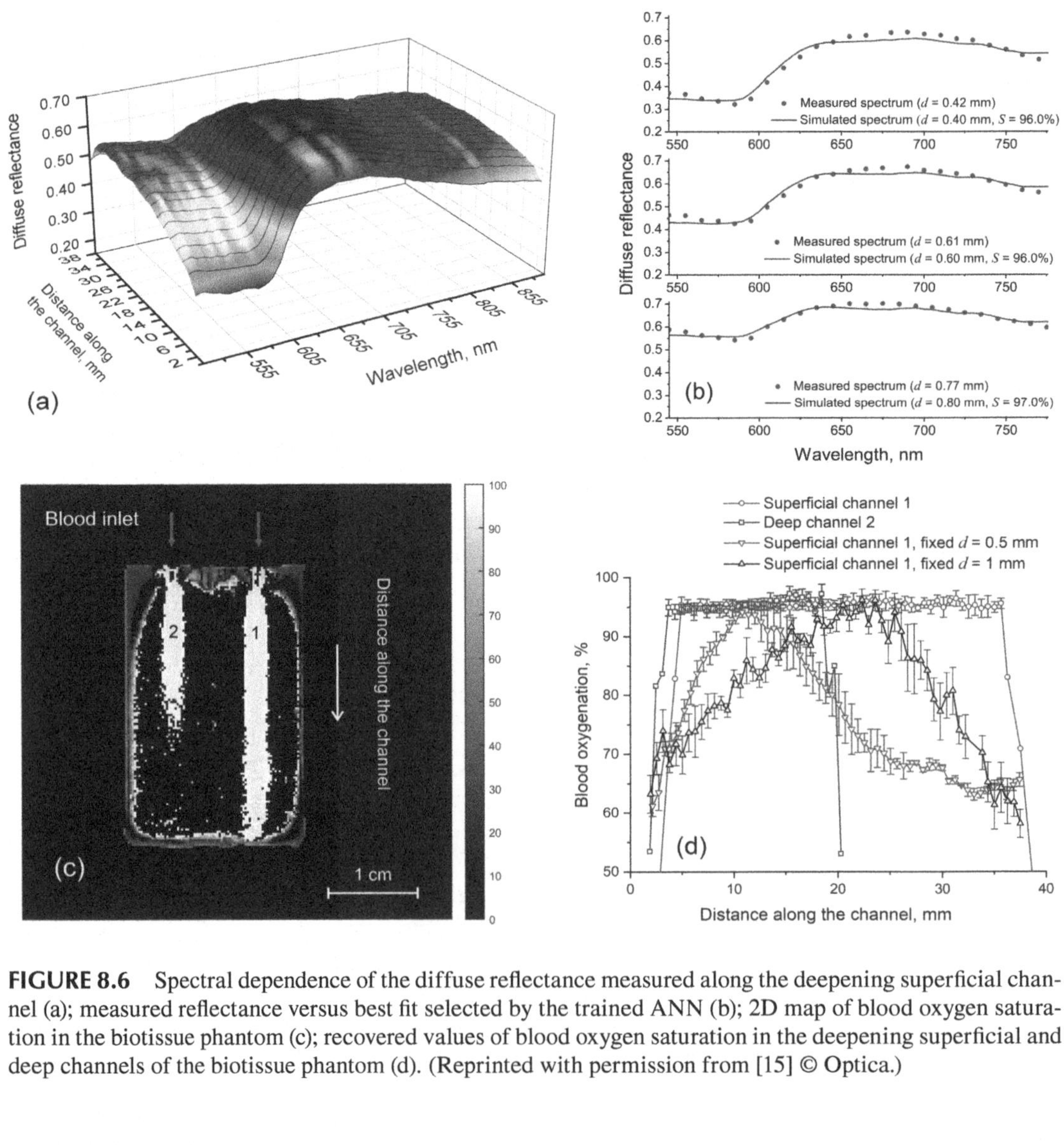

FIGURE 8.6 Spectral dependence of the diffuse reflectance measured along the deepening superficial channel (a); measured reflectance versus best fit selected by the trained ANN (b); 2D map of blood oxygen saturation in the biotissue phantom (c); recovered values of blood oxygen saturation in the deepening superficial and deep channels of the biotissue phantom (d). (Reprinted with permission from [15] © Optica.)

The phantom studies also allow the investigation of the effect of the chosen tissue model on the accuracy of blood oxygen level recovery. Figure 8.6d shows the recovered oxygen level in the deepening superficial channel of the phantom, using a biotissue model with a fixed embedding depth of the blood layer (d=0.5 and 1 mm, correspondingly). Thus, it is seen that the application of the model with a fixed embedding depth of the blood layer may lead to significant (up to 40%) underestimation of the recovered blood oxygen saturation. In this case, the best fit performed by the neural network is obtained for a reflectance spectrum corresponding to lower oxygen saturation, which is causing the underestimation in the recovered oxygen level. The correct value is obtained only for the embedding depth corresponding to that in the biotissue model.

8.4 *IN VIVO* FUNCTIONAL SKIN IMAGING

The proposed imaging approach was applied to human skin *in vivo*. The developed HSI system was used to perform trial measurements with healthy volunteers. Typical dependencies of the simulated skin reflectance spectra on the variable parameters (K_b, S, K_{mel}, and K_{epi}) obtained for the seven-layer skin model used for ANN training are presented in Figure 8.7a–d.

The increase in BVF in the skin (shown in Figure 8.7a) causes the decrease of skin reflectance within the entire spectral range of 500–900 nm. However, the most significant changes are observed in the region of strong blood absorption (500–600 nm). The most prominent changes in the skin reflectance caused by the variation in blood oxygen saturation are observed in the range of 600–800 nm (see Figure 8.7b). The influence of the melanin content in the epidermis is shown in Figure 8.7c. It is seen that an increase in melanin content decreases the reflectance in the entire spectral range considered. However, the change in the shape of spectra is significantly different compared to the increase in BVF. Increase in the epidermal thickness (see Figure 8.7d) acts differently for the spectral region 500–600 nm and 600–900 nm. Below 600 nm, the increase in the epidermal thickness causes an increase in skin reflectance; however, for the longer wavelengths, a similar increase causes a decrease in reflectance. This effect can be explained by the significantly lower (up to two orders of magnitude) absorption of oxyhemoglobin in the red-IR spectral region relative to absorption in the green region. Thus, the oxygenated blood turns into a moderate absorber, and the decrease in the reflectance in the red-IR region is caused by the increased optical density of the epidermis.

The developed HSI system was used to perform trial measurements and the arterial occlusion tests with healthy volunteers. According to the University of Oulu Ethics Committee regulations, informed consent was obtained from all tested subjects. First, Caucasian and Indian male skin corresponding to the type II and V of Fitzpatrick scale (Figure 8.8a and b) was considered and compared.

During the occlusion test, the middle or ring finger of the right hand was occluded with a rubber band. The pressure applied was above both systolic/diastolic pressures to ensure the suppression of

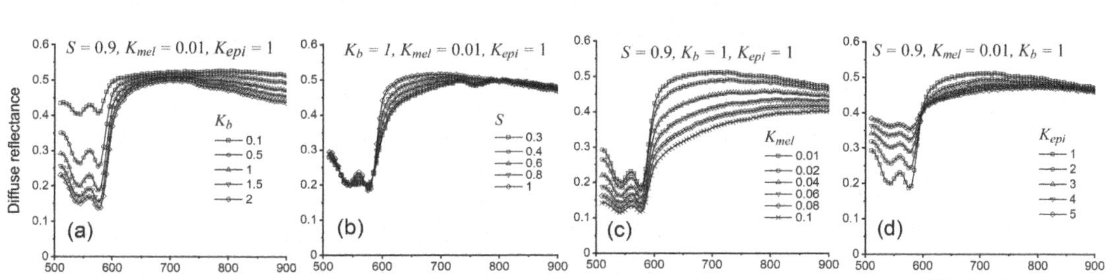

FIGURE 8.7 Simulated reflectance spectra for human skin at different values of blood volume coefficient K_b – (a), blood oxygen saturation S – (b), melanin concentration K_{mel} – (c) and epidermis thickness coefficient K_{epi} – (d). (Adapted with permission from [15] © Optica.)

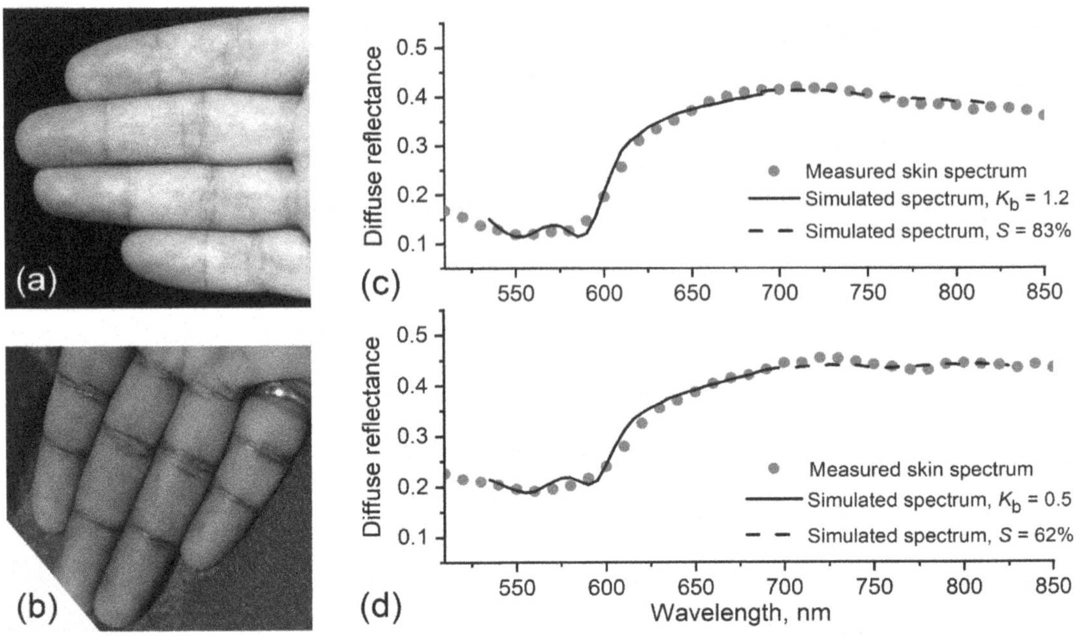

FIGURE 8.8 Photographs of the Caucasian (a) and Indian (b) skin *in vivo* used for the measurements. Measured skin diffuse reflectance versus best fit selected by the trained ANN for Caucasian skin at different values of BVF and SBO before (c) and during (d) the occlusion. The measurements were taken at a distal phalanx of a ring finger. Left and right simulated results indicate the parts of the spectrum used for the estimation of BVF and SBO, correspondingly. (Reprinted with permission from [15] © Optica.)

blood circulation within the finger and, hence, its oxygenation. Hyperspectral images of the intact skin were obtained before the occlusion, after 3 minutes of continuous occlusion, and 1 minute after the release of the occlusion ring. Finally, 2D maps of BVF and SBO were reconstructed. Spatial variations in the thickness of the epidermis and melanin content of the skin were also taken into account. Figure 8.8c and d shows a comparison of the measured skin reflectance spectra with those selected as the best fit by the trained ANN. A good general agreement of measured and simulated spectra is demonstrated. The measurements were performed for Caucasian skin at a distal phalanx of a ring finger before (Figure 8.8c) and during the occlusion (Figure 8.8d), resulting in a simultaneous decrease of BVF and SBO.

The reconstructed values of epidermal thickness and melanin content for the considered skin types are presented in Figure 8.9a and i and Figure 8.9b and j, respectively. The average value of epidermal thickness in both cases is within 140–180 μm, which is typical for palmar skin. The uneven distribution of the epidermal thickness across the measured area can be explained by the intraindividual variability related to uneven loads to the different parts of the palm during the lifetime. The average recovered value of melanin content is 1.3% for Caucasian skin and 8% for Indian skin. Figure 8.9c–e and k–m presents the reconstructed maps of BVF during the occlusion test for both considered skin types. It is seen that, in the case shown in Figure 8.9c, the blood content for the distal phalanges is up to three times higher than for the middle or proximal ones. This increase could also be noticed from the photograph presented in Figure 8.8a. For the second case (see Figure 8.9k), the corresponding increase is about two times. During the occlusion (see Figure 8.9d and l), in both cases, a significant decrease in blood content is observed. The release of the occlusion ring causes the appearance of reactive hyperemia, resulting in the increase of BVF (see Figure 8.9e and m), which can be reliably detected by the developed system in cases of both low and high melanin content of the skin. Figure 8.9f–h and n–p presents the maps of the SBO obtained before, during, and after the occlusion for both of skin types. The average blood oxygen level is 81%–84% for

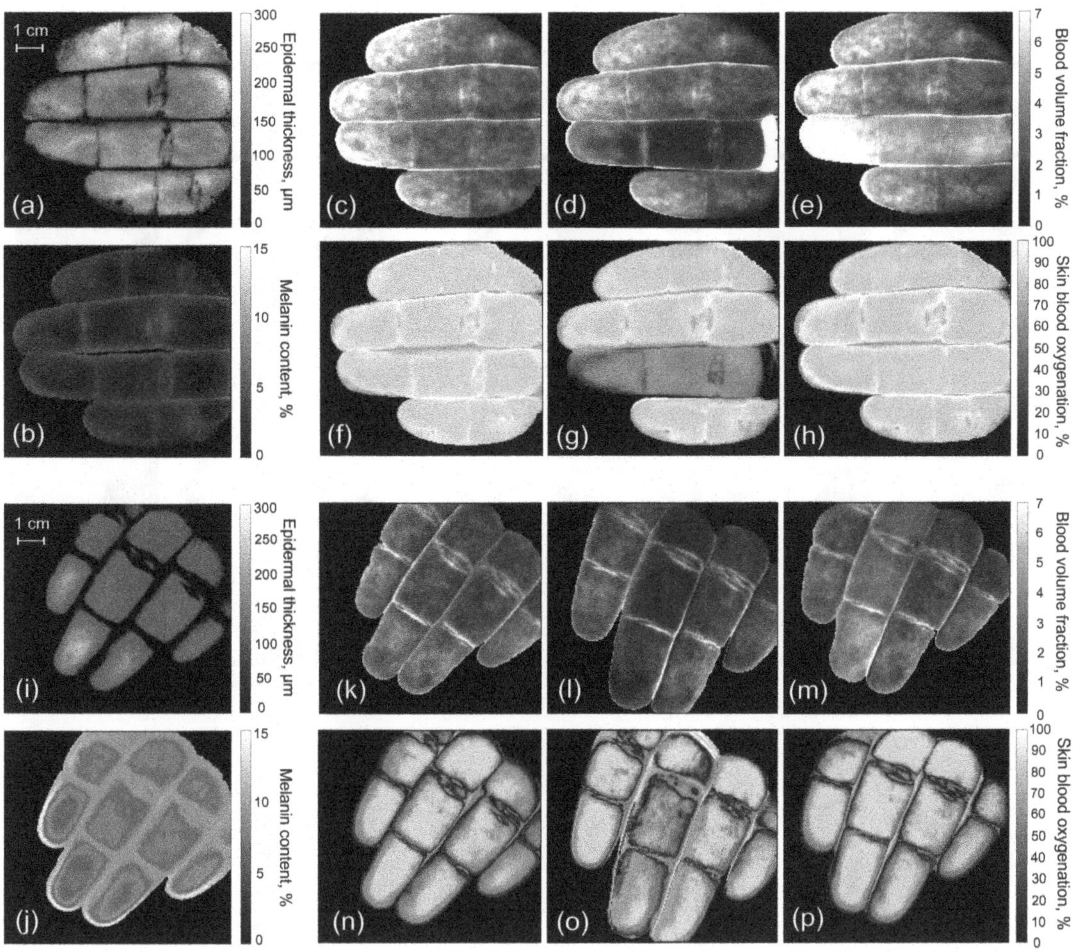

FIGURE 8.9 Reconstructed values of the epidermal thickness (a, i) and melanin content (b, j) for Caucasian and Indian skin, correspondingly. Retrieved maps of BVF before (c, k), during (d, l), and 1 minute after (e, m) the ring/middle finger occlusion for the considered skin types. Corresponding maps of SBO before (f, n), during (g, o), and 1 minute after (h, p) the occlusion. (Reprinted with permission from [15] © Optica.)

not occluded tissues. The obtained parameter represents tissue oxygenation and corresponds to the average between the oxygen saturation of the arterial (95%–99%) and venous ~75%) blood [33]. Three-minute occlusion lowers the blood oxygen content down to the level of 60%–65% (see Figure 8.9g and o). After the release of the occlusion ring, the oxygenation rapidly recovers to its initial values (see Figure 8.9h and p).

Next, the developed HSI system was used to perform different occlusion tests in healthy volunteers with Caucasian skin. Additional arterial occlusion and venous occlusion tests were performed on the upper arm, using a sphygmomanometer air cuff with a pressure of 200–220 mmHg and 80–100 mmHg, correspondingly. Hyperspectral images of the skin of the palm and fingers were obtained before the occlusion, after 3 min of continuous occlusion, and 1 min after the release of the occluded arm. Each occlusion test was a separate measurement, with a significant time interval in between to enable full tissue recovery. Here, the epidermis thickness was fixed at the typical average thickness [34] of 100 µm. The average value of melanin content is 1%, which is typical for Caucasian skin.

Figure 8.10a–d presents the reconstructed maps of *BVF* of the palm and fingers during the different occlusion tests performed. During the arterial occlusion of the palm and fingers (Figure 8.10a

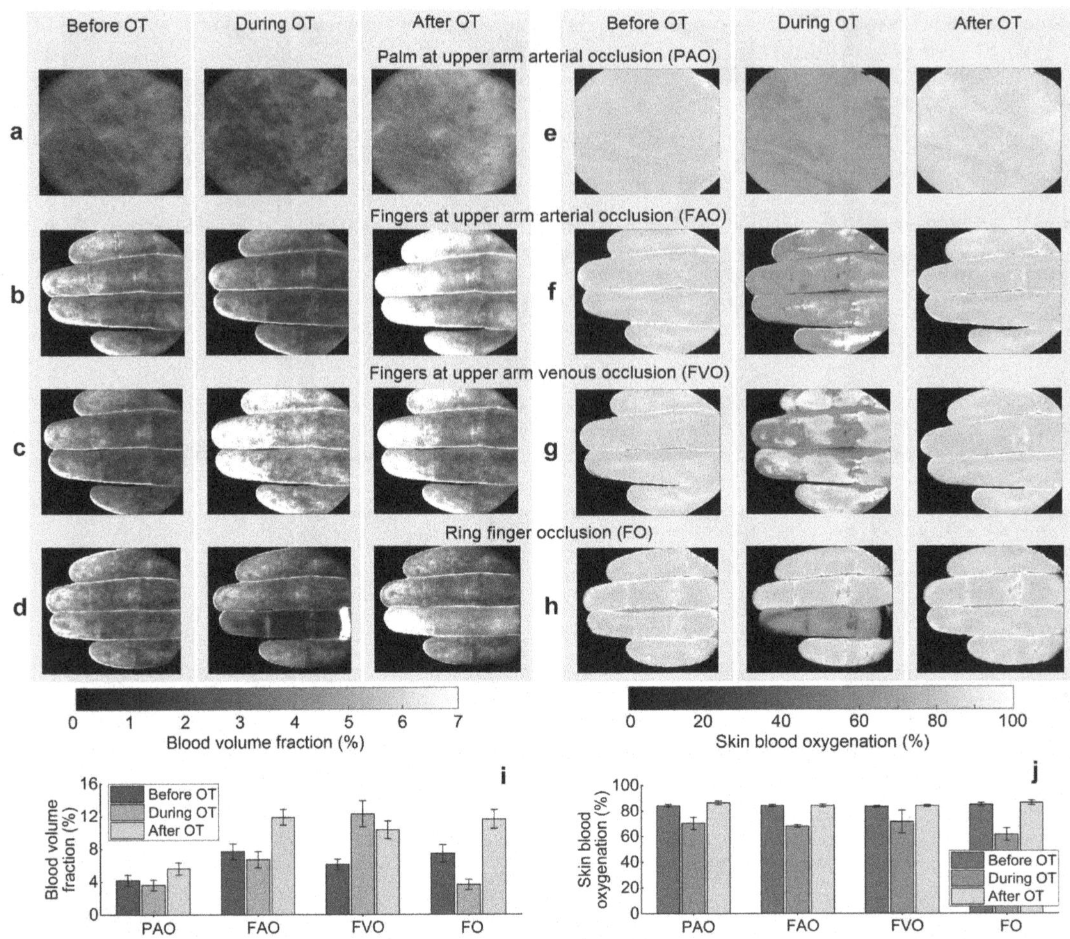

FIGURE 8.10 Retrieved maps of BVF before, during, and 1 minute after different occlusion tests (a–d). Palm with upper arm arterial occlusion (a). Fingers with upper arm arterial occlusion (b). Fingers with upper arm venous occlusion (c). Fingers with arterial occlusion of one finger (d). Corresponding maps of SBO before, during, and 1 minute after different occlusion tests for the considered skin zones (e–h). Mean values and standard deviation of BVF for each stage of occlusion (i). Mean values and standard deviation of SBO for each stage of occlusion (j) [16].

and b) and finger occlusion by elastic band (Figure 8.10d), a significant decrease in blood content is observed. Venous occlusion (Figure 8.10c) greatly increases the blood content in the fingers, which is associated with the cessation of blood outflow. When the pressure in the cuff is low, only the veins are squeezed, while maintaining the arterial blood flow. The release of the occlusion cuff or occlusion ring causes reactive hyperemia, which leads to an increase in the *BVF* parameter.

Figure 8.10e–h shows the distribution maps of the *SBO* parameter before, during, and after the different occlusion tests for the considered skin zones. Gentler arterial and venous occlusions (Figure 8.10e–g) than finger occlusion (Figure 8.10h) lower the oxygenation values to 70%–75%. It can be seen that the decrease in oxygenation in venous occlusion is more local. The mean values and standard deviation of *BVF* and *SBO* for each stage of different occlusion tests are presented in Figure 8.10i and j, correspondingly.

The system was also adapted to measure the degree of linear polarization (DOLP) of radiation reflected from a biological tissue and was equipped with a rotating linear polarizer, which can be placed either parallel or perpendicular to the initial polarization of the light source. Accounting for

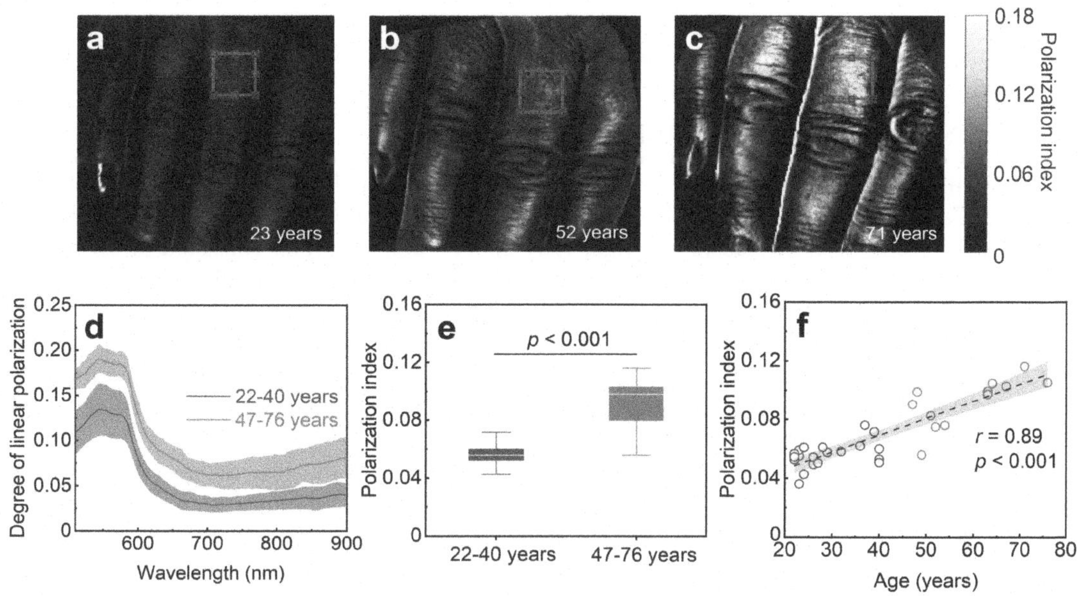

FIGURE 8.11 Polarization-sensitive hyperspectral system allows us to identify age-specific changes. Retrieved PI maps for healthy female volunteers: 23 years. (a); 52 years. (b); 71 years. (c). Region of interest (rectangle, 8×8 mm) indicates the averaging area to obtain the DOLP spectra. DOLP spectra for two age groups: 22–40 and 47–76 years. (d). (e) Comparison of PI among two age groups (e). Correlation between PI and age (confidence level is 95%) (f) [16].

the DOLP of the light scattered from a sample significantly improves the imaging contrast between slightly different tissue structures and offers, in consequence, a sensitive parameter for the discriminant analysis [35]. In addition, the normalization of the difference image by the sum image cancels the variation in the illumination and the melanin effect [36]. As a diagnostic criterion, the polarization index PI is proposed. PI represents the integrated value of the DOLP spectrum normalized by the number of bands of the hyperspectral camera. When measuring BVF and SBO, the rotating polarizer was fixed in the crossed position in respect to the polarizer on the illumination ring for the reduction of specular reflection from the measured object.

Preliminary studies of the capacity of the proposed PI parameter for monitoring age-related human skin changes were conducted. For this purpose, 32 healthy volunteers (9 males and 23 females, aged 22–76 years.) were recruited. Hyperspectral images of the skin at the parallel and perpendicular orientation of the polarizers were recorded on the palm dorsal surface. Figure 8.11a–c shows examples of the PI images for three female volunteers of different ages. A general increase of the PI parameter is observed with age. For the detailed analysis, the volunteers were divided into two age groups: 22–40 and 47–76 years.

Figure 8.11d shows the spectral distributions of the DOLP of radiation backscattered from skin at different ages. The spectral dependences of the DOLP are affected by the presence of skin absorbers (e.g., blood, melanin) and scatterers (e.g., collagen, elastin). A statistically significant increase in the PI parameter for the elderly group (47–76 years) is observed (Figure 8.11e). In particular, the average value of PI for this group is 76% higher than for younger volunteers (22–40 year.). Figure 8.11f shows the relationship between the PI and the age (r=0.89, p<0.001, Pearson's correlation). The increase can be explained by the negative correlation between the skin scattering coefficient and human age demonstrated in Ref. [37]. Specifically, a decrease in the concentration of skin collagen during aging [38] causes a decrease in skin scattering. However, for polarized light, multiple scattering results in a loss of polarization. Thus, low values of the scattering coefficient in older age lead to lower depolarization and, as a consequence, to higher values of DOLP. The proposed approach shows promise for *in vivo* noninvasive real-time assessment of age-related skin changes and can be also extended to monitor changes associated with the development of DM.

8.5 CLINICAL STUDY OF DIABETIC PATIENTS

The experimental clinical study involved 20 patients (10 men and 10 women, mean age 54±11 yrs) with Caucasian skin type and with type 2 DM at the Medical Center Plavnieki and the Pauls Stradins Clinical University Hospital (Riga, Latvia). The control group consisted of 20 healthy volunteers (mean age 48±14 years). Thirteen participants (65%) were women. The compared groups were statistically homogeneous with differences in fasting glucose level and body mass index ($p<0.01$, Mann-Whitney *U*-test). The measurements were taken on the dorsal surface of the foot. All studies were performed in the sitting position. The study participants refrained from food and drink 2 hours before the study to exclude the influence of these factors on microcirculation. Room temperature was maintained steadily at 24°C–25°C.

Hyperspectral images were obtained of the dorsal surface of the foot of patients with DM. Moreover, 2D distribution maps of BVF and SBO were reconstructed, taking into account the melanin content and the epidermal thickness of the skin.

Figure 8.12a and b shows the measurement area (FOV of HSI system) with the calculated parameters of BVF and SBO, respectively. Typical distribution maps of the calculated parameters for the control and the diabetic groups are presented in Figure 8.12c–h.

All parameters were averaged over the FOV of the system. Experimental studies have shown that diabetic patients have elevated values of BVF and PI, as well as a lower SBO level (see Figure 8.13a).

The results of the computation of BVF show that high blood content is observed in DM patients. This can be explained by a number of factors. When venous stasis is associated with venous insufficiency characteristic of diabetic patients [39], all functioning veins become wider, and those venous vessels that had not previously functioned are open. Capillaries also expand, mainly in the venous parts. The elevated values of this parameter can also be accounted for by a blood flow to the lower limbs due to inflammation and healing processes, which can be observed in patients with DM. For patients with DM, it is also necessary to take into account the presence of complex cardiovascular pathology and nephropathy, which can lead to variations in blood content.

The obtained data on understated tissue saturation in diabetics in comparison with the control group match those previously obtained by other studies [2,40,41]. It is known that hyperglycemia may increase oxygen consumption in mitochondria, resulting in cellular hypoxia [42,43]. In this regard, the retrieval of blood content and oxygen saturation can be useful in the study of blood stasis, leading to edema and trophic disorders, as well as in monitoring therapeutic procedures aimed at healing trophic ulcers.

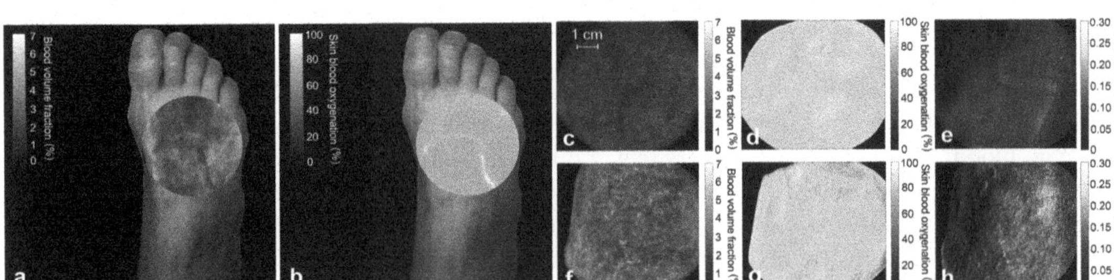

FIGURE 8.12 The developed hyperspectral system is capable of reliably visualizing the BVF and SBO parameters in the dorsal surface of feet. FOV of the HSI system (8×8 cm^2) with the calculated parameters of BVF (a) and SBO (b). Typical reconstructed maps of the BVF (c, f), SBO (d, g) and PI (e, h) for the control (upper row, c–e) and diabetic (lower row, f–h) groups, correspondingly [16].

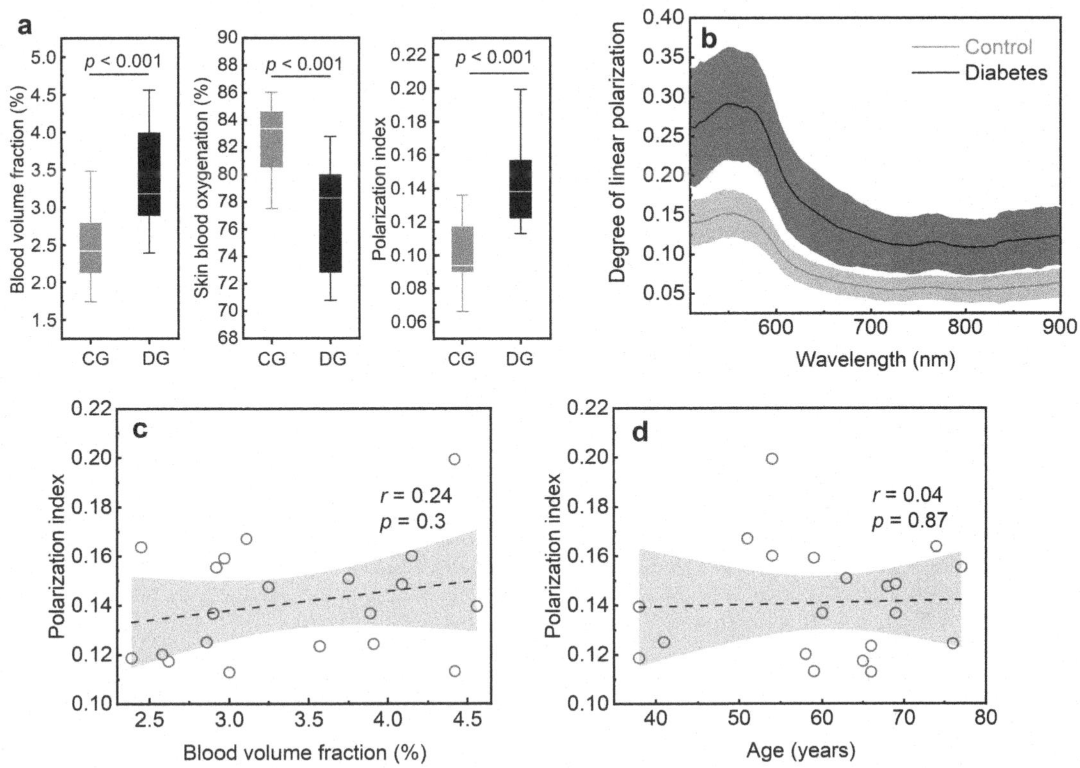

FIGURE 8.13 Comparison of parameters between control and diabetic groups (a). The box central line is the median of the group, and the edges are the 25th and 75th percentiles. Mean spectral dependences of the DOLP (solid line) and standard deviation for two groups: control (lower curve) and diabetes (upper curve) (b). Correlation between PI and BVF (c) and between PI and age (d) for diabetic group (confidence level is 95%) [16].

Figure 8.13b shows the averaged spectra of DOLP for control and diabetic groups with the difference in the entire wavelength range. It should be noted that the spectral dependences of the DOLP reflect the presence of blood in the skin tissue (an increase in the DOLP in the spectral bands of hemoglobin absorption) [36]. In this regard, to assess the influence of blood content on the difference in the spectra, the dependence of PI from BVF was plotted. As can be seen from Figure 8.13c, there is no statistically significant correlation between these parameters. We assume that it is the presence of diabetes and associated skin changes that affect the difference in DOLP spectra.

Considering the connection of diabetic changes and aging processes [44], we assume that the presence of DM, as well as aging, reduces the loss of polarization. Additionally, we checked the relationship between PI and age for the diabetic group. As can be seen from Figure 8.13d, there is also no correlation between these parameters. This confirms the assumption that the presence of diabetes is the main factor affecting the polarization properties of patients' skin.

The analyzed parameters (BVF, SBO, and PI) were used for the synthesis of the decision rule. These parameters satisfy the principles of statistical independence and the significance of the differences of their values, calculated for the patients' and control groups (see Figure 8.13a). The combination of all three parameters provides the highest values of sensitivity and specificity.

Figure 8.14a shows the 3D scatter plot of experimental data with the applied discriminant surface that divides the experimental points into two groups with a better combination of sensitivity and specificity. The phenomenological diagnostic criterion, in the form of discriminant functions,

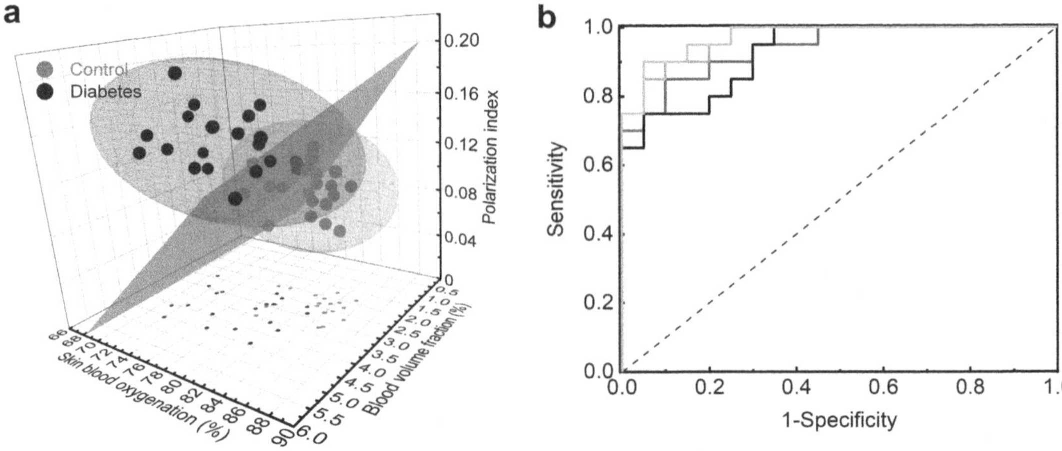

FIGURE 8.14 The simultaneously measured values of BVF, SBO, and PI underlies the set of developed diagnostic classifiers of high sensitivity and specificity. The 3D scatter plot and the decision surface for a three-dimensional LDA classifier (a). 3D ellipsoids are confidence regions (confidence level is 95%). ROC curves for assessing the effectiveness of the classifiers (b). For the ROC analysis: 1st line from the top denotes BVF, SBO, PI; 2nd line – SBO, PI; 3rd line – BVF, SBO; 4th line – BVF, PI [16].

allows for attributing a newly measured subject to one of the two groups. The ROC curves were calculated for the obtained discriminant functions (see Figure 8.14b).

8.6 SUMMARY

The developed hyperspectral system has shown the ability to sense the alterations of blood content, blood oxygenation, and collagen structure in human skin *in vivo*. The implementation of ANN-based processing allows fast recovery of the considered skin parameters. Our concept was successfully demonstrated in preclinical experiments and clinical studies and satisfied the major requirements for modern medical imaging techniques by providing high FOV, high spectral resolution, close to real-time data acquisition and processing, and high level of sensitivity and specificity.

This is the first *in vivo* human clinical study to report a label-free, noninvasive polarization-sensitive approach that can discriminate between the skin of DM patients and control participants. We showed that polarization-enhanced hyperspectral analysis can differentiate the skin tissue with age-related and DM changes. We assume that the differences shown in the PI parameter, characterizing the age-related collagen changes, can form the basis for the diagnosis of DM complications. Prediction of ulcers can also be based on the detection of changes in collagen structure. Figure 8.13d shows the absence of correlation between PI and the age of patients with diabetes. At the same time, a significant correlation is observed for the control group of healthy volunteers. This can be explained by the fact that the presence of diabetes can distort the collagen structure at any age and thus hinders the dependence of PI on age for the diabetic group. That, in particular, supports our suggestion of the sensitivity of PI to the changes in collagen structure that occur in the presence of diabetes (Figure 8.13a). It should be borne in mind that the blood content in the biotissue, epidermis thickness, and some other factors may specifically affect residual polarization spectra. Thus, hyperspectral polarization imaging, in contrast to the standard approach with the registration of reflected light, allows for more detailed tissue analysis, including selectively tracking changes in the concentration of various absorbers and scatterers in the biotissue.

Nonetheless, for future clinical applications, our technique can be improved and expanded. The developed data processing algorithm allows almost real-time processing of the hyperspectral images; however, the time of the hypercube acquisition and the duration of the data transfer via

USB port impose the main limitations on system performance. The hyperspectral sensor with specific spectral resolution and wavelength range could be flexibly optimized according to application requirements. In this study, a spectral band of 510–900 nm was utilized. However, the application of a sensor with an extended spectral range up to 1000 nm may provide an additional possibility to analyze the content of water and fat in tissues. Additionally, the use of UV or blue sources in the lighting channel will allow the HSI method to be used for the fluorescence analysis of AGEs, as well as the respiratory chain coenzymes NADH and FAD [45,46,47]. There is also further potential to expand the number of clinical investigations that will eventually allow the methods in question to be introduced into the practice of the attending physician.

In summary, we have developed a compact HSI system utilizing an ANN-based processing approach for the calculation of distribution maps of skin blood content, blood oxygenation, and the polarization index of the reflected radiation. The proposed system is capable of close to real-time hyperspectral image processing. The use of polarization-sensitive HSI and parameters of BVF, SBO, and PI in combination or individually can have important clinical applications, since they can help to identify patients with different degrees of microcirculation impairments in the lower extremities. The proposed parameters could serve as biomarkers of diabetic complications. They can also be used to evaluate therapeutic procedures aimed at preventing or reversing diabetic complications. Employment of the PI parameter increases the sensitivity and specificity of the method up to the level of 95% and 85%, respectively, compared to the approach that relies only on BVF and SBO (sensitivity 85%, specificity 80%). Our results can facilitate the development of the HIS technique and give a new direction in the studies of age-related diseases.

ACKNOWLEDGMENTS

Viktor Dremin thanks the support of the European Union's Horizon 2020 research and innovation program under the Marie Skłodowska-Curie Grant Agreement No. 839888. The authors are grateful for the support from the Academy of Finland (Grant Nos. 314369 and 318281).

REFERENCES

1. P. Zimmet, K. G. M. M. Alberti, and J. Shaw, "Global and societal implications of the diabetes epidemic," *Nature* **414**(6865), 782–787 (2001).
2. R. L. Greenman, S. Panasyuk, X. Wang, et al., "Early changes in the skin microcirculation and muscle metabolism of the diabetic foot," *Lancet* **366**(9498), 1711–1717 (2005).
3. O. S. Zhernovaya, V. V. Tuchin, and I. V. Meglinski, "Monitoring of blood proteins glycation by refractive index and spectral measurements," *Laser Phys. Lett.* **5**(6), 460–464 (2008).
4. E. Zharkikh, V. Dremin, E. Zherebtsov, A. Dunaev, and I. Meglinski, "Biophotonics methods for functional monitoring of complications of diabetes mellitus," *J. Biophotonics* **13**(10), 202000203 (2020).
5. D. Huang, E. A. Swanson, C. P. Lin, et al., "Optical coherence tomography," *Science* **254**(5035), 1178–1181 (1991).
6. M. Bonesi, S. Matcher, and I. Meglinski, "Doppler optical coherence tomography in cardiovascular applications," *Laser Phys.* **20**(6), 1491–1499 (2010).
7. M. Bonesi, S. G. Proskurin, and I. V. Meglinski, "Imaging of subcutaneous blood vessels and flow velocity profiles by optical coherence tomography," *Laser Phys.* **20**(4), 891–899 (2010).
8. R. Argarini, R. A. McLaughlin, S. Z. Joseph, et al., "Optical coherence tomography: A novel imaging approach to visualize and quantify cutaneous microvascular structure and function in patients with diabetes," *BMJ Open Diabetes Res. Care* **8**(1), 1479 (2020).
9. B. E. Sumpio, R. O. Forsythe, K. R. Ziegler, et al., "Clinical implications of the angiosome model in peripheral vascular disease," *J. Vasc. Surg.* **58**(3), 814–826 (2013).
10. O. A. Mennes, J. J. van Netten, R. H. J. A. Slart, and W. Steenbergen, "Novel optical techniques for imaging microcirculation in the diabetic foot," *Curr. Pharm. Des.* **24**(12), 1304–1316 (2018).
11. R. J. Hinchliffe, J. R. W. Brownrigg, J. Apelqvist, et al., "IWGDF guidance on the diagnosis, prognosis and management of peripheral artery disease in patients with foot ulcers in diabetes," *Diabetes. Metab. Res. Rev.* **32**, 37–44 (2016).

12. B. Chance, "Spectrophotometry of intracellular respiratory pigments," *Science* **120**(3124), 767–775 (1954).
13. S. J. Matcher, "Signal quantification and localization in tissue near-infrared spectroscopy," in *Handbook of Optical Biomedical Diagnostics, Second Edition, Volume 1: Light-Tissue Interaction*, pp. 585–687, SPIE Press, Bellingham, WA (2017).
14. I. V. Meglinski and S. J. Matcher, "Computer simulation of the skin reflectance spectra," *Comput. Methods Programs Biomed.* **70**(2), 179–186 (2003).
15. E. Zherebtsov, V. Dremin, A. Popov, et al., "Hyperspectral imaging of human skin aided by artificial neural networks," *Biomed. Opt. Express* **10**(7), 3545 (2019).
16. V. Dremin, Z. Marcinkevics, E. Zherebtsov, et al., "Skin complications of diabetes mellitus revealed by polarized hyperspectral imaging and machine learning," *IEEE Trans. Med. Imaging* **40**(4), 1207–1216 (2021).
17. D. W. Paul, P. Ghassemi, J. C. Ramella-Roman, et al., "Noninvasive imaging technologies for cutaneous wound assessment: A review," *Wound Repair Regen.* **23**(2), 149–162 (2015).
18. D. Yudovsky, A. Nouvong, K. Schomacker, and L. Pilon, "Assessing diabetic foot ulcer development risk with hyperspectral tissue oximetry," *J. Biomed. Opt.* **16**(2), 026009 (2011).
19. L. Khaodhiar, T. Dinh, K. T. Schomacker, et al., "The use of medical hyperspectral technology to evaluate microcirculatory changes in diabetic foot ulcers and to predict clinical outcomes," *Diabetes Care* **30**(4), 903 (2007).
20. A. Nouvong, B. Hoogwerf, E. Mohler, et al., "Evaluation of diabetic foot ulcer healing with hyperspectral imaging of oxyhemoglobin and deoxyhemoglobin," *Diabetes Care* **32**(11), 2056–2061 (2009).
21. V. Dremin, E. Zherebtsov, A. Bykov, et al., "Influence of blood pulsation on diagnostic volume in pulse oximetry and photoplethysmography measurements," *Appl. Opt.* **58**(34), 9398–9405 (2019).
22. A. Doronin and I. Meglinski, "Online object oriented Monte Carlo computational tool for the needs of biomedical optics," *Biomed. Opt. Express* **2**(9), 2461–2469 (2011).
23. A. V. Moco, S. Stuijk, and G. de Haan, "New insights into the origin of remote PPG signals in visible light and infrared.," *Sci. Rep.* **8**(1), 8501 (2018).
24. G. I. Petrov, A. Doronin, H. T. Whelan, I. Meglinski, and V. V. Yakovlev, "Human tissue color as viewed in high dynamic range optical spectral transmission measurements," *Biomed. Opt. Express* **3**(9), 2154–2161 (2012).
25. A. V. Bykov, A. P. Popov, A. V. Priezzhev, and R. Myllyla, "Multilayer tissue phantoms with embedded capillary system for OCT and DOCT imaging," in *Optics InfoBase Conference Papers* **8091**, R. A. Leitgeb and B. E. Bouma, Eds., p. 80911R, International Society for Optics and Photonics (2011).
26. M. S. Wróbel, A. P. Popov, A. V. Bykov, et al., "Measurements of fundamental properties of homogeneous tissue phantoms," *J. Biomed. Opt.* **20**(4), 045004 (2015).
27. M. S. Wróbel, A. P. Popov, A. V. Bykov, et al., "Multi-layered tissue head phantoms for noninvasive optical diagnostics," *J. Innov. Opt. Health Sci.* **8**(3), 8 (2015).
28. D. Loginova, E. Sergeeva, I. Fiks, and M. Kirillin, "Probing depth in diffuse optical spectroscopy and structured illumination imaging: A Monte Carlo study," *J. Biomed. Photonics Eng.* **3**(1), 010303 (2017).
29. N. Bosschaart, G. J. Edelman, M. C. G. Aalders, T. G. van Leeuwen, and D. J. Faber, "A literature review and novel theoretical approach on the optical properties of whole blood," *Lasers Med. Sci.* **29**(2), 453–479 (2014).
30. E. Salomatina, B. Jiang, J. Novak, and A. N. Yaroslavsky, "Optical properties of normal and cancerous human skin in the visible and near-infrared spectral range," *J. Biomed. Opt.* **11**(6), 64026 (2006).
31. X. Ma, J. Q. Lu, H. Ding, and X.-H. Hu, "Bulk optical parameters of porcine skin dermis at eight wavelengths from 325 to 1557 nm," *Opt. Lett.* **30**(4), 412 (2005).
32. S. L. Jacques, "Optical properties of biological tissues: A review," *Phys. Med. Biol.* **58**(11), R37 (2013).
33. K. Akons, E. J. Dann, and D. Yelin, "Measuring blood oxygen saturation along a capillary vessel in human," *Biomed. Opt. Express* **8**(11), 5342 (2017).
34. T. Gambichler, R. Matip, G. Moussa, P. Altmeyer, and K. Hoffmann, "*In vivo* data of epidermal thickness evaluated by optical coherence tomography: Effects of age, gender, skin type, and anatomic site," *J. Dermatol. Sci.* **44**(3), 145–152 (2006).
35. A. Pierangelo, A. Benali, M.-R. Antonelli, et al., "*Ex-vivo* characterization of human colon cancer by Mueller polarimetric imaging," *Opt. Express* **19**(2), 1582 (2011).
36. V. V. Tuchin, "Polarized light interaction with tissues," *J. Biomed. Opt.* **21**(7), 071114 (2016).
37. H. Jonasson, I. Fredriksson, S. Bergstrand, et al., "*In vivo* characterization of light scattering properties of human skin in the 475- to 850-nm wavelength range in a Swedish cohort," *J. Biomed. Opt.* **23**(12), 121608 (2018).

38. S. Shuster, M. M. Black, and E. McVitie, "The influence of age and sex on skin thickness, skin collagen and density," *Br. J. Dermatol.* **93**(6), 639–643 (1975).
39. V. Fejfarová, K. Roztočil, A. Svědínková, et al., "The relationship between chronic venous insufficiency and diabetes mellitus," *Int. Angiol.* **36**(1), 90–91 (2017).
40. L. Xi, C.-M. Chow, and X. Kong, "Role of tissue and systemic hypoxia in obesity and type 2 diabetes," *J. Diabetes Res.* **2016**, 1527852 (2016).
41. J. Ditzel and E. Standl, "The problem of tissue oxygenation in diabetes mellitus," *Acta Med. Scand. Suppl.* **578**, 59–68 (1975).
42. J. R. Nyengaard, Y. Ido, C. Kilo, and J. R. Williamson, "Interactions between hyperglycemia and hypoxia: Implications for diabetic retinopathy," *Diabetes* **53**(11), 2931–2938 (2004).
43. K. Sada, T. Nishikawa, D. Kukidome, et al., "Hyperglycemia induces cellular hypoxia through production of mitochondrial ROS followed by suppression of aquaporin-1," *PLoS One* **11**(7), e0158619 (2016).
44. P. Gkogkolou and M. Bohm, "Advanced glycation end products: Key players in skin aging?," *Dermatoendocrinol.* **4**(3), 259–270 (2012).
45. V. V. Dremin, E. A. Zherebtsov, V. V. Sidorov, et al., "Multimodal optical measurement for study of lower limb tissue viability in patients with diabetes mellitus," *J. Biomed. Opt.* **22**(8), 085003 (2017).
46. F. Bartolome and A. Y. Abramov, "Measurement of mitochondrial NADH and FAD autofluorescence in live cells," *Methods Mol. Biol.* **1264**, 263–270 (2015).
47. A. Vinokurov, V. Dremin, G. Piavchenko, et al., "Assessment of mitochondrial membrane potential and NADH redox state in acute brain slices," *Methods Mol. Biol.* **2276**, 193–202 (2021).

9 Fluorescent Technology in the Assessment of Metabolic Disorders in Diabetes

Elena V. Zharkikh, Viktor V. Dremin, and Andrey V. Dunaev

CONTENTS

9.1 INTRODUCTION

Noninvasive optical methods are increasingly used in biomedical diagnostics. Fluorescence spectroscopy, in particular, has found its application in chemistry, biology, and various fields of medicine. This method is highly sensitive and allows us to study various pathological changes of biological tissues in the development of socially significant diseases. Fluorescence spectroscopy and imaging techniques are probably the most common biomedical photonics methods used in skin research. By analyzing fluorescence data, one can extract information about the structure and component composition of the biological tissue and its functional state. Fluorescence provides insight into both the conformation of fluorescent molecules and their binding as well as their interactions within biological tissues.

Speaking about the application of fluorescence technologies in biomedical research, one can distinguish two large groups of methods: (1) those based on the study of endogenous fluorescence of biological tissue molecules (autofluorescence) and (2) methods based on the use of exogenous fluorophores or photosensitizers. Registration of biological tissues' intrinsic fluorescence is more difficult compared to the use of fluorescent dyes, but recently in the imaging and diagnosis of living tissues, a growing interest is focused on intrinsic fluorescence. This trend is closely related to recent advances in high-throughput spectroscopic and microscopic techniques.

Biological tissues are characterized by a complex organization and contain a large amount of intrinsic endogenous fluorophores, which have different spectral regions of absorption and fluorescence, different fluorescence quantum yields, and different fluorescence lifetime. Most of the intrinsic fluorophores of biological tissues are associated with the structural matrix of tissues and components involved in cellular metabolic pathways. The former include such elements as collagen and elastin. Autofluorescence of collagen and elastin has an excitation peak in the range from 300 to 400 nm and shows broad emission bands from 400 to 600 nm with a maximum around 400, 430, and 460 nm [1–3]. By analyzing the fluorescence of collagen and elastin, it is possible to distinguish different types of tissues, for example, to separate epithelial tissues from connective ones [4].

DOI: 10.1201/9781003112099-9

The second group includes some of the most intense fluorophores that exist in biological tissues, involved in cellular metabolic processes – the reduced form of nicotinamide adenine dinucleotide (NADH), flavins (FAD, etc.), and strongly fluorescent lipopigments (lipofuscin). NADH is excited in a wavelength range between 330 and 370 nm. Flavin mononucleotide (FMN) and flavin adenine dinucleotide (FAD) with excitation maxima around 380 and 450 nm also contribute to tissue autofluorescence [5,6]. The fluorescence of the coenzymes NADH and FAD separately or together is used in the assessment of mitochondrial respiratory chain function.

In addition to the above-mentioned molecules, there are many other endogenous fluorophores possessing emission of different strengths and covering different spectral ranges in the ultraviolet (UV) and visible regions of the electromagnetic spectrum. These fluorophores include aromatic compounds, amino acids (tryptophan, tyrosine, phenylalanine), various porphyrins (hemoglobin, myoglobin) [7], etc. The absorption and fluorescence spectra of various tissue endogenous fluorophores are shown in Figure 9.1.

A large amount of diagnostic information is provided not only by fluorescence spectra and two-dimensional image registration, but also by time-resolved measurements of fluorescence lifetime. Fluorescence lifetime imaging (FLIM) provides additional diagnostic information by measuring tissue autofluorescence lifetime, helping to separate fluorophores with close overlapping absorption and emission spectra. One of the significant applications of FLIM is the *in vivo* assessment of the metabolic status of tissues based on the measurement of the fluorescence lifetime of free and protein-bound NADH [9,10].

The area most benefiting from the development of fluorescent methods is oncology. Studies show that fluorescence spectroscopy demonstrates high informativity in detecting malignant formations of the oral cavity [11] and gastrointestinal tract, bladder [12], etc. Currently, a large amount of research is also being conducted toward the development of fluorescence-guided surgical oncology, in particular in minimally invasive procedures [13].

Another area of wide implementation of fluorescence spectroscopy and imaging technologies in medicine is research related to diabetes mellitus (DM) [14]. According to the International Diabetes Federation (IDF), the challenges of early-stage diagnosis and monitoring of the treatment effectiveness in diabetes is currently one of the highest priorities in modern healthcare [15]. The medical, social, and economic significance of diabetes is primarily determined by the high prevalence of this disease and the frequency of debilitating effects suffered by affected individuals. The IDF report for 2019 indicated that there are 463 million diabetic patients worldwide, with this figure projected to grow to 700 million by 2045. Experts indicate that almost every patient with DM will develop at least one complication during his or her lifetime [16,17]. Complications of DM range from acute to chronic and include a number of conditions such as severe hypoglycemia, ketoacidosis, polyneuropathy, and diabetic nephropathy. Chronic complications of diabetes are mainly related to vascular and nervous function disorders and in the most severe cases can lead to vision loss, limb amputation,

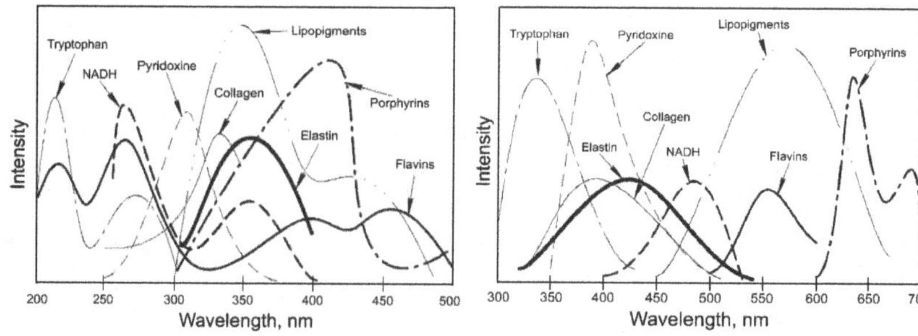

FIGURE 9.1 Absorption (a) and fluorescence (b) spectra of major biological tissue fluorophores. (Adapted from [8] with the permission of John Wiley & Sons.)

and end-stage renal disease leading to the need for dialysis or kidney transplantation and even death. Currently, it is an undeniable fact that life expectancy and quality of life in patients with diabetes depend primarily on the timely detection and competent treatment of its accompanying complications [18,19]. Multiple studies have indicated that a timely diagnosis and treatment, including an increased level of patient monitoring, reduce the manifestation of various complications, potentially even reversing them at early preclinical stages [20]. To this effect, modern diagnosis of feet in DM patients can be done using an array of various methodologies. These methodologies offer both unique advantages and disadvantages. Today, the use of various optical noninvasive methods, which include the methods of fluorescence spectroscopy and imaging, is promising and informative for the diagnosis of DM complications.

In the field of DM complications diagnostics, much attention is paid to the assessment of advanced glycation end products (AGE) accumulation in patient tissues based on their fluorescence properties, as well as to the study of other intrinsic biotissue fluorophores, since changes in these parameters may be associated with the development of diabetic complications. Methods of fluorescence studies of biotissues in combination with other methods of biophotonics also play an important role in the development of multimodal optical noninvasive diagnostics. In the following sections of this chapter, we will take a closer look at the listed research areas.

9.2 ASSESSMENT OF AGE ACCUMULATION IN DIABETES USING SKIN AUTOFLUORESCENCE

One of the mechanisms responsible for the development of diabetic complications is the increased intracellular formation of AGE [21,22]. AGE are the product of a series of non-enzymatic reactions between reducing sugars and amino groups of proteins, lipids, or nucleic acids. AGE can be formed in the body in various ways [23], and the most common way of AGE production is called the Maillard reaction. During this reaction, early glycation products, Schiff bases, are first formed, from which Amadori products (intermediate glycation products) are formed. Subsequently, Amadori products are undergoing oxidation, dehydration, and crosslinking, resulting in the formation of compounds that demonstrate a significantly increased reactivity toward arginine and lysine residues on proteins, which leads to the formation of AGE [24].

Previous studies have shown that AGE production in the extracellular matrix is much greater with increasing age and in diabetic patients compared to healthy individuals [25]. An increased production of AGE precursors has various negative effects on the patient's body: it affects the changes in the functions of intracellular proteins, altered under the influence of AGE; changes the nature of the interaction of extracellular matrix proteins exposed to AGE with each other and with matrix receptors; causes the binding of altered plasma proteins to the receptors of AGE, which in turn leads to an increase in the production of reactive oxygen species, causing the development of numerous pathologies [23].

Evidence that collagen-linked fluorescence may be associated with the development of micro- and macrovascular complications in diabetes was first published in 1986 by Monnier et al. [26]. The authors found that when collagen was incubated with glucose under physiological conditions, after a prolonged time it was possible to register fluorescence excited by UV light. Later, it was shown that this parameter increases significantly with aging and diabetes [27]. The first studies to assess the accumulation of AGE using fluorescence were conducted on blood and urine samples and tissue biopsies.

Complicated diabetes leads to an even greater increase in the fluorescence of cross-linked proteins. The study by Yanagisawa et al. showed a significant increase in this parameter in patients with complications of diabetes, particularly kidney disease, compared to those without complications [28].

For many years, studies on the accumulation of AGE in patients with DM have been performed on samples of blood, urine, or biopsy material. However, such studies have a number of drawbacks, as it has been shown that AGE levels in urine and blood samples do not necessarily reflect actual

AGE concentrations in patients' tissues [29–31], and evaluation of skin biopsy specimens is invasive and painful. Invasive and painful procedure of biopsy sampling for subsequent AGE fluorescence analysis prevents wider use of this method for screening of prediabetic conditions and complications of DM. Also, some studies have shown the relationship of increased lens and cornea fluorescence with the development of DM complications [32,33]. Such studies are also characterized by some methodological difficulties. In this aspect, the skin is of the greatest interest for research due to its easy accessibility.

The first study on noninvasive assessment of AGE accumulation by skin fluorescence levels was published in 2004 by Meerwaldt et al. [31]. To validate skin fluorescence as a tool capable of assessing AGE accumulation in DM, the authors performed noninvasive measurements in conditionally healthy volunteers and patients with DM, comparing results with AGE concentration in biopsy specimens (of both fluorescent and nonfluorescent AGE) from the same patients. The study showed a high correlation of skin autofluorescence measurement results with collagen-linked fluorescence in biopsy specimens of the same subjects. Figure 9.2 shows the correlation analysis of skin autofluorescence with collagen-linked (a) and pentosidine (b) fluorescence of skin biopsy specimens [31].

Subsequent studies confirmed a significant correlation of AGE fluorescence with long-term glycated hemoglobin (HbA1c) levels [34,35], while no significant relationship with current levels of this parameter was found. The authors suggest that skin fluorescence measurements can provide insight into long-term glycemic control, thus representing the so-called glycemic memory. These findings are particularly important when working with patients whose long-term glycemic control data are not available for whatever reason.

Several prospective studies have also investigated the possibility of using skin autofluorescence as a marker of the development of micro- and macrocirculatory complications of DM. For example, Meerwaldt et al. studied a group of 117 patients with DM (48 – type 1 and 69 – type 2) and 43 healthy controls over a 5-year follow-up period [36]. Gerrits et al. studied 881 patients with type 2 DM to determine the feasibility of skin fluorescence assessment as a marker for the development of diabetic complications [37]. The study lasted 5 years during which patients underwent measurement at least once a year. Both studies confirmed a significant correlation between increased skin autofluorescence parameters and the presence of DM complications, such as neuropathy and diabetic (micro)albuminuria, as well as the development of cardiovascular disease and cardiovascular

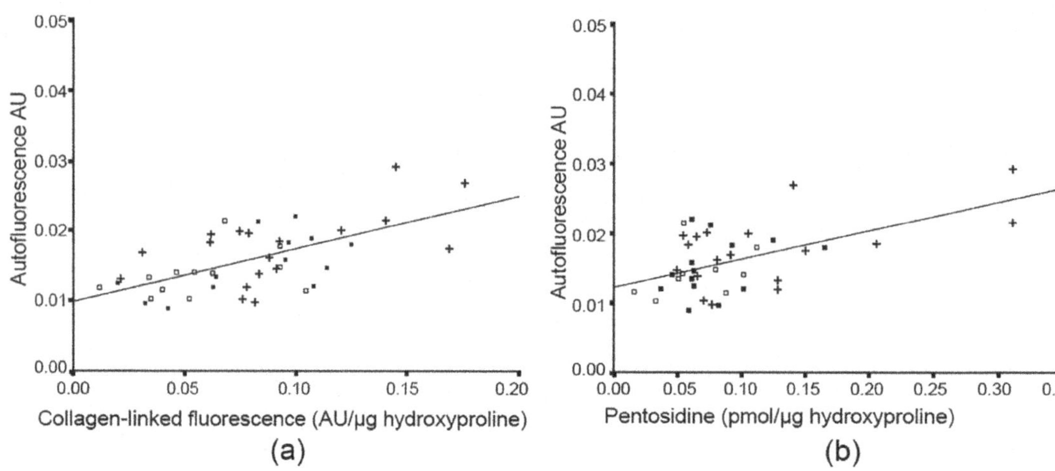

FIGURE 9.2 The relationship between noninvasive skin autofluorescence measurements and (a) skin collagen-linked fluorescence and (b) pentosidine levels in skin biopsies: black squares denote the results from type 1 diabetic patients; plus – results from type 2 diabetic patients; white squares – results from control subjects. (Reproduced from [31] with the permission of Springer Nature.)

mortality. In a further multicenter study, it was confirmed that skin autofluorescence is associated with the risk of micro- and macrovascular complications [38]. For the aforementioned reasons, many studies have indicated that the level of AGE fluorescence can be considered as an independent predicator of microvascular complications, neuropathy, and renal disease.

A review and meta-analysis published in 2011 by Bos et al. based on the example of several studies showed that measurements of skin autofluorescence correlate with the development of one or more complications of diabetes, including all-cause mortality, cardiovascular mortality, micro- and macroangiopathies, nephropathy, neuropathy etc., except for retinopathy [39]. However, the authors point out that caution should be taken in interpreting the results of such studies, as they are very heterogeneous, only a small percentage of the studies included in the analysis were prospective studies, and several studies in the analyzed sample were published by the same research group.

Later Yasuda et al. investigated the correlation between skin autofluorescence and the severity of diabetic retinopathy [40]. They showed a gradual increase in this parameter during the development of retinopathy (from patients without complications to patients with non-proliferative and proliferative forms of retinopathy). At the same time, the HbA1c parameter, traditionally used to assess the severity of diabetic complications, did not show significant changes corresponding to the changing retinopathy severity. The authors conclude that skin fluorescence may be a promising parameter for predicting the development of diabetic retinopathy.

The widespread use and acceptance of AGE content evaluation by measuring skin autofluorescence has led to the development and implementation of devices for such diagnosis in medical practice. The first commercially available device was released by the Dutch company DiagnOptics (Groningen, the Netherlands), called the AGE-Reader (formerly called Autofluorescence Reader, AFR). A photograph of the first modification of the device is shown in Figure 9.3. VeraLight Inc. (Albuquerque, New Mexico) also manufactures the SCOUT DS, a noninvasive skin fluorescence measurement device for type 2 diabetes screening (Albuquerque, New Mexico). Measurements in both devices are suggested to be taken on the volar side of the forearm. The AGE Reader measures skin fluorescence when exposed to a UV light source with a peak at about 350–370nm, while the SCOUT DC excites fluorescence with light-emitting diodes centered at 375, 405, 417, 435, and 456nm [41].

In addition to detecting the presence of complications in patients with DM, assessment of AGE fluorescence in the skin is used to assess risks and predict cardiovascular events, cardiovascular morbidity and mortality, as well as all-cause mortality [36,42,43]. A recent review and meta-analysis indicated that elevated levels of AGE fluorescence could be a predictor of cardiovascular and all-cause mortality in patients with diabetes-related cardiovascular and renal diseases [43]. This parameter was also demonstrated to effectively assess the 5-year risk of

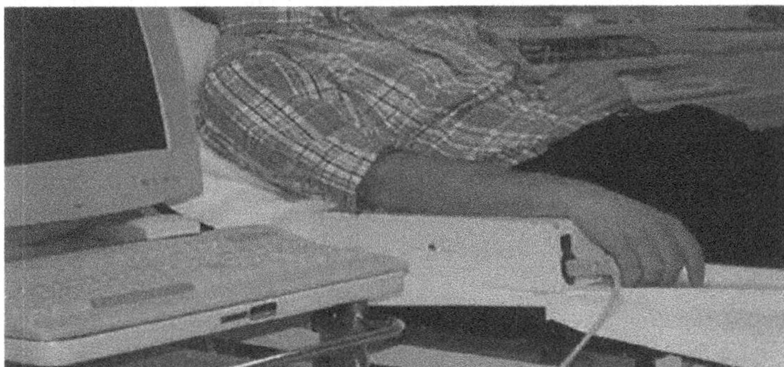

FIGURE 9.3 A photo of the first modification of the AGE reader during the study. (Reproduced from [31] with the permission of Springer Nature.)

amputation in patients with peripheral arterial disease [44]. Another study involving nearly 73,000 subjects showed a significant increase in basal skin fluorescence levels in subjects who had developed DM, cardiovascular disease, or died by the end of the 4-year follow-up period [45]. Current studies assessing AGE accumulation by measuring skin fluorescence are aimed at predicting the risk of diabetes development among healthy population and patients with prediabetes [46,47].

Measuring the fluorescence lifetime of fluorophores of the biological tissues provides additional information for researchers, including in the diagnosis of complications associated with DM. Several studies have shown an increase in the lifetime of AGE-associated fluorescence in the eyes, skin, and dentin of patients with diabetes [48–50]. Using two- and three-term exponential fitting with estimation of parameters related to the fluorescence relaxation time constant as well as to the intensity of the corresponding fluorescence components, a gradual decrease in the average lifetime parameters of the first and second decay components with increasing AGE was recorded after incubation of collagen gel and dentin in ribose [51]. Remarkably, no significant differences in lifetime parameters were found in measurements on the skin of diabetic patients and healthy volunteers [48], whereas an increase in cutaneous fluorescence is indicative of the development of diabetic ulcers [52,53].

9.3 STUDIES OF OTHER INTRINSIC BIOLOGICAL FLUOROPHORES IN RESPECT TO DIABETES

In addition to studies related to fluorescence of collagen and AGE, other fluorophores are also of interest in regard to the clinical picture of diabetic complications. These substances include, in particular, redox-regulated fluorophores, participants in the Krebs cycle NADH and FAD. The redox status of tissues can be determined by assessing the fluorescence ratio of these coenzymes.

Redox studies at the level of cells and biological tissues are often used, in particular, to assess wound healing processes in diabetic and normal conditions. Several studies conducted in rodents have shown the usefulness of such method for assessing the wound healing ability [54,55]. Some studies have analyzed the difference in the AGE/NADH ratio between diabetics and healthy controls [56]. They reported a significant increase in this ratio among the former group. Evaluation of the NADH/FAD ratio may have an important diagnostic value in the diagnosis of tissue oxygen metabolism disorder in DM [57]. However, it is worth noting the difficulty of using such an analysis for *in vivo* assessment of the risks of developing complications at the organ level. Of particular interest is the noninvasive measurement of the fluorescence of NADH and FAD in the skin, hindered by the presence of high concentrations of collagen in the skin tissue, which has its own fluorescence in the same spectral range.

In order to determine the possibility of noninvasive measurement of NADH fluorescence in the skin, preliminary test experiments were conducted with the participation of healthy volunteers in the nail bed area. This area was chosen for reasons of visual control of blood flow parameters and gas exchange zones (perivascular zones) using the video capillaroscopy method.

In this evaluation study, the area of the middle finger nail bed was illuminated by a LED radiation source with a wavelength of 365 nm (power ~ 2–3 mW) and a broadband halogen radiation source HL-2000 (Ocean Optics, USA, 360–2400 nm, ~ 5–7 mW). The high-aperture micro lens with an aperture of 0.12 and a projection long-focus lens formed an image on a monochrome CCD video camera. The fluorescence image was filtered using a long wavelength cut filter. Figure 9.4 shows a schematic representation of the experimental setup.

For a dynamic observation of changes in fluorescence, an occlusion test was chosen as a provocative effect (cuff placement on the brachial artery). Thus, a state of artificial ischemia was created. Depending on oxygen availability, NADH molecules formed in the sixth glycolysis reaction have

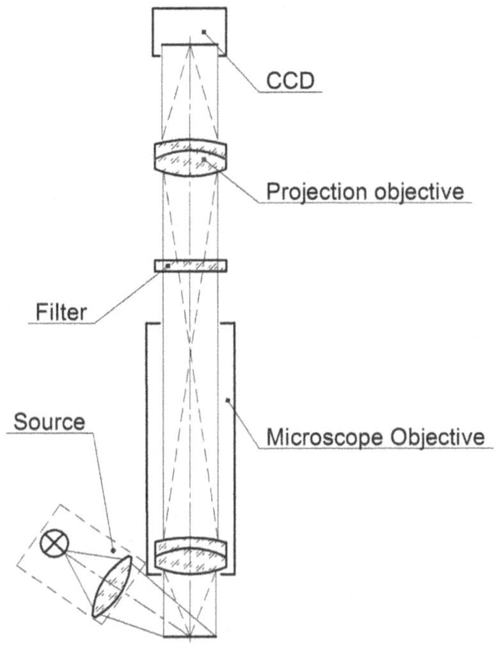

FIGURE 9.4 The experimental setup. (Reproduced from [58] with the permission of SPIE.)

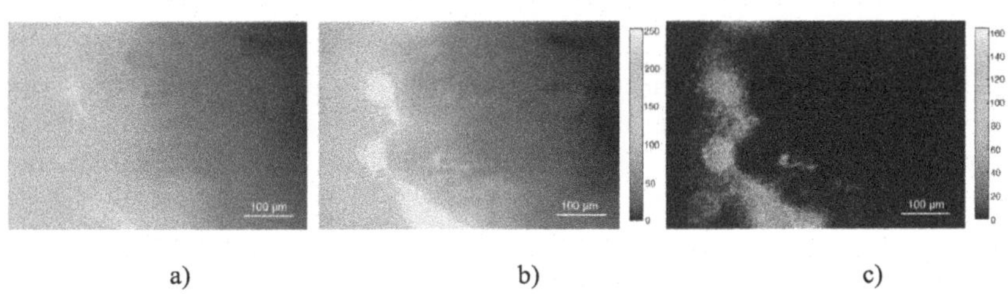

FIGURE 9.5 Fluorescent images before (a) and at the end (b) of occlusion test; difference of two images (c). (Reproduced from [58] with the permission of SPIE.)

two ways of their further transformation: either stay in the cytosol and enter the eleventh glycolysis reaction (anaerobic conditions), or penetrate into the mitochondria and oxidize in the Krebs cycle respiratory chain (aerobic conditions). When hypoxia occurs, the oxidation of NADH in mitochondria slows down, and the glycolysis (anaerobic) pathway of NADH formation is also activated. In this regard, it can be assumed that the recorded level of fluorescence during such an experiment should increase [59].

The study included occlusion with a cuff pressure of 220 mmHg. within 1.5 minutes. A pair of images (fluorescence and diffuse reflection) were recorded before and at the end of the occlusion test. Further stabilization and the imposition of frames were made.

Images of diffuse reflection, as well as for the case with spectral data, were calculated using the formulas presented in Ref. [58]. The fluorescence spectra were corrected by dividing to the images of diffuse reflection. As a result of experimental studies, images of fluorescence of the precapillary zones normalized to diffuse reflection images were obtained (Figure 9.5).

As can be seen from the data obtained, by the end of the occlusion test, an increase in fluorescence occurs. This is especially pronounced in the precapillary zones, which may indicate the accumulation of NADH due to tissue hypoxia. Thus, skin fluorescence measurements can also be used to study the dynamics of changes in NADH concentrations.

9.4 MULTIMODAL OPTICAL APPROACH WITH FLUORESCENCE

One of the promising directions in the development of optical noninvasive diagnostic methods at present is the joint application of several biophotonics methods for the simultaneous and comprehensive assessment of the parameters of the object under the study. Such an approach is called multimodal or multiparametric; it can include different methods of biophotonics, depending on the purpose of the study. This approach has found its application in various fields of medicine, including diabetes [60], minimally invasive surgery [61], and cancer diagnosis [62].

An example of a multimodal approach in the diabetes clinic using fluorescence is the 1996 study by Yu et al. in which an approach was used to simultaneously record both fluorescence and Rayleigh components of light to examine the human lens in aging and diabetes [63]. Rayleigh scattering data were used to normalize fluorescence, as this normalization has previously been shown to provide greater separation between the lenses of control and DM patients [64]. The authors proposed a fluorescence/Rayleigh scattering ratio estimation method to screen the population for latent diabetes.

Another popular combination of methods in multimodal diabetes research is fluorescence and Raman spectroscopy. Raman spectroscopy has previously shown the possibility of estimating nonenzymatic glycation of proteins. Studies evaluating AGE accumulation in human eye tissues associated with aging or DM have been conducted [65,66]. Studies have shown a high ability of the method to detect changes correlated with the accumulation of fluorescent AGEs. In 2015, Shi et al. jointly analyzed AGE fluorescence and Raman bands of skin collagen scaffolds in animal models of diabetes [67]. The study conducted by Paolillo et al. for the first time simultaneously utilized Raman and fluorescence techniques for noninvasive analysis of AGE accumulation in the skin of diabetic patients [68]. In this study, the degradation of type I collagen and increased glycation was shown in diabetics, and changes in the collagen hydration state were noted for diabetic and chronological aging groups, according to Raman spectroscopy data. Thus, the combination of fluorescence and Raman spectroscopy techniques was proposed for noninvasive diabetes diagnostics or diabetes risk prediction.

It seems promising to apply together the methods of fluorescence and microcirculation analysis in diabetes practice, as it will allow assessing both metabolic and vascular disorders in the patient's body. However, to the best of our knowledge, there is a limited amount of research being done in this area. A recent animal study assessing wound healing in diabetes used multimodal optical imaging based on the combined use of multiphoton microscopy, second harmonic generation, optical coherence tomography, angiography, and FLIM [69]. The developed method of multimodal study has proved its effectiveness in assessing the pharmacodynamics of the wound region in different therapy approaches and seems promising for further use in research, since it allows us to assess metabolic and structural changes in the biological tissue, as well as to monitor the process of angiogenesis. The method of laser Doppler flowmetry is also of some interest when used together with fluorescent methods to study metabolic and microcirculatory disorders in diabetes.

9.5 COMBINED USE OF FLUORESCENCE SPECTROSCOPY AND LASER DOPPLER FLOWMETRY

To determine the prospects for the joint application of fluorescence spectroscopy and laser Doppler flowmetry methods in identifying the presence and severity of diabetic complications, experimental studies were carried out. The method of laser Doppler flowmetry allows noninvasive diagnostics of the state of blood microcirculation and has been used for many years in various fields of medicine. The choice of this method as an adjunct to fluorescence in this study was due to the fact that microcirculation disorders in diabetes are closely associated with metabolic disorders.

The experimental study involved 76 patients with type 2 DM. Patients were recruited at the Endocrinology Department of the Orel Regional Clinical Hospital (Orel, Russia). Fourteen people from the patients' group were considered to have a more severe course of the disease (diabetic group 2), and the remaining 62 patients were considered to have a normal course of the disease

TABLE 9.1
The Main Characteristics of the Groups under Study

Characteristics	Diabetic Group 1	Diabetic Group 2	Control Group	*p* value
Sex, M/F	20/42	8/6	30/18	0.08
Age, y	54±10	53±13	46±6	0.89
Body mass index, kg/m²	31.9±6.3	32.0±6.2	128±9	0.29
Diabetes duration, y	11±7	18±12	80±5	0.12
Fasting glucose (mmol/l)	9.3±4.8	9.1±2.5	24.2±2.5	<0.001[a]
HbA1c, %	8.6±1.5	8.6±0.7	4.8±0.4	<0.001[a]
HbA1c, mmol/mol	70±9	70±4		0.03[b]
Total cholesterol, mmol/l	4.9±1.2	5.1±1.3		0.28[b]
Creatinine, umol/l	81.5±28.7	90.3±16.9		0.31[b]
Urea, mmol/l	7.7±6.8	8.2±2.3		0.76[b]
ALT, IU/L	34.2±20.3	28.4±19.3		0.02[b]
AST, IU/L	31.0±24.8	24.1±13.4		0.09[b]
Systolic BP, mmHg	136±16	144±22		0.25[b]
Diastolic BP, mmHg	83±8	85±8		0.13[b]

Reference values of the laboratory: HbA1c 4.0%–6.0%, total cholesterol 3.5–5.0 mmol/l, urea 2.5–7.5 mmol/l, creatinine 70-110 μmol/l, ALT 10-38 IU/L, AST 10-40 IU/L.

[a] Control group versus diabetic group 1 and diabetic group 2.

[b] Diabetic group 1 versus diabetic group 2. Nonparametric data of the three studied groups were analyzed by a Kruskal–Wallis analysis of variance (ANOVA). A Mann–Whitney U-test was used to identify differences between the two groups.

(diabetic group 1). The decision on the degree of severity in each case was taken individually. This was based on a combination of elevated laboratory parameters, presence of trophic disorders in the form of ulcers, and in consultation with the attending physician. The control group consisted of 48 conditionally healthy volunteers who were recruited through internal university advertising. Laboratory parameters were measured according to standard laboratory procedures. The parameters of the groups under the study are presented in Table 9.1.

The study included the skin fluorescence spectra recording upon excitation with 365- and 450-nm light, as well as recording the LDF signal under thermoneutral conditions and during a series of temperature tests with temperature modes of 25°C, 35°C, and 42°C, respectively. A multi-optical fiber probe was used for delivery of probing radiation and registration of back-reflected secondary radiation from the tissue. The probe was secured on the dorsal surface of the foot to a point located on a plateau between the 1st and 2nd metatarsal bones. Figure 9.6 shows the location of the optical sensor on the patient's foot. To register these parameters, the "LAZMA-D" (SPE "LAZMA" Ltd, Russia) system was used, which consists of a "LAZMA-MC" multichannel laser analyzer and a "LAZMA-TEST" unit for providing functional tests.

After registration of the signals, the fluorescence intensity was estimated, normalized to the intensity of back-reflected radiation (AF_{460} and AF_{525} – for the excitation by the 365- and 450-nm light source, respectively); according to the LDF signal, the main recorded parameter (I_m – index of microcirculation) was analyzed at the stages of heating up to 35°C and 42°C.

Experimental studies have shown that diabetic patients have elevated values of normalized fluorescence amplitudes (AF_{460} and AF_{525}). At the same time, the values of the I_m were significantly reduced in patients at both stages of heating. The results of the study revealed not only the differences in the parameter values between patients with diagnosed diabetes and healthy controls, but also showed a gradual deterioration of the parameters with the progression of complications of the

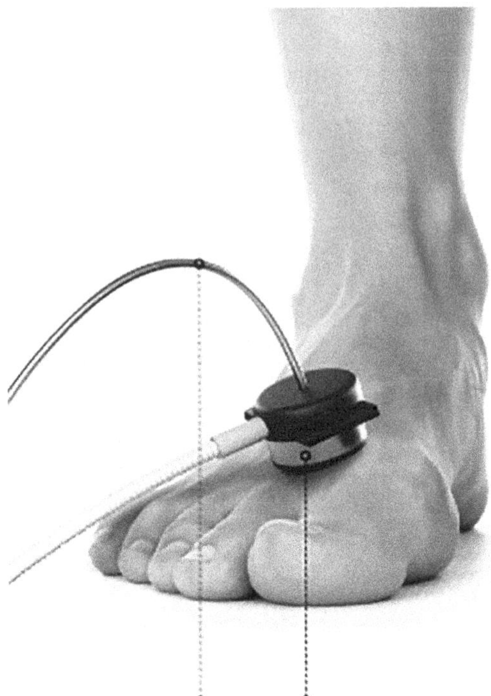

FIGURE 9.6 Location of the fiber optic probe (1) with Peltier thermal-pad (2) during the experimental study. (Reproduced from [53] with the permission of Frontiers Media.)

disease. The results of the analysis of fluorescence intensity and LDF parameters during the study are presented in Figure 9.7.

The results obtained in the study can be attributed to a number of reasons. As mentioned earlier, it has been established that long-existing diabetes with hyperglycemia results in an increase in protein glycation levels, which eventually leads to the accumulation of AGE [70,71]. AGEs can induce cross-linking of collagen, which results in the increased collagen-linked fluorescence. It should be noted that the level of skin fluorescence in the present study corresponded to the level of the conventional parameter for assessing the level of glycemia and diabetic complications HbA1c determined *in vitro*. It is important to note, however, that the standard measure of glycation using HbA1c characterizes the glycation processes that occur only in the short term (around 3 months), while the assessment of the AGE accumulation allows us to assess the so-called metabolic memory, characterizing the processes occurring in tissues over a longer period of time.

Diabetes also causes tissue hypoxia, leading to a violation of the cells' aerobic respiration pathway [72–74]. In this case, the mitochondrial oxidation of NADH slows down, while the glycolysis pathway of NADH formation activates. Therefore, NADH accumulation may be considered as a sign of tissue hypoxia, and its contribution to the total fluorescence signal is a marker of general oxygen deficiency in tissues.

The reduced increase in perfusion during heating can also be explained by a number of reasons. In particular, it is known that in diabetes, long-term accumulation of AGE in tissues causes vascular stiffening as well as activation of inflammatory pathways in vascular endothelial cells [22,75], which can lead to the development of endothelial dysfunction.

When analyzing the data obtained during the study, the registered values of fluorescence and microcirculation index were used for the synthesis of the classification rule. These parameters

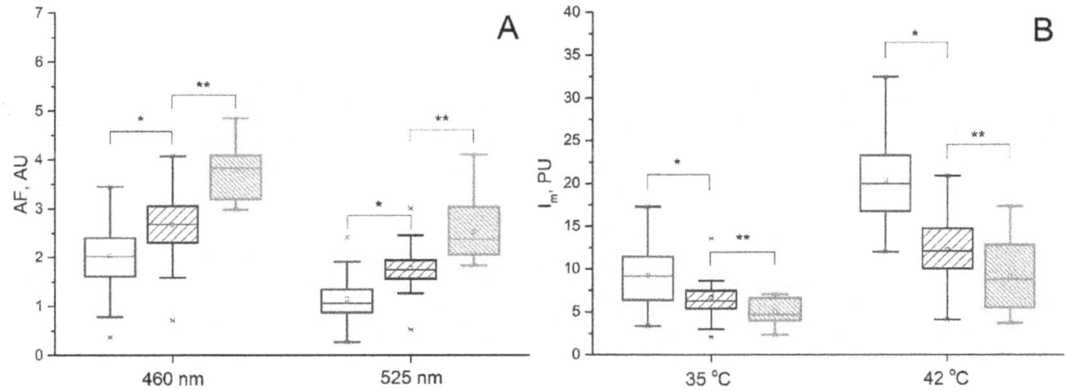

FIGURE 9.7 Comparison of the normalized fluorescence amplitude (a) and index of microcirculation (b) among control (empty bars), diabetic (loose shading), and diabetic with ulcers (tight shading) groups. In each box, the central line is the median of the group, while the edges are the 25th and 75th percentiles. (* – Confirmed statistically significant differences between control group and diabetic group 1 ($p < 0.01$). ** – Confirmed statistically significant differences between diabetic group 1 and diabetic group 2 ($p < 0.01$). Reproduced from [52] with the permission of SPIE.)

TABLE 9.2
Sensitivity and Specificity for the First Classification Rule

Parameter	AF_{460}	AF_{525}	$I_m^{35°C}$	$I_m^{42°C}$	AF_{460}, $I_m^{35°C}$	AF_{460}, $I_m^{42°C}$	AF_{525}, $I_m^{35°C}$	AF_{525}, $I_m^{42°C}$
Sensitivity	0.64	0.77	0.77	0.88	0.75	0.92	0.87	0.90
Specificity	0.7	0.79	0.62	0.81	0.77	0.90	0.77	0.83

satisfy the principles of statistical independence, as well as the significance of the differences of their values, calculated for the patients' and control groups, as well as for groups of patients with different severity of the disease. The discriminant function is synthesized in such a way as to provide high sensitivity while providing good specificity. Table 9.2 summarizes the sensitivity and specificity for the first classification rule (the control group is compared with the diabetic group 1) for a different combination of measured parameters. As can be seen from the table, the lowest level of error is obtained with the combination of fluorescence intensity at an excitation wavelength of 365 nm and the level of stimulation of microcirculatory perfusion at 42°C.

For the second classification rule (the diabetic group 1 is compared with the diabetic group 2), a sensitivity and specificity of 0.86 and 0.85, respectively, were obtained.

Figure 9.8a demonstrates the scatter plot of the experimental data with applied discriminant lines for the better combination of sensitivity and specificity. The immediate diagnostic criterion is the classification model as discriminant functions (D1 and D2) that allow to relate the newly emerging object to one of the three groups:

$$\begin{cases} D1 = -2.55 - 0.45AF_{460} + 0.23I_m^{42°C}, \\ D2 = -1.8 + 1.16AF_{460} - 0.18I_m^{42°C} \end{cases} . \quad (9.1)$$

It can be seen from Figure 9.8a that the shift toward the top-left characterizes the deterioration of the patient's condition and an increased risk of foot ulcers development.

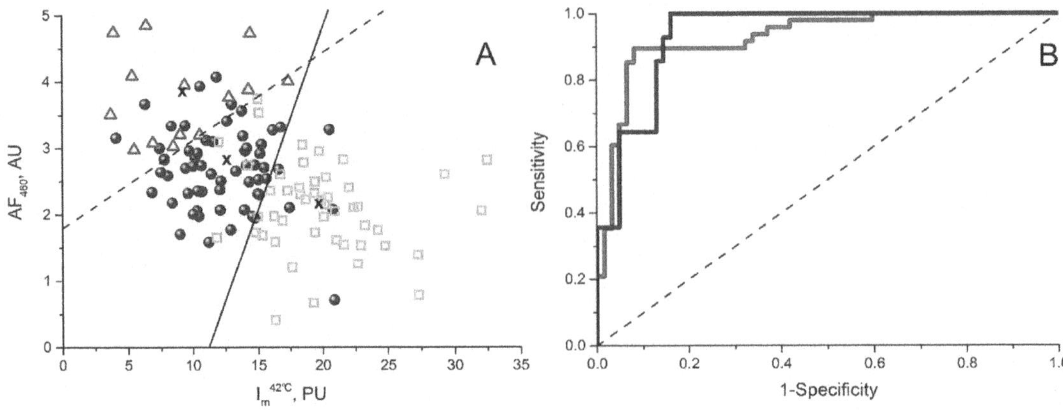

FIGURE 9.8 The scatter plot with applied discriminant lines, obtained by linear discriminant analysis method (a) and ROC curves for assessing the effectiveness of the classifiers (b). A healthy group is shown by squares, diabetic group – circles, diabetic group with ulcers – triangles. (Reproduced from [52] with the permission of SPIE.)

In Figure 9.8b, the ROC curves calculated for the obtained discriminant functions are demonstrated. To compare the quality of different classifying rules, the integral characteristic AUC – area under curve – was used. In this particular case, AUC=0.93 for both classification rules. This indicates a high level of the classification rule quality.

Based on the above observations, we can conclude that the skin fluorescence and level of tissue blood perfusion during a heating test measured together can act as markers for various stages of diabetes, beginning with early disease, to the development of secondary complications.

9.6 SUMMARY

This chapter reviews and analyzes the prospects for the use of fluorescence techniques in the diagnosis and assessment of the severity of diabetic complications. Methods of fluorescence spectroscopy and imaging are considered to be promising modern methods of optical diagnostics. Noninvasive assessment of fluorescence of biological tissues allows us to draw conclusions about structural and metabolic changes that occur in various pathological processes. Fluorescence spectroscopy and imaging are widely used in various medical fields, including oncology and skin diagnostics.

The vast majority of the studies in the field of DM is conducted in the field of noninvasive assessment of accumulation of AGE in the skin of patients with diabetes. These studies demonstrate a high correlation of increased cutaneous autofluorescence due to AGE accumulation with the development of end-organ complications. The success of research in this area is also confirmed by innovative developments, including the development of commercial AGE readers.

Studies on the noninvasive assessment of other biotissue fluorophores contributing to the diagnosis of disorders occurring in DM, in particular the Krebs cycle coenzymes NADH and FAD, were also reviewed. Finally, the advantages of using fluorescence methods in multimodal optical noninvasive studies in various combinations with other biophotonics methods are evaluated. The results of joint application of fluorescence with laser Doppler flowmetry to improve diagnosis of the presence and severity of microcirculatory and metabolic disorders in DM are presented.

ACKNOWLEDGMENTS

The authors acknowledge the support of the grant of the Russian Science Foundation under project No. 22-75-10088 and the grant of the President of the Russian Federation for state support of young Russian scientists No. MK-398.2021.4. E.Z. acknowledges the support of the Finnish National Agency for Education (EDUFI, fellowship TM-20-11476).

REFERENCES

1. S. Andersson-Engels, J. Johansson, K. Svanberg, and S. Svanberg, "Fluorescence imaging and point measurements of tissue: Applications to the demarcation of malignant tumors and atherosclerotic lesions from normal tissue," *Photochem. Photobiol.* **53**(6), 807–814 (1991).
2. E. Fujimori, "Cross-linking and fluorescence changes of collagen by glycation and oxidation," *Biochim. Biophys. Acta - Protein Struct. Mol. Enzymol.* **998**(2), 105–110 (1989).
3. H. Schneckenburger, M. H. Gschwend, R. J. Paul, et al., "Time-gated spectroscopy of intrinsic fluorophores in cells and tissues," *Proc. SPIE* **2324**, 187–195 (1995).
4. T. Galeotti, G. D. V. van Rossum, D. H. Mayer, and B. Chance, "On the fluorescence of NAD(P)H in whole-cell preparations of tumours and normal tissues," *Eur. J. Biochem.* **17**(3), 485–496 (1970).
5. J.-M. Salmon, E. Kohen, P. Viallet, et al., "Microspectrofluorometric approach to the study of free/bound nad(p)h ratio as metabolic indicator in various cell types," *Photochem. Photobiol.* **36**(5), 585–593 (1982).
6. P. Galland and H. Senger, "New trends in photobiology the role of flavins as photoreceptors," *J. Photochem. Photobiol. B Biol.* **1**(3), 277–294 (1988).
7. R. Richards-Kortum and E. Sevick-Muraca, "Quantitative optical spectroscopy for tissue diagnosis," *Annu. Rev. Phys. Chem.* **47**(1), 555–606 (1996).
8. G. A. Wagnieres, W. M. Star, and B. C. Wilson, "In vivo fluorescence spectroscopy and imaging for oncological applications," *Photochem. Photobiol.* **68**(5), 603–632 (1998).
9. J. R. Lakowicz, H. Szmacinski, K. Nowaczyk, and M. L. Johnson, "Fluorescence lifetime imaging of free and protein-bound NADH," *Proc. Natl. Acad. Sci.* **89**(4), 1271 (1992).
10. T. S. Blacker, Z. F. Mann, J. E. Gale, et al., "Separating NADH and NADPH fluorescence in live cells and tissues using FLIM," *Nat. Commun.* **5**(1), 3936 (2014).
11. L. Tiwari, O. Kujan, and C. S. Farah, "Optical fluorescence imaging in oral cancer and potentially malignant disorders: A systematic review," *Oral Dis.* **26**(3), 491–510 (2020).
12. S. Palmer, K. Litvinova, E. U. Rafailov, and G. Nabi, "Detection of urinary bladder cancer cells using redox ratio and double excitation wavelengths autofluorescence," *Biomed. Opt. Express* **6**(3), 977–986 (2015).
13. V. Dremin, E. Potapova, E. Zherebtsov, et al., "Optical percutaneous needle biopsy of the liver: a pilot animal and clinical study," *Sci. Rep.* **10**(1) (2020).
14. E. Zharkikh, V. Dremin, E. Zherebtsov, A. Dunaev, and I. Meglinski, "Biophotonics methods for functional monitoring of complications of diabetes mellitus," *J. Biophotonics* **13**(10), e202000203 (2020).
15. "International Diabetes Federation. IDF Diabetes Atlas, 9th edn.," 2019, <https://www.diabetesatlas.org> (accessed 8 April 2021).
16. H. D. Nickerson and S. Dutta, "Diabetic complications: Current challenges and opportunities," *J. Cardiovasc. Transl. Res.* **5**(4), 375–379 (2012).
17. J. M. Forbes and M. E. Cooper, "Mechanisms of diabetic complications," *Physiol. Rev.* **93**(1), 137–188 (2013).
18. D. J. Wexler, R. W. Grant, E. Wittenberg, et al., "Correlates of health-related quality of life in type 2 diabetes," *Diabetologia* **49**(7), 1489–1497 (2006).
19. A. M. Jacobson, B. H. Braffett, P. A. Cleary, et al., "The long-term effects of type 1 diabetes treatment and complications on health-related quality of life: A 23-year follow-up of the diabetes control and complications/epidemiology of diabetes interventions and complications cohort," *Diabetes Care* **36**(10), 3131–3138, Am Diabetes Assoc (2013).
20. J. C. Schramm, T. Dinh, and A. Veves, "Microvascular changes in the diabetic foot," *Int. J. Low. Extrem. Wounds* **5**(3), 149–159 (2006).
21. M. Brownlee, "The pathobiology of diabetic complications," *Diabetes* **54**(6), 1615–1625 (2005).

22. G. Ferdinando, B. Michael, and S. A. Marie, "Oxidative stress and diabetic complications," *Circ. Res.* **107**(9), 1058–1070 (2010).
23. R. Singh, A. Barden, T. Mori, and L. Beilin, "Advanced glycation end-products: A review," *Diabetologia* **44**(2), 129–146 (2001).
24. N. Ahmed, "Advanced glycation endproducts—role in pathology of diabetic complications," *Diabetes Res. Clin. Pract.* **67**(1), 3–21 (2005).
25. A. W. Stitt, J. E. Moore, J. A. Sharkey, et al., "Advanced glycation end products in vitreous: Structural and functional implications for diabetic vitreopathy," *Investig. Ophthalmol. Vis. Sci.* **39**(13), 2517–2523 (1998).
26. V. M. Monnier, V. Vishwanath, K. E. Frank, et al., "Relation between complications of type I diabetes mellitus and collagen-linked fluorescence," *N. Engl. J. Med.* **314**(7), 403–408 (1986).
27. M. H. Dominiczak, J. Bell, N. H. Cox, et al., "Increased collagen-linked fluorescence in skin of young patients with type I diabetes mellitus," *Diabetes Care* **13**(5), 468 LP–472 (1990).
28. K. Yanagisawa, Z. Makita, K. Shiroshita, et al., "Specific fluorescence assay for advanced glycation end products in blood and urine of diabetic patients," *Metabolism* **47**(11), 1348–1353 (1998).
29. D. E. Hricik, Y. C. Wu, A. Schulak, and M. A. Friedlander, "Disparate changes in plasma and tissue pentosidine levels after kidney and kidney-pancreas transplantation," *Clin. Transplant.* **10**(6 Pt 1), 568–573 (1996).
30. C. A. Dorrian, S. Cathcart, J. Clausen, D. Shapiro, and M. H. Dominiczak, "Factors in human serum interfere with the measurement of advanced glycation endproducts," *Cell. Mol. Biol. (Noisy-le-grand).* **44**(7), 1069–1079 (1998).
31. R. Meerwaldt, R. Graaff, P. H. N. Oomen, et al., "Simple non-invasive assessment of advanced glycation endproduct accumulation," *Diabetologia* **47**(7), 1324–1330 (2004).
32. T. Abiko, A. Abiko, S. Ishiko, et al., "Relationship between autofluorescence and advanced glycation end products in diabetic lenses," *Exp. Eye Res.* **68**(3), 361–366 (1999).
33. E. Sato, F. Mori, S. Igarashi, et al., "Corneal advanced glycation end products increase in patients with proliferative diabetic retinopathy," *Diabetes Care* **24**(3), 479 LP–482 (2001).
34. V. R. Aroda, B. N. Conway, S. J. Fernandez, et al., "Cross-sectional evaluation of noninvasively detected skin intrinsic fluorescence and mean hemoglobin A1C in type 1 diabetes," *Diabetes Technol. Ther.* **15**(2), 117–123 (2013).
35. E. Sugisawa, J. Miura, Y. Iwamoto, and Y. Uchigata, "Skin autofluorescence reflects integration of past long-term glycemic control in patients with type 1 diabetes," *Diabetes Care* **36**(8), 2339–2345 (2013).
36. R. Meerwaldt, H. L. Lutgers, T. P. Links, et al., "Skin autofluorescence is a strong predictor of cardiac mortality in diabetes," *Diabetes Care* **30**(1), 107–112 (2007).
37. E. G. Gerrits, H. L. Lutgers, N. Kleefstra, et al., "Skin autofluorescence," *Diabetes Care* **31**(3), 517–521 (2008).
38. M. J. Noordzij, D. J. Mulder, P. H. N. Oomen, et al., "Skin autofluorescence and risk of micro- and macrovascular complications in patients with Type 2 diabetes mellitus—a multi-centre study," *Diabet. Med.* **29**(12), 1556–1561 (2012).
39. D. C. Bos, W. L. de Ranitz-Greven, and H. W. de Valk, "Advanced glycation end products, measured as skin autofluorescence and diabetes complications: A systematic review," *Diabetes Technol. Ther.* **13**(7), 773–779 (2011).
40. M. Yasuda, M. Shimura, H. Kunikata, et al., "Relationship of skin autofluorescence to severity of retinopathy in type 2 diabetes," *Curr. Eye Res.* **40**(3), 338–345 (2015).
41. B. P. Olson, N. I. Matter, M. N. Ediger, E. L. Hull, and J. D. Maynard, "Noninvasive skin fluorescence spectroscopy is comparable to hemoglobin A1c and fasting plasma glucose for detection of abnormal glucose tolerance," *J. Diabetes Sci. Technol.* **7**(4), 990–1000 (2013).
42. H. L. Lutgers, E. G. Gerrits, R. Graaff, et al., "Skin autofluorescence provides additional information to the UK Prospective Diabetes Study (UKPDS) risk score for the estimation of cardiovascular prognosis in type 2 diabetes mellitus," *Diabetologia* **52**(5), 789 (2009).
43. C. Ivan, S. Alba, Á. Celia, et al., "Skin autofluorescence–indicated advanced glycation end products as predictors of cardiovascular and all-cause mortality in high-risk subjects: A systematic review and meta-analysis," *J. Am. Heart Assoc.* **7**(18), e009833 (2018).
44. L. C. de Vos, D. J. Mulder, A. J. Smit, et al., "Skin autofluorescence is associated with 5-year mortality and cardiovascular events in patients with peripheral artery disease," *Arterioscler. Thromb. Vasc. Biol.* **34**(4), 933–938 (2014).
45. R. P. van Waateringe, B. T. Fokkens, S. N. Slagter, et al., "Skin autofluorescence predicts incident type 2 diabetes, cardiovascular disease and mortality in the general population," *Diabetologia* **62**(2), 269–280 (2019).

46. A. Kakar, S. Tripathi, A. Gogia, N. Gupta, and S. P. Byotra, "Stratification of risk groups for developing diabetes among healthy nondiabetic population using skin autofluoroscence spectroscopic screening," *Curr. Med. Res. Pract.* **10**(2), 41–43 (2020).
47. L. La Sala, E. Tagliabue, P. de Candia, F. Prattichizzo, and A. Ceriello, "One-hour plasma glucose combined with skin autofluorescence identifies subjects with pre-diabetes: the DIAPASON study," *BMJ Open Diabetes Res. Care* **8**(1), e001331 (2020).
48. J. Blackwell, K. M. Katika, L. Pilon, et al., "In vivo time-resolved autofluorescence measurements to test for glycation of human skin," *J. Biomed. Opt.* **13**(1), 1–15 (2008).
49. D. Schweitzer, L. Deutsch, M. Klemm, et al., "Fluorescence lifetime imaging ophthalmoscopy in type 2 diabetic patients who have no signs of diabetic retinopathy," *J. Biomed. Opt.* **20**(6), 1–13 (2015).
50. J. Schmidt, S. Peters, L. Sauer, et al., "Fundus autofluorescence lifetimes are increased in non-proliferative diabetic retinopathy," *Acta Ophthalmol.* **95**(1), 33–40 (2017).
51. S. Fukushima, M. Shimizu, J. Miura, et al., "Decrease in fluorescence lifetime by glycation of collagen and its application in determining advanced glycation end-products in human dentin," *Biomed. Opt. Express* **6**(5), 1844–1856 (2015).
52. V. V Dremin, E. A. Zherebtsov, V. V Sidorov, et al., "Multimodal optical measurement for study of lower limb tissue viability in patients with diabetes mellitus," *J. Biomed. Opt.* **22**(8), 085003 (2017).
53. A. I. Zherebtsova, V. V Dremin, I. N. Makovik, et al., "Multimodal optical diagnostics of the microhaemodynamics in upper and lower limbs," *Front. Physiol.* **10**, 416 (2019).
54. K. P. Quinn, E. C. Leal, A. Tellechea, et al., "Diabetic wounds exhibit distinct microstructural and metabolic heterogeneity through label-free multiphoton microscopy," *J. Invest. Dermatol.* **136**(1), 342–344 (2016).
55. S. Mehrvar, K. T. Rymut, F. H. Foomani, et al., "Fluorescence imaging of mitochondrial redox state to assess diabetic wounds," *IEEE J. Transl. Eng. Heal. Med.* **7**, 1–9 (2019).
56. V. V Dremin, V. V Sidorov, A. I. Krupatkin, et al., "The blood perfusion and NADH/FAD content combined analysis in patients with diabetes foot," *Proc. SPIE* **9698**, 969810 (2016).
57. D. M. Ciobanu, L. E. Olar, R. Stefan, et al., "Fluorophores advanced glycation end products (AGEs)-to-NADH ratio is predictor for diabetic chronic kidney and cardiovascular disease," *J. Diabetes Complications* **29**(7), 893–897 (2015).
58. V. V Dremin, N. B. Margaryants, M. V Volkov, et al., "Assessment of tissue ischemia of nail fold precapillary zones using a fluorescence capillaroscopy," *Proc. SPIE* **10412**, 104120W (2017).
59. A. Mayevsky and B. Chance, "Oxidation-reduction states of NADH in vivo: From animals to clinical use," *Mitochondrion* **7**(5), 330–339 (2007).
60. O. A. Smolyanskaya, E. N. Lazareva, S. S. Nalegaev, et al., "Multimodal optical diagnostics of glycated biological tissues," *Biochem.* **84**(1), 124–143 (2019).
61. E. Potapova, V. Dremin, E. Zherebtsov, A. Mamoshin, and A. Dunaev, "Multimodal optical diagnostic in minimally invasive surgery," in *Multimodal Optical Diagnostics of Cancer*, p. 597, Springer International Publishing, Cham (2020).
62. V. V Tuchin, J. Popp, and V. Zakharov, *Multimodal Optical Diagnostics of Cancer*, Springer, Cham (2020).
63. N.-T. Yu, B. S. Krantz, J. A. Eppstein, et al., "Development of a noninvasive diabetes screening device using the ratio of fluorescence to Rayleigh scattered light," *J. Biomed. Opt.* **1**(3), 280–288 (1996).
64. M. A. Samuels, S. W. Patterson, J. A. Eppstein, T. Y. Nai, and S.-E. Bursell, "Apparatus and methods for quantitatively measuring molecular changes in the ocular lens," p. U.S. Patent No. 5,203,328, U.S. Patent and Trademark Office, Washington, DC (1993).
65. J. V Glenn, J. R. Beattie, L. Barrett, et al., "Confocal Raman microscopy can quantify advanced glycation end product (AGE) modifications in Bruch's membrane leading to accurate, nondestructive prediction of ocular aging," *FASEB J.* **21**(13), 3542–3552 (2007).
66. J. Sebag, S. Nie, K. Reiser, M. A. Charles, and N. T. Yu, "Raman spectroscopy of human vitreous in proliferative diabetic retinopathy," *Invest. Ophthalmol. Vis. Sci.* **35**(7), 2976–2980 (1994).
67. P. Shi, H. Liu, X. Deng, et al., "Label-free nonenzymatic glycation monitoring of collagen scaffolds in type 2 diabetic mice by confocal Raman microspectroscopy," *J. Biomed. Opt.* **20**(2), 1–9 (2015).
68. F. R. Paolillo, V. S. Mattos, A. O. de Oliveira, et al., "Noninvasive assessments of skin glycated proteins by fluorescence and Raman techniques in diabetics and nondiabetics," *J. Biophotonics* **12**(1), e201800162 (2019).
69. J. Rico-Jimenez, J. H. Lee, A. Alex, et al., "Non-invasive monitoring of pharmacodynamics during the skin wound healing process using multimodal optical microscopy," *BMJ Open Diabetes Res. Care* **8**(1), e000974 (2020).

70. K. C. B. Tan, W.-S. Chow, V. H. G. Ai, et al., "Advanced glycation end products and endothelial dysfunction in type 2 diabetes," *Diabetes Care* **25**(6), 1055–1059 (2002).
71. P. Gkogkolou and M. Böhm, "Advanced glycation end products," *Dermatoendocrinol.* **4**(3), 259–270 (2012).
72. J. Ditzel and E. Standl, "The problem of tissue oxygenation in diabetes mellitus: Its relation to the early functional changes in the microcirculation of diabetic subjects," *Acta Med. Scand.* **197**(S578), 49–58, Wiley Online Library (1975).
73. H. Thangarajah, D. Yao, E. I. Chang, et al., "The molecular basis for impaired hypoxia-induced VEGF expression in diabetic tissues," *Proc. Natl. Acad. Sci.* **106**(32), 13505–13510 (2009).
74. L. Xi, C.-M. Chow, and X. Kong, "Role of tissue and systemic hypoxia in obesity and type 2 diabetes," Hindawi (2016).
75. R. Medzhitov, "Origin and physiological roles of inflammation," *Nature* **454**(7203), 428–435 (2008).

10 Terahertz Time-Domain Spectroscopy in the Assessment of Diabetic Complications

Sviatoslav I. Gusev, Ravshanjon Kh. Nazarov, Petr S. Demchenko, Tianmiao Zhang, Olga P. Cherkasova, and Mikhail K. Khodzitsky

CONTENTS

10.1 INTRODUCTION

Despite the fact that diabetes is one of the most common diseases, its effective diagnosis and therapy is an urgent task. In this regard, over the past decades, noninvasive methods have been developed [1,2]: reverse iontophoresis [3], infrared spectroscopy [4], Raman spectroscopy [5], impedance spectroscopy, etc.; however, the department for sanitary supervision of quality Food and Drug Administration approved only the GlucoWatch device, which works based on the phenomenon of reverse iontophoresis. The main obstacle to the development of a noninvasive glucometer is a low signal level from glucose molecules, which leads to low sensitivity and specificity in measuring sucrose content [6]. Today, there are various diagnostic methodologies, but one of the most promising of them is the optical diagnostic method. The composition and structure of blood are important indicators of diseases, including diabetes. For example, with an increase in glucose level, creatinine level can change simultaneously. A deviation of glucose from the normal level is considered as a symptom of impaired carbohydrate metabolism. In addition, one of the complications of diabetes is diabetic nephropathy. It can cause an increase in creatinine levels. An increase in bilirubin level is less common in patients with diabetes mellitus, but it may occur with concomitant liver disease [7]. The bilirubin level varies in a narrower range in patients that are not affected by liver disease than in patients with organ pathology. However, measuring the level of bilirubin, as well as other biochemical parameters, after a meal is difficult for current research methods used in routine clinical

DOI: 10.1201/9781003112099-10

practice [7]. This is due to the fats that are obtained from food. Fats may cause an increase in the level of triglycerides in the blood, which leads to the errors in the analysis of blood composition [8].

Terahertz spectroscopy, which is rapidly developing, opens new possibilities in various noninvasive methods of diagnosing diseases [4–6]. The THz frequency range is noteworthy in biomedicine because it covers the resonance frequencies of each vibration and rotation mode of biomolecules. By using THz radiation, we can evaluate and determine the state of biological molecular bonds [9–13]. In addition, THz radiation is very sensitive to different types of H_2O molecule conformations with other molecules contained in biological tissues [11–13]. Thus, the states of water molecules, i.e., free water and bound water (including the hydration shell), can be distinguished. For example, six H_2O molecules are bound to one glucose molecule [14], and a transport protein albumin binds about 300 water molecules [15]. In addition, water molecules constantly change their conformations of free water and bound water [16]. The changes in the proportions of free and bound water, as well as the changes in relaxation time of each state, are represented in the spectral characteristics and optical properties of biological samples. Also, the concentration of substances in biological fluids affects the conformations of water molecules, and correspondingly, the optical properties of these liquids are also influenced [17]. THz spectroscopy makes it possible to characterize H_2O molecules in detail: determine their relaxation times, understand conformation exchange, and measure what happens during the interaction between proteins and other substrates in solutions [9–12]. The absorption spectra of polar liquids have wide spectral features at frequencies below 3.0 THz [11]; therefore, all changes in the solution will be expressed in the dispersions of its refractive index and absorption coefficient. This feature gives a great promise that THz radiation can be used to develop optical methods for blood composition analysis. Currently, terahertz time-domain spectroscopy (THz TDS) and imaging techniques are applied in various fields such as astronomy, security screening, communications, genetic engineering, pharmaceutical quality control, medical imaging, and biomedical engineering [18–21]. Several unique features make THz-based technology very suitable for medical applications. The main feature of THz spectroscopy, which is really important for biophotonics, is that the characteristic energies of molecules' rotational and vibrational motions lay in THz frequency range, so many chemical and biological molecules can be identified by their characteristic resonant peaks [22]. Also, THz radiation is very sensitive to water [23], so it could provide a great contrast of samples with different water content. In addition, THz radiation is non-ionizing because it has very low photon energy. As a result, it might be safely applied to *in vivo* diagnostics. This fact is a significant advantage for biological structural investigation because there does not appear to be any tissue damage during the procedure [24,25]. Delayed diagnostics of diabetes mellitus and insufficient control on patient are the general reasons of complications, which lead to invalidation and higher mortality. It should be noted that a large number of patients have macrovascular complications due to abnormal blood glucose level: hyperglycemia, hypoglycemia, variability of glycemia. Statistics shows that growth of glycated hemoglobin (HbA1c) index from 6% to 9% leads to a fourfold increased risk of microvascular complications and twice the risk of myocardial infarction [26]. Hypoglycemia (even in subclinical situations) is associated with a 2- to 2.5-fold increased risk of mortality from cardiovascular complications and a fourfold increase in the number of cardiovascular cases in Ref. [27]. Variability of glycemia is another risk factor, as variability of glucose level by more than 60 mg/dl is a predictor of coronary atherosclerosis (2.6-fold increased risk) even more accurate than HbA1c [28]. In addition, glycemia variability on an empty stomach is an independent predictor of cardiovascular lethal cases for senior patients with type 2 diabetes mellitus [29], urgent patients [30], and patients with myocardial infarction [31]. The present clinical methods of glucose-level sensing include the following: opto-chemical (Accu-Chek Active) and electrochemical (Ascensia Contour TS, Abbott Freestyle, OneTouch Select) glucometers, continuous glucose-monitoring sensors (Medtronic MiniMed, Freestyle Libre), and electro-chemistry for labs (EKF Biosen C), all of which require fresh blood sampling. At the same time, there is still lack of noninvasive methods of glucose measuring: some are out of production due to drawbacks [32], while other are only at experimental stages [33,34], or slowly coming to market [35,36].

10.2 STUDY OF GLUCOSE CONCENTRATION INFLUENCE ON BLOOD OPTICAL PROPERTIES IN THz FREQUENCY RANGE

This section consists of a series of works [37-40], the goal of which is to use the benefits of THz radiation for noninvasive glucose measuring technique. In this section, the possibility of transmission and reflection spectroscopic methods in THz frequency range to analyze glucose level in blood samples is considered. The transmission mode was used to make *in vitro* analysis of blood and nail samples. The collected data were used for modeling of *in vivo* noninvasive experiment in CST Microwave Studio software.

The optical properties of blood samples were studied in the frequency range of 0.3–0.5 THz using THz TDS in transmission mode [41]. The scheme of the setup is shown in Figure 10.1.

THz radiation is generated by means of an InAs semiconductor in a 2.0 T magnetic field irradiated with a Yb:KYW femtosecond laser (the wavelength of 1040 nm, the pulse duration of 120 fs, the pulse repetition frequency of 75 MHz, and the power of 1W). THz radiation has the following output characteristics: the pulse duration is 2.7 ps, and the main power is concentrated in the frequency range of 0.3–0.5 THz with an average power up to 30 μW, which was measured by a Golay cell. The typical transmission spectrum of air for this THz TDS is shown in Figure 10.2 (spectral resolution up to 1 GHz). THz radiation passes through a Teflon filter (which blocks wavelengths shorter than 50 μm). After that, the THz radiation passes through the sample fixed at the focal plane perpendicular to the beam. The existence of the THz field changes the birefringence of the electro-optical crystal, i.e., causing the refractive index difference along different axes of the crystal. The electric field-induced birefringence changes the probe beam's polarization [43]. As a result, the THz pulse induces the birefringence of the probe beam in the crystal due to the electro-optical effect. The birefringence magnitude is directly proportional to the electric field amplitude of terahertz radiation E(t) in the time point. These data are required to calculate E(ω) using the Fourier transform.

In the experiment, eight blood samples were used, each with a different glucose concentration. The samples were obtained from the same person over a short period of time (about 2 hours). The patient had no significant diseases other than diabetes mellitus. The patient had an

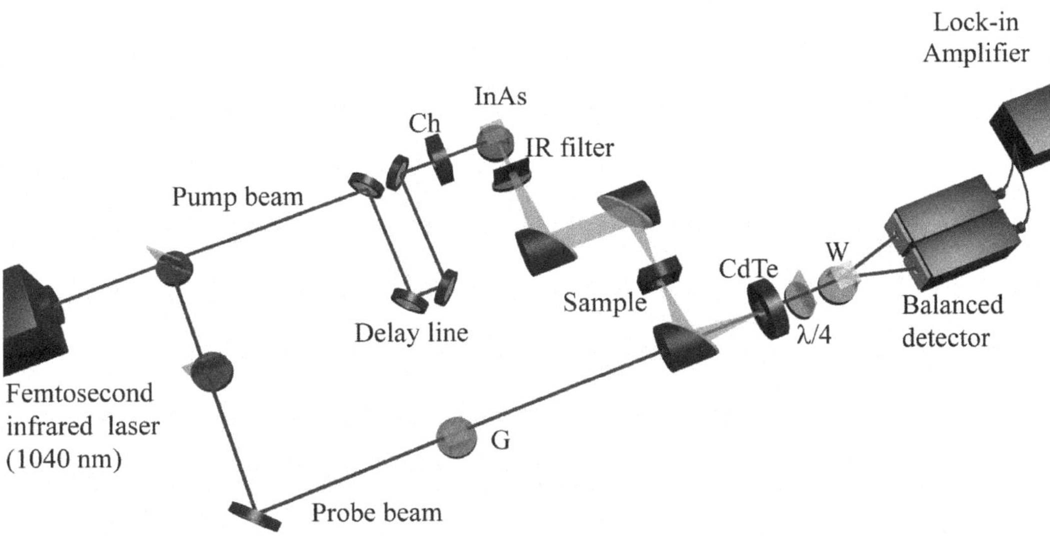

FIGURE 10.1 The diagram of THz TDS system. λ/2 – half-wave plate, G – Glan prism, Ch – chopper, λ/4 – quarter-wave plate, W – Wollaston prism. (Reprinted with permission from [42].)

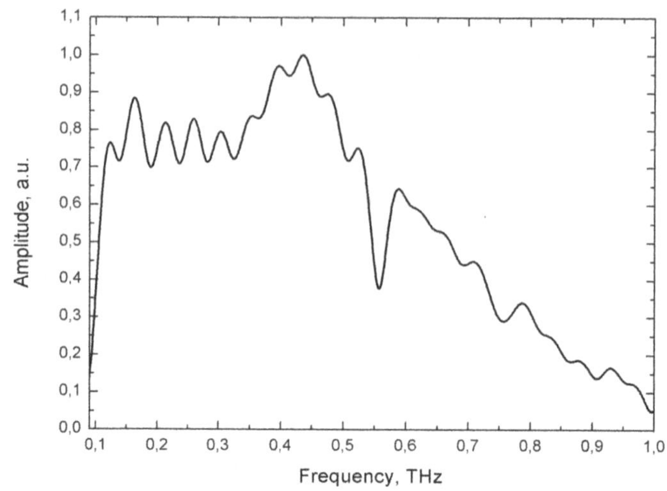

FIGURE 10.2 Transmission spectrum of air. (Reprinted with permission from [44].)

TABLE 10.1
The Glucose Concentration in Samples used in Experiment

Sample #	mmol/l	mg/dl
1	3.0	54.0
2	3.8	68.4
3	6.2	111.6
4	9.2	165.6
5	11.0	198.0
6	14.9	268.2
7	18.0	324.0
8	19.0	342.0

Data from [44].

increased level of HbA1c (7%, 9%); LD cholesterol (4.60 mmol/l); uric acid (9.1 mmol/l). Other blood component levels were normal. This fact helped us to provide the stable concentration of blood components (excluding glucose concentration) to avoid multiple dependences within the time of experiment. The study was performed in accordance with Good Clinical Practice (GCP) and with the 1964 Helsinki declaration and its later amendments. All measurements were taken with the assistance of the staffs from the Institute of Endocrinology of Federal Almazov North-West Medical Research Center (St. Petersburg, Russia). The first step of sample preparation was increasing the glucose level of the diabetic patient up to hyperglycemia level. It is quite a fast process, so samples were obtained during the time of decreasing glucose level after insulin injection. The glucose concentration in the samples is shown in Table 10.1.

The glucose concentrations in the blood were measured twice per a sample using Abbott Freestyle Optium glucometer. This instrument shows the concentration in values of mmol/l which can be converted to mg/dl by multiplying 18. This model of electrochemical glucometer model is widely used in clinical research [45]. The glucose concentration measurement and THz test happened

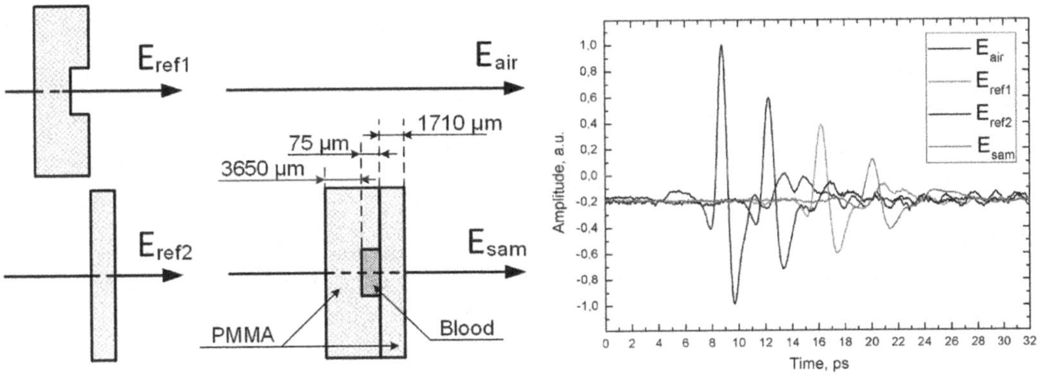

FIGURE 10.3 The scheme of sample preparation and experimental THz waveforms. (Reprinted with permission from [44].)

FIGURE 10.4 Polymethyl methacrylate (PMMA) container with blood. (Reprinted with permission from [44].)

simultaneously. The blood drops were located inside a special polymethyl methacrylate (PMMA) container (Figures 10.3 and 10.4). There was a recess with a depth of 75 μm in the middle. This type of blood holder provides stable sample thickness and protects biological fluid from drying.

For each sample, a time-amplitude transmission waveform was taken 100 times and averaged for each timepoint. In addition, transmission waveforms of air, bottom and top parts of container were measured. All the acquired waveforms were converted to the frequency domain $\hat{E}(\omega)$ using Fourier transform. Then, the THz electric field is:

$$\hat{E}(\omega)=\frac{1}{\sqrt{2\pi}}\int_{-\infty}^{+\infty} f(t)\exp(-\mathrm{i}\omega t)\mathrm{d}t = E_0(\omega)\exp(-\mathrm{i}\varphi(\omega)), \tag{10.1}$$

where $f(t)$ is the time-amplitude waveform, ω is the angular frequency, $E_0(\omega)$ is the amplitude data, and $\varphi(\omega)$ is the phase data.

At the first step, $E_{ref1}(t)$, $E_{ref2}(t)$, THz waveforms were used to obtain the phase delay dispersions for the bottom part and the top part of container, $\Delta\varphi_1(f)$ and $\Delta\varphi_2(f)$ correspondingly, where $\Delta\varphi_1(f)=\varphi_{ref1}(f)-\varphi_{air}(f), \cdot \Delta\varphi_2(f)=\varphi_{ref2}(f)-\varphi_{air}(f)$ At the second step, $E_{sam}(t)$ THz waveform was used to obtain the phase dispersion for the container with blood $\varphi_{sam}(f)$. Then, the phase delay dispersion for blood was calculated as: $\Delta\varphi_{blood}(f)=\varphi_{sam}(f)-\varphi_{ref}(f)$, where

$\varphi_{\text{ref}}(f)=\varphi_{\text{air}}(f)+\Delta\varphi_1(f)+\Delta\varphi_2(f)$. The phase delay dispersion for blood $\Delta\varphi_{\text{blood}}(f)$ was used to calculate the complex refractive index and complex permittivity.

As a result, the refractive index n of a sample in the container can be calculated. The real part of the blood refractive index $n_{\text{real blood}}$ is calculated as:

$$n_{\text{real blood}}(f)=1+\frac{c\cdot\Delta\varphi_{\text{blood}}(f)}{2\pi f d_{\text{blood}}}, \tag{10.2}$$

where c is the speed of light in vacuum, d_{blood} is the thickness of blood layer, f is a frequency, and *ref* and *sam* indices the mean values attitude to reference and sample signals correspondingly.

The absorption coefficients α are calculated using the electric field amplitude data:

$$d_{\text{sam}}=d_{\text{blood}}+d_{\text{ref1}}+d_{\text{ref2}}, \tag{10.3}$$

$$\alpha(f)_{\text{ref1}}=\frac{1}{d_{\text{ref1}}}\ln\left(\frac{E_{\text{air}}(f)}{E_{\text{ref1}}(f)}\right)^2, \tag{10.4}$$

$$\alpha(f)_{\text{ref2}}=\frac{1}{d_{\text{ref1}}}\ln\left(\frac{E_{\text{air}}(f)}{E_{\text{ref2}}(f)}\right)^2, \tag{10.5}$$

$$\alpha(f)_{\text{sam}}=\frac{1}{d_{\text{sam}}}\ln\left(\frac{E_{\text{air}}(f)}{E_{\text{sam}}(f)}\right)^2, \tag{10.6}$$

$$\alpha_{\text{blood}}(f)=\frac{d_{\text{sam}}\alpha(f)_{\text{sam}}-d_{\text{ref1}}\alpha(f)_{\text{ref1}}-d_{\text{ref2}}\alpha(f)_{\text{ref2}}}{d_{\text{blood}}}. \tag{10.7}$$

In addition, the blood penetration depth L_{blood} is the reverse function to the blood absorption coefficient:

$$L_{\text{blood}}(f)=\frac{1}{\alpha_{\text{blood}}(f)}. \tag{10.8}$$

The imaginary part of the refractive index $n_{\text{imag blood}}$ requires data about the blood absorption coefficient α_{blood}:

$$n_{\text{imag blood}}(f)=\frac{\alpha_{\text{blood}}(f)\cdot c}{4\pi f}, \tag{10.9}$$

Both parts of the blood permittivity $\varepsilon_{\text{blood}}$ use both parts of the complex refractive index n_{blood}:

$$\varepsilon_{\text{real blood}}(f)=n^2_{\text{real blood}}(f)-n^2_{\text{imag blood}}(f), \tag{10.10}$$

$$\varepsilon_{\text{imag blood}}(f)=2\cdot n_{\text{real blood}}(f)\cdot n_{\text{imag blood}}(f). \tag{10.11}$$

All of these optical properties are available as results of Spectrina software [38], to obtain the dispersion of optical properties of materials (complex refractive index, complex permittivity, complex conductivity) and the spectral characteristics (transmission, reflection, absorption spectra) in transmission and reflection modes. Based on eight biosamples, we have investigated the frequency dispersions of n_{real}, α, $\varepsilon_{\text{real}}$, and $\varepsilon_{\text{imag}}$ in the THz frequency range (Figure 10.5).

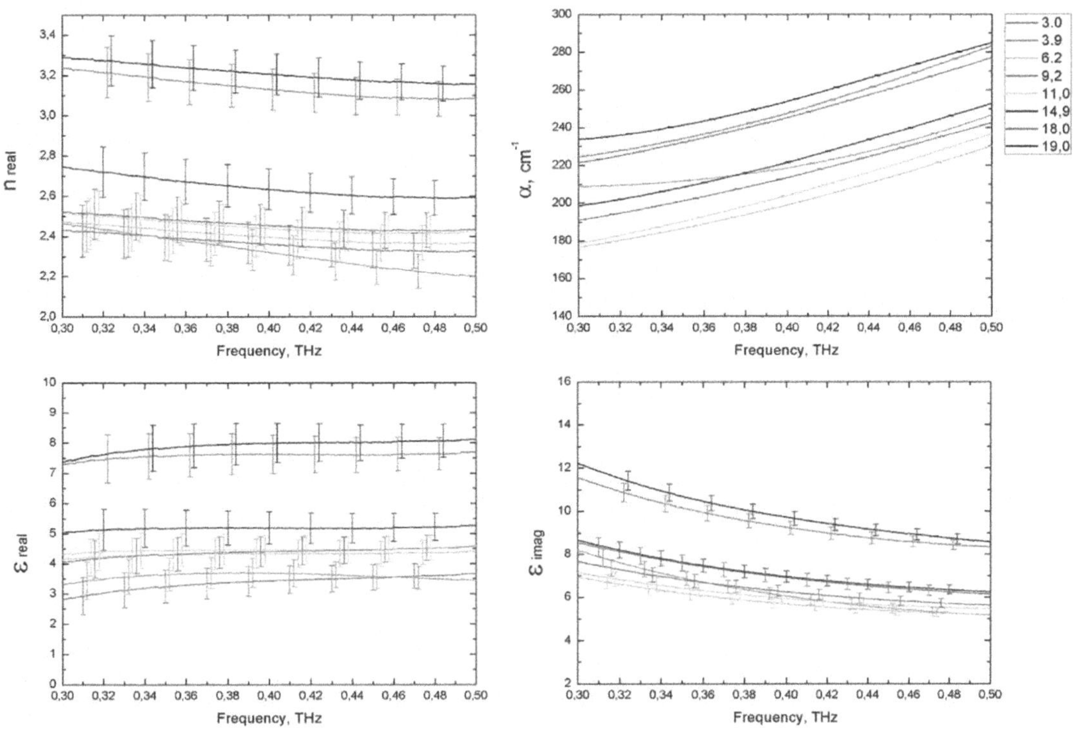

FIGURE 10.5 Dispersions of optical parameters of samples with different glucose concentration $C_{glucose}$. (Reprinted with permission from [44].)

After data acquisition, the next step is finding dependencies between glucose concentrations and optical properties. To determine the dependence of blood optical properties upon glucose concentration, we selected the experimental data at 0.30, 0.35 0.40, and 0.45 THz. These frequencies were chosen due to maximal THz electric field amplitude and minimal absorption of water vapor [46]. Figure 10.6 shows the dependences of glucose concentration on the real part of the refractive index and on the real part of the permittivity. The dependence is not linear with blood glucose concentrations above 16 mmol/l. This may be due to the change in physicochemical properties of blood components with high glucose concentration [47,48]. As known, hyperglycemia caused by insulin deficiency is accompanied by a large loss of electrolytes, dehydration of tissues, and osmolality increase of blood plasma. In a week, we repeated these experiments to check the dependence of glucose level on the blood optical properties. The comparison of data of two experiments is shown in Figure 10.6. We observe the shift of the calibration curves, but the shape of the curves was similar. This shift may be caused by the variation of the concentrations of the patient blood components (cholesterol, uric acid, etc.) in another experiment.

The dependences of glucose concentration on the real part of the refractive index and the real part of the permittivity were approximated by all OriginPro embedded functions to find the most suitable one. The Gompertz function (Eq. 10.12) fitted these dependences better than other ones.

$$y = A\exp\left(-\exp\left(-C(x-B)\right)\right). \tag{10.12}$$

These dependencies indicate that quantitative blood glucose-level analysis is feasible by using TDS in the THz frequency range. According to these data, we can simulate glucose-level analysis by performing numerical modeling in the reflective mode. Often, during the process of mixing, analysis of the model of effective medium is used. It means the following case: if we mix few media

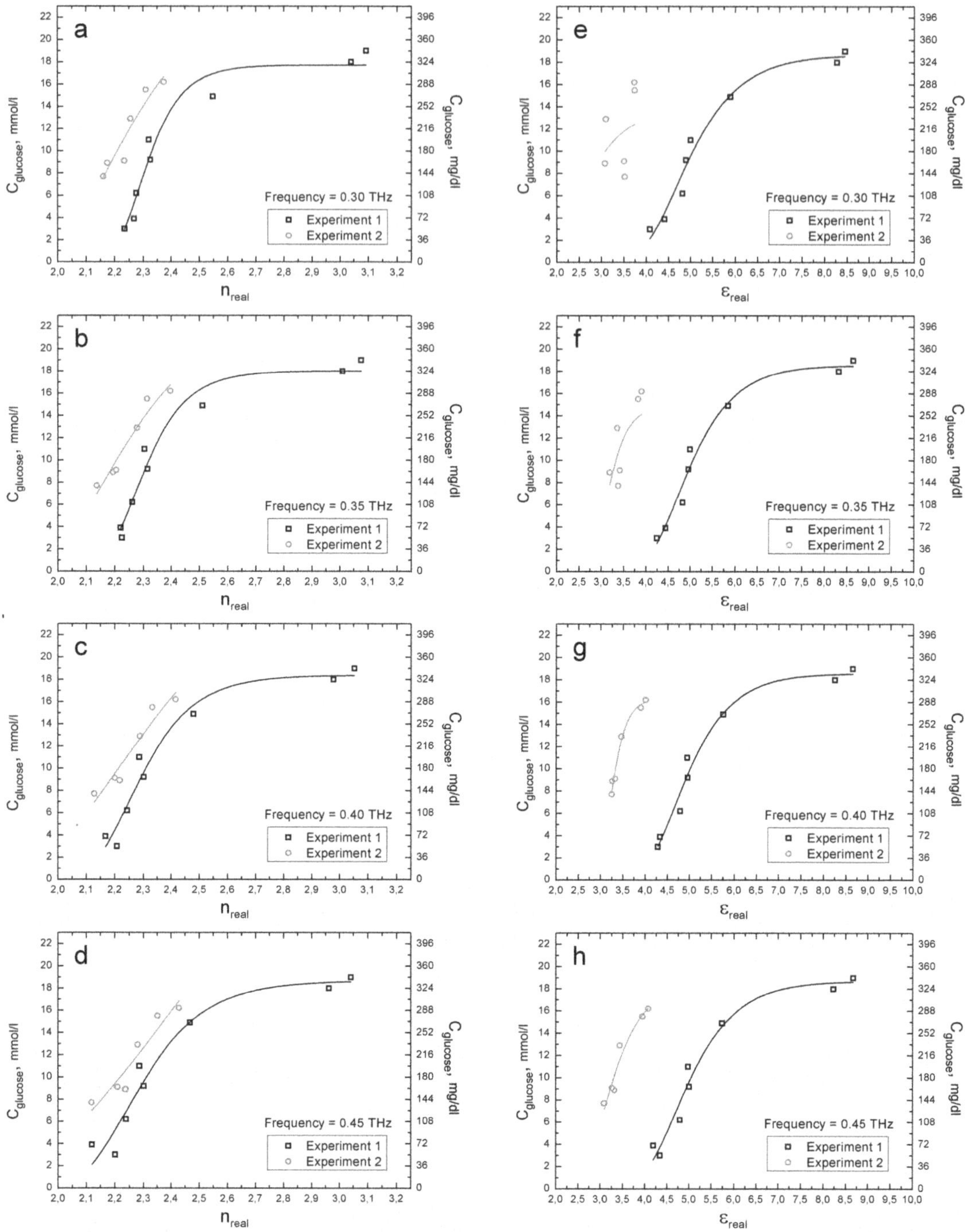

FIGURE 10.6 Dependencies of glucose concentration on the real part of refractive index $C_{\text{glucose}}(n_{\text{real}})$ (a,b,c,d) and real part of permittivity $C_{\text{glucose}}(\varepsilon_{\text{real}})$ (e,f,g,h) at 0.30, 0.35, 0.40, 0.45 THz. (Reprinted with permission from [44]. Comparison of results of two experiments.)

TABLE 10.2
The Coefficients of Gompertz Fitting for Dependence of Glucose Concentration $C_{glucose}$ on Real Parts of Refractive Index (1st tab) and Permittivity (2nd tab) at 0.30, 0.35, 0.40, and 0.45 THz

	Coefficient A		Coefficient B		Coefficient C	
Frequency, THz	**Ex.1**	**Ex.2**	**Ex.1**	**Ex.2**	**Ex.1**	**Ex.2**
0.30	19.0	16.2	2.49	2.17	4.14	16.98
0.35	19.0	16.2	2.44	2.13	4.21	14.59
0.40	19.0	16.2	2.40	2.12	4.24	12.71
0.45	19.0	16.2	2.36	2.11	4.23	11.65
	Coefficient A		**Coefficient B**		**Coefficient C**	
Frequency, THz	**Ex.1**	**Ex.2**	**Ex.1**	**Ex.2**	**Ex.1**	**Ex.2**
0.30	19.0	16.2	3.73	2.73	0.84	1.86
0.35	19.0	16.2	3.99	3.17	0.85	4.54
0.40	19.0	16.2	4.07	3.21	0.87	4.30
0.45	19.0	16.2	4.07	3.18	0.87	3.52

Data from [44].

with known optical parameters and proportions, we can calculate optical parameters as the sum of multiplication products of medium concentrations and its optical parameters. In our case, the result cannot be fitted into a model of effective medium theory due to the model being developed for non-interacting components; thus, the biochemical interaction of components cannot be fitted into this model [49] (Table 10.2).

Terahertz radiation is highly absorbed by water and biological tissues, so signal transmission through a biotissue for noninvasive measurements is not possible practically at THz frequencies. At the same time, it is possible to use THz time domain spectroscopy in the reflective mode, which allows studying a layered structure [50]. Figure 10.7 shows a sketch of THz spectrometer for this case. The nail may be used as a reference sample for noninvasive glucose-level sensing. The time domain of the resulting pulse train detected in the reflection measurements consists of a reference pulse, a pulse reflected from the front of the nail plate, and a sample pulse, and a delayed pulse reflected from the nail plate/nail bed interface. The optical properties of the nail bed may be retrieved from the ratio between the sample and reference electric fields obtained experimentally.

So the nail is a type of layered structure, which has quasi-homogenous layers with defined boundaries: nail plate and nail bed. Moreover, the nail bed contains significant part of capillaries, the optical properties of which strongly depend on the glucose level in blood. Therefore, such structure may be simply simulated using the finite difference time domain (FDTD) solver by CST. The simulated nail is shown in Figure 10.8.

A waveguide port simulated a TE plane wave propagating through a medium. The port works as an emitter and detector. The THz bipolar waveform of the source was experimentally recorded and was fitted by the formula for a THz surface field emitter:

$$E(t) = \frac{2A}{\tau^2}\exp\left(-\frac{(t+\Delta t)^2}{\tau^2}\right) - 4A\frac{(t+\Delta t)^2}{\tau^4}\exp\left(-\frac{(t+\Delta t)^2}{\tau^2}\right), \qquad (10.13)$$

where τ is 0.765 ps, A is 0.41, and Δt is –5.8 ps.

The experimental and theoretical ones are shown in Figure 10.9.

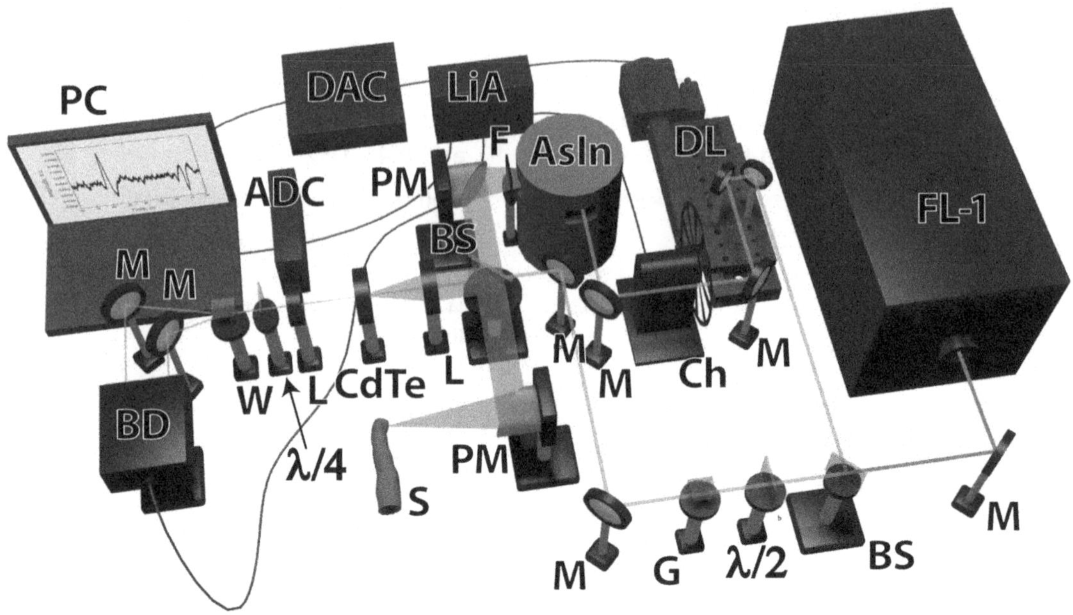

FIGURE 10.7 Sketch of THz TDS spectrometer in reflection mode for glucose-level sensing (FL-1 – femtosecond laser based on potassium-yttrium tungstate crystal activated with ytterbium (Yb: KYW), generating femtosecond pulses; F – a set of Teflon filters for IR wavelength range cutting off, BS – beamsplitter, DL – optical delay line, M – mirrors, S – investigated sample, W – Wollaston prism, CdTe – electro-optical cadmium-telluric crystal, BD – balanced detector, LiA – lock-in amplifier, PC – personal computer, G – Glan-Taylor prism, PM – parabolic mirrors, Ch – chopper, DAC – digital-to-analog converter, ADC – analog-to-digital converter). (Reprinted with permission from [44].)

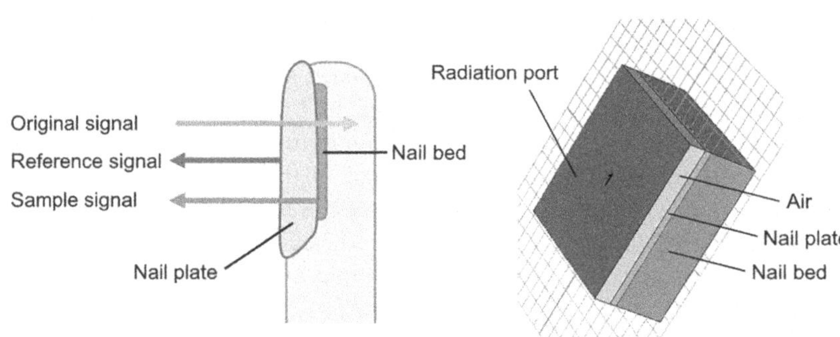

FIGURE 10.8 Principal scheme of the reflecting nail structure (left) and its realization in CST Microwave Studio (right). (Reprinted with permission from [44].)

In the model, we used the experimental data for the human nails [40] (Figure 10.10). The thickness of the nail plate was 0.58 mm.

The experimental dispersions of the real and imaginary parts of the nail plate permittivity are shown in Figure 10.11. The thickness of the nail bed layer was chosen as 3 mm due to the fact that we observed only the reflective signal from the structure. It was assumed that the dispersions of nail bed permittivity for the different glucose concentrations can be replaced by the dispersions of the permittivity of blood samples with the different glucose levels. The assumption was made due to the nail bed containing blood-filled capillaries [51] and the change

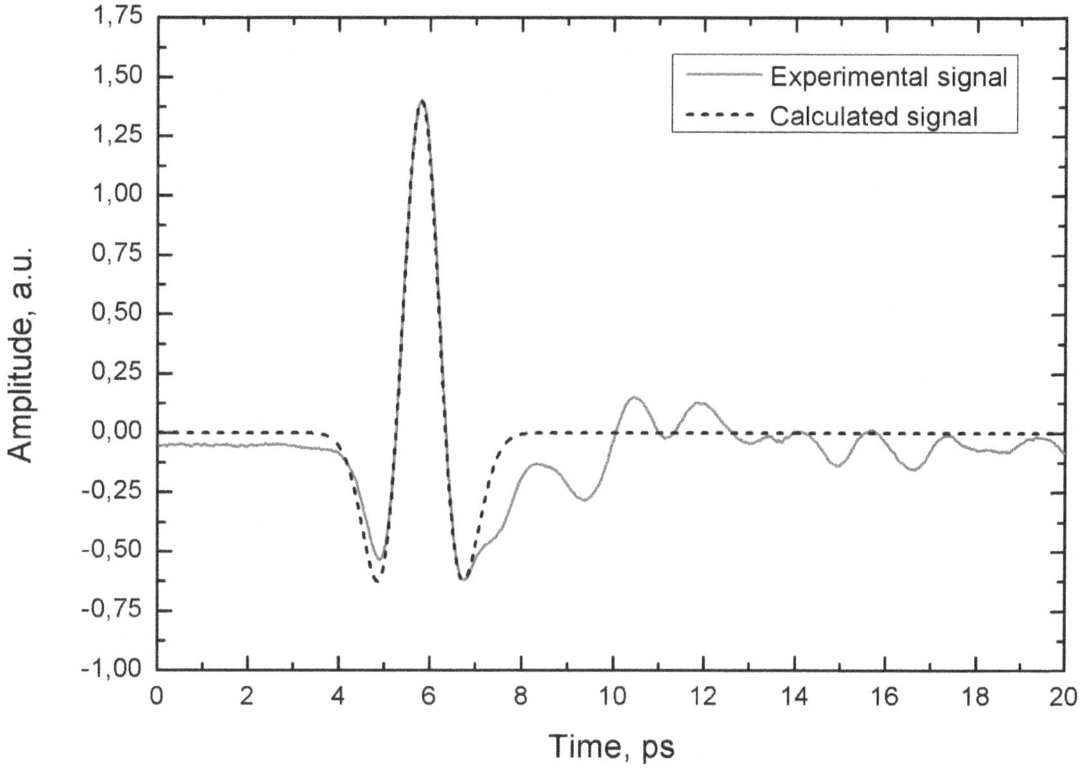

FIGURE 10.9 THz source signal: experimental (solid line) and theoretical (dotted line). (Reprinted with permission from [44].)

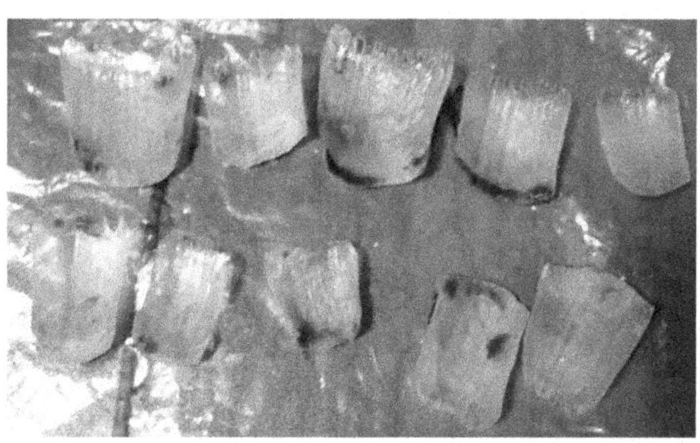

FIGURE 10.10 The nail samples under study. (Reprinted with permission from [44].)

in nail bed's optical properties is due to varying blood glucose concentrations. Modeled THz waveforms of reflected signals from the nail for different glucose levels are shown in Figure 10.12. As seen from the figure, the peak-to-peak amplitude of the sample pulse reflected from the nail plate/nail bed interface changes with varying glucose levels. The increasing blood glucose level results in an increase in the amplitude of the pulse reflected from the nail bed (Figure 10.13). So, the reflection mode of THz TDS may be used as noninvasive technique for glucose-level control.

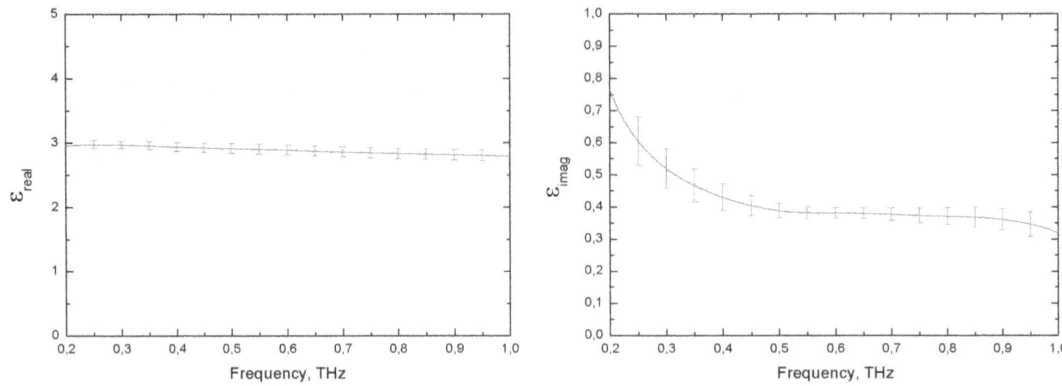

FIGURE 10.11 The dispersion of the real and imaginary parts of the permittivity of the nails. Spectral resolution is 1 GHz. (Reprinted with permission from [44].)

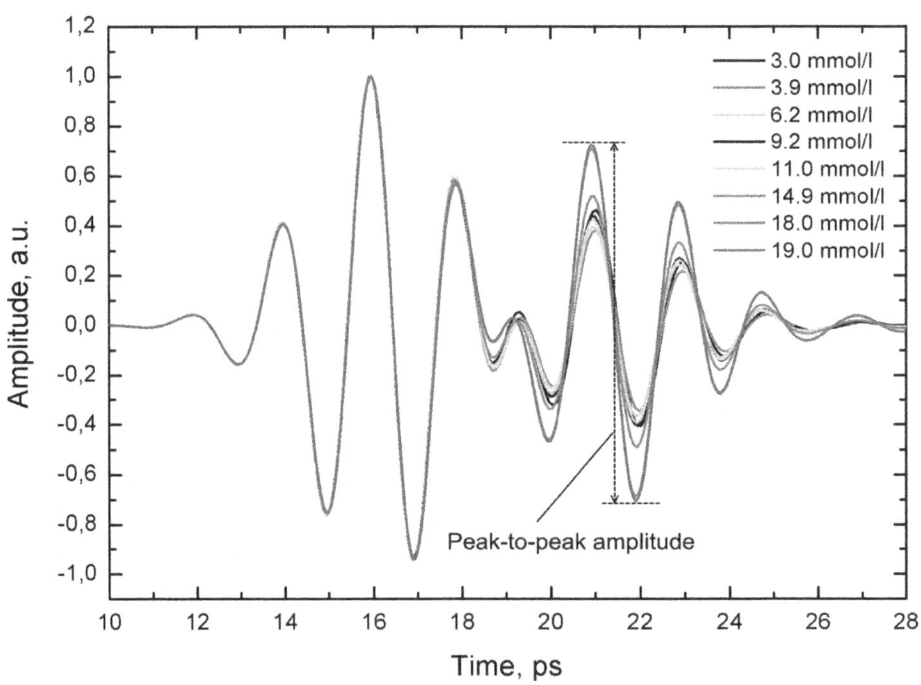

FIGURE 10.12 THz waveforms for the different glucose levels in blood. (Reprinted with permission from [44].)

10.3 THE USE OF BIOSENSOR BASED ON METAFILM TO DETERMINE THE CONCENTRATION OF GLUCOSE IN HUMAN BLOOD

Sensitivity to changes in absorption makes it possible to develop various biosensors based on metafilm [52]. Bandpass filters are often used for biosensors; their spectral characteristics can vary at the changing of the environment or material properties [53]. In this section, the possibility of using the metafilm as a highly sensitive sensor for determining the glucose concentration in human blood in the frequency range of 0.1–1 THz is discussed.

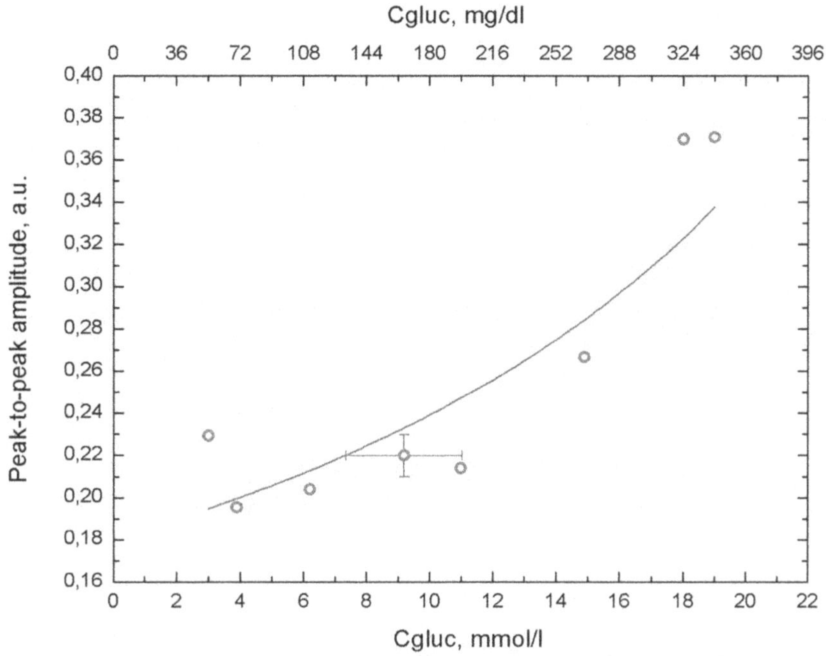

FIGURE 10.13 The dependence of the peak-to-peak amplitude of THz reflected signal on the glucose level in blood. (Reprinted with permission from [44].)

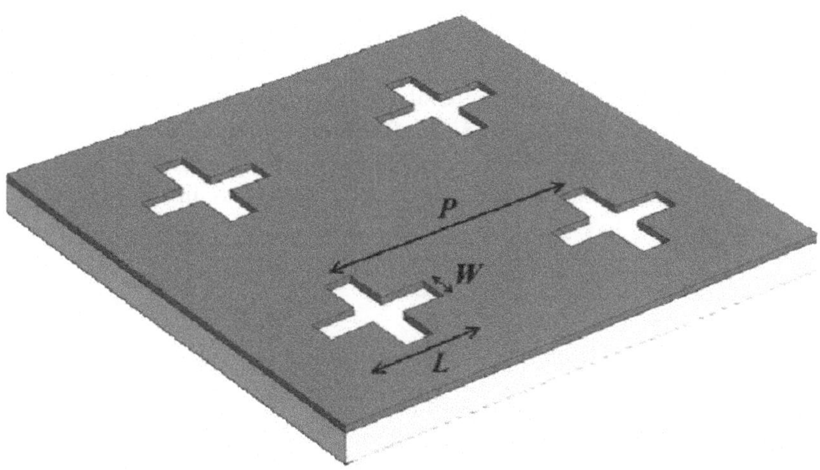

FIGURE 10.14 Sketch and geometric parameters of a bandpass filter based on cross-shaped resonators. (Reprinted with permission from [54].)

A metafilm is a film with periodic cuts of any geometry, the unit cells of which are comparable to the wavelength of the radiation. In the submillimeter wavelength range, the dimensions of the unit cells will also be on the order of tenths of a millimeter; therefore, such metamaterials are easy to manufacture. As already mentioned, for biosensors it is convenient to use bandpass filters, and their spectral characteristics are strongly influenced by the optical properties of the environment; for example, the transmission peak can shift toward lower frequencies due to a change in the refractive index [15].

A bandpass filter based on cross-shaped resonators is considered as a biosensor (Figure 10.14). The shape of the cross is symmetric, and the spectral characteristics of such a filter do not depend on the polarization of the incident radiation. Cross-shaped resonators are characterized by three geometric parameters – a period P, a length L, and a width W. By changing these parameters, we can change the spectral properties of the filter. For biosensors, it is convenient to use flexible filters; therefore, in this work, we use a filter on a polyethylene terephthalate substrate, which is transparent for the terahertz frequency range, with the dielectric constant of $\varepsilon=3$ and the thickness of $h=68\,\mu m$. The substrate is coated with the aluminum coating of 0.5 μm thick with cut-out elementary cells.

The spectral characteristics of the filter are described mainly by the resonance transmission frequency f_r (the frequency at which the maximum transmission is observed) and the quality factor Q (the ratio of the resonance frequency to the bandwidth Δf). The increase in the period at the constant size of the cross strongly influences the passband of filter and, therefore, on the filter Q-factor. Using numerical simulations in COMSOL Multiphysics, it is shown that at an increase in the period P, the bandwidth decreases, and therefore, the Q-factor increases (Figure 10.15a), and it corresponds to a nonlinear law (Figure 10.15b). The dependence shown in Figure 10.15b can be used only with a constant ratio $L/W = 4$.

The selection of the resonator parameters for a given resonance frequency is carried out by simple scaling [55]. For this, it is necessary to have a filter with known geometric parameters P, L, and W and known resonance frequency. Then, to change the resonance wavelength by a factor of k, it is necessary to change all parameters also by a factor of k. In addition, the Q-factor of the filter does not change at the scaling. Since a filter on a substrate is used, it is also necessary to know the effect of the substrate on the spectrum. With the addition of a substrate, the resonance frequency and the passband of the filter decrease by a factor of k_s [56]:

$$k_s = \left(\frac{n_{\text{sub}}^2 + 1}{2}\right)^{\frac{1}{2}}, \tag{10.14}$$

which depends on the refractive index of the substrate n_{sub}. Since both the resonance frequency and the bandwidth decrease by a factor of k_s, the Q-factor will not change. Based on the above-mentioned factors, it is possible to formulate a general algorithm for selecting the geometric parameters of cross-shaped resonators for a given resonance frequency and quality factor: (1) first of all, the influence of the substrate on the resonance frequency position is taken into account, i.e., to compensate for this shift, the parameters for the resonance frequency are initially calculated, multiplied

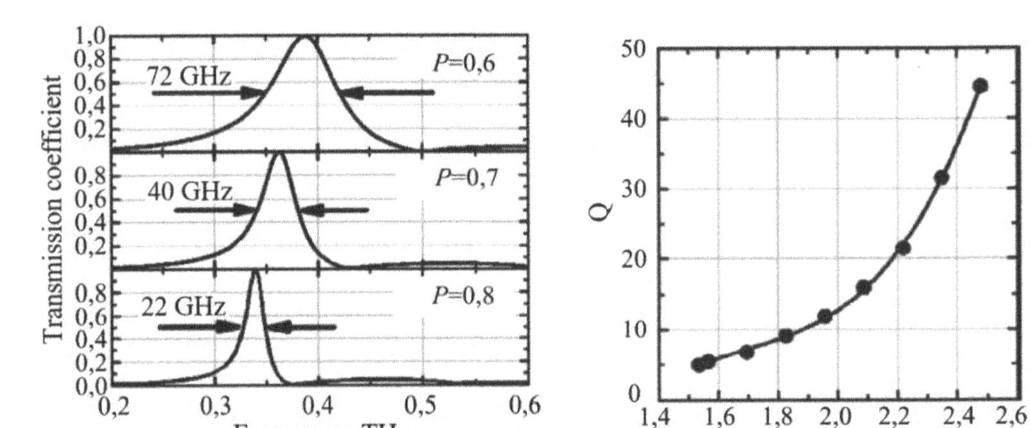

FIGURE 10.15 Transmission spectra of the filters obtained by numerical simulation with the increase of the period P at the constant dimensions of the cross L and W (a); increasing Q-factor of the filter with increasing the ratio of the period P to the cross size L (b). (Reprinted with permission from [54].)

by the factor k_s, calculated by Eq. (10.14); (2) the specified value of the quality factor Q is realized, which, as mentioned earlier, unambiguously depends on the ratio of the period P to the cross size L (Figure 10.15b); (3) using scaling, the parameters of a filter with known parameters and spectral characteristics are recalculated for the resonance frequency multiplied by the factor k_s. The result is the filter without a substrate with the Q-factor and the resonance frequency, but after adding a substrate, the peak is shifted to the given resonance frequency f_r. It was previously shown that the permittivity of whole blood in the terahertz frequency range strongly depends on the glucose concentration [57]. The studies were carried out in the frequency range of 0.2–0.7 THz with a change in glucose concentration from 3 to 19 mmol/l in human whole blood (Figure 10.16a) by the method of time-domain terahertz spectroscopy. In numerical modeling, a simplified version of human whole blood without absorption was considered and only the real part of the permittivity was used, since it makes the main contribution to the shift of the resonance frequency of the filter. To study biosensors with different characteristics, three different frequencies were chosen−0.3, 0.4, and 0.5 THz. For each case, the blood permittivity monotonically increases with an increase in glucose concentration (Figure 10.16b).

For these selected frequencies, using the algorithm described above, the geometric parameters of the filters on the substrate were selected, which in this case act as biosensors, with resonance frequencies of 0.3, 0.4, and 0.5 THz. The transmission spectra of the filters with the selected parameters were obtained numerically by the finite element method in the COMSOL Multiphysics. Their transmission spectra, spectral characteristics (resonance frequency f_r, bandwidth Δf, and quality factor Q), and selected geometric parameters of cross-shaped resonators (length L, width W, and period P) are shown in Figure 10.17. Using the calculated biosensors (Figure 10.17) and the real part of the permittivity of human blood with different glucose concentrations (Figure 10.16a), the numerical experiment can be carried out: if the model medium with the permittivity (Figure 10.16a) is irradiated at the resonance frequency (0.3, 0.4, or 0.5 THz) and the measurements are taken from the side of the substrate, then, depending on the permittivity of the blood, the different shifts of the filter resonance frequency will be observed in the reflection spectrum. Moreover, depending on the value of the blood permittivity, the shift in the resonance frequency will be more (Figure 10.18).

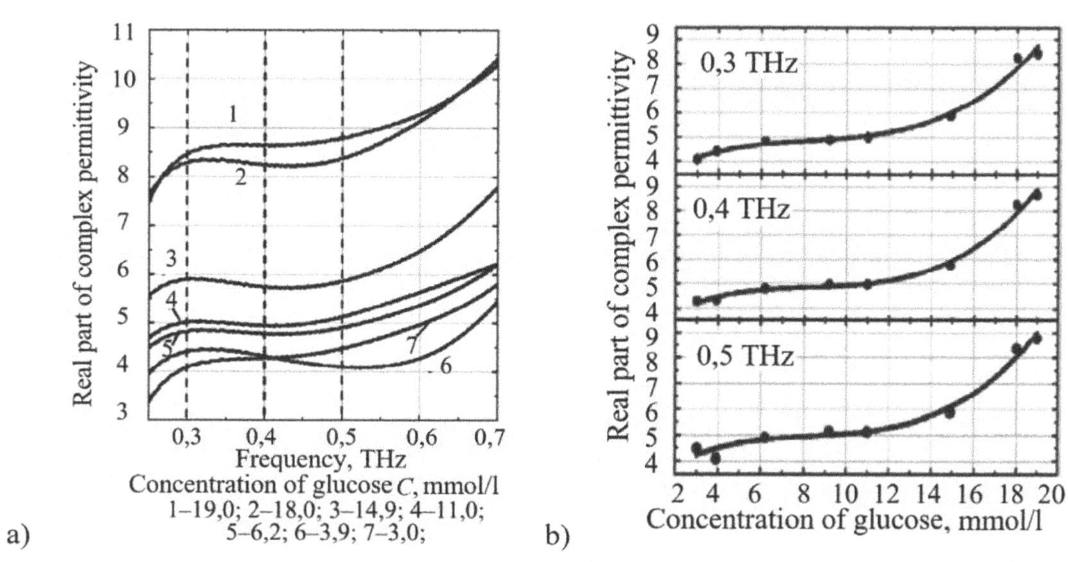

FIGURE 10.16 (Dispersion of the real part of the blood permittivity with glucose concentrations C 3.0–19.0 mmol/l (a); increase in the permittivity with an increase in the concentration of glucose in the blood at frequencies 0.3, 0.4, and 0.5 THz (b). (Reprinted with permission from [54].)

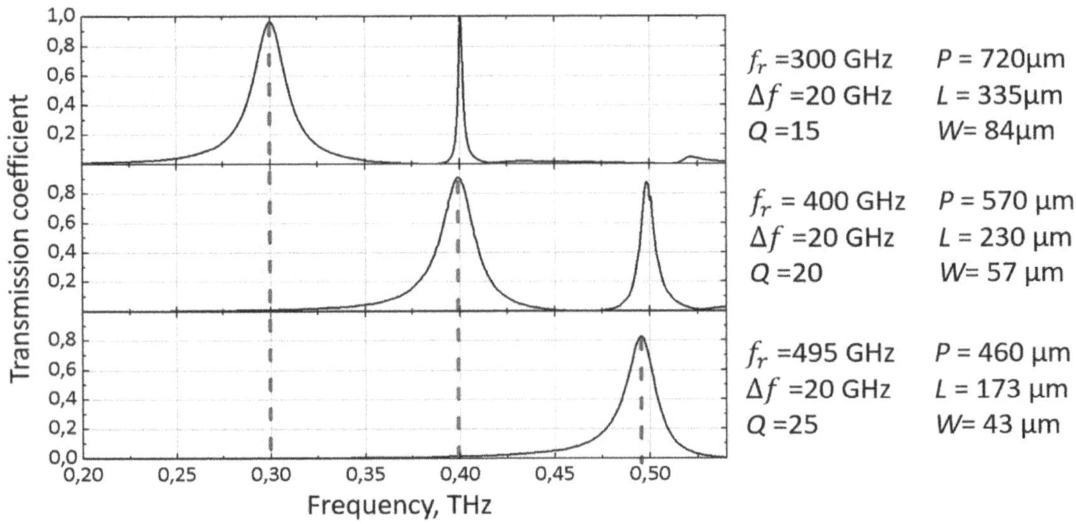

FIGURE 10.17 Transmission spectra of three different biosensors obtained by numerical simulation and their spectral characteristics. (Reprinted with permission from [54].)

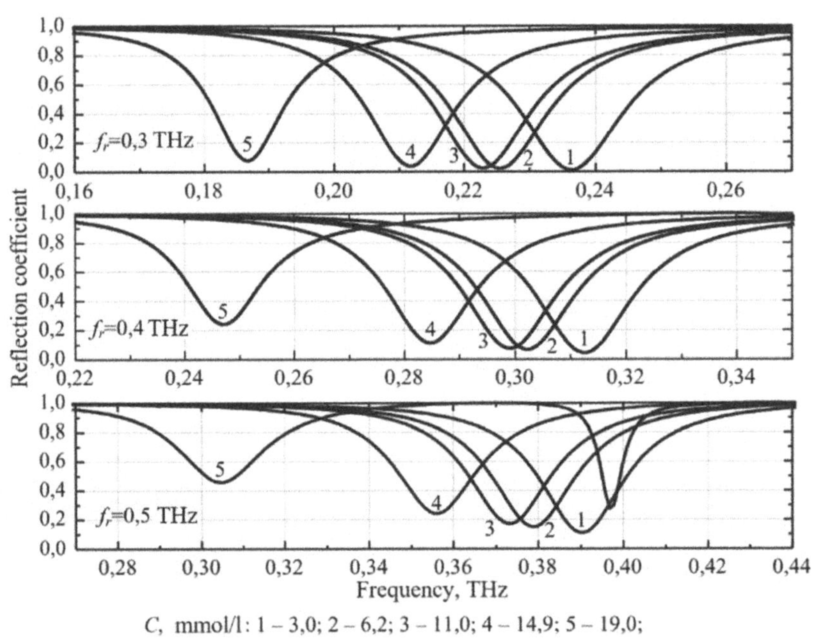

FIGURE 10.18 Resonance frequency shift of biosensors with different initial resonance frequencies with an increase in the concentration of glucose in the blood. (Reprinted with permission from [54].)

The shift is linearly dependent on the blood permittivity value. For biosensors with different spectral characteristics, the different slope of the dependence of resonant frequency shift on the real part of permittivity was observed, i.e., the different sensitivity: the filter with the initial resonance frequency of 0.5 THz has the highest sensitivity (Figure 10.19). With the increasing in the concentration of glucose in the blood, the permittivity and the resonance shift increase accordingly (Figure 10.19).

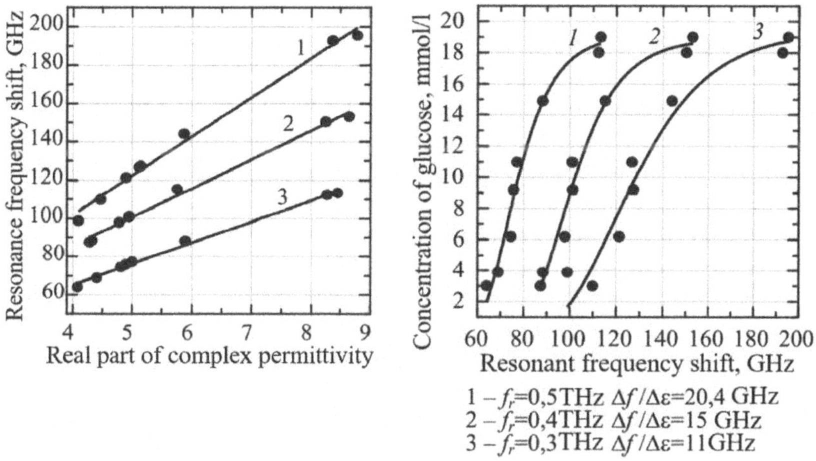

(a) (b)

FIGURE 10.19 Dependence of the biosensor resonance frequency shift on the real part of permittivity of the model medium like blood (a); dependence of the blood glucose concentration on the resonance frequency shift (b). (Reprinted with permission from [54].)

The possibility of using a bandpass filter based on cross-shaped resonators as a highly sensitive sensor to determine the concentration of glucose in human blood was shown. Using the dispersions of the blood permittivity in the terahertz frequency range for different concentrations of glucose in the blood, the parameters of cross-shaped resonators for resonance frequencies of 0.3, 0.4, and 0.5 THz were calculated. By using numerical simulation by the finite element method in COMSOL Multiphysics, the transmission spectra of the calculated filters were obtained, and it is shown that when the permittivity of the blood changes, the resonance in the filter's reflection spectrum shifts. The sensitivity of the method to changes in glucose concentration was calculated based on the sensitivity of the filter to changes in the permittivity of medium. The highest sensitivity, which is a derivative of $dC/d(\Delta f)$, is possessed by a filter with a resonance frequency of 0.5 THz: in the range 3–12 mmol/l, the sensitivity is about 2 mmol/l; in the range 12–19 mmol/l, the sensitivity is about 0,5 mmol/l. Increasing the resonance frequency increases the sensitivity of the method. On the other hand, the sensitivity of the method is limited by the used scheme of time-domain terahertz spectrometer, in which the spectral resolution of 5 GHz is associated with a limitation in the scanning time interval. If time-domain THz spectrometer with a better spectral resolution (for example, 1 GHz) or a continuous terahertz spectrometer with a spectral resolution of 1 kHz is used, then the sensitivity of the method will significantly increase.

This result is important for further development of a noninvasive glucometer, since a change in the concentration of glucose in the blood entails the diffusion of glucose into the skin and, in parallel, the dehydration of the surrounding tissue. This, in turn, leads to a change in the refractive index of the skin [58]. Due to the developed highly sensitive biosensors, it became possible to register these small changes in the optical properties of the skin [59].

10.4 RESEARCH OF HUMAN BLOOD OPTICAL PROPERTIES WITH CONCENTRATION CHANGES OF BLOOD COMPONENTS IN TERAHERTZ FREQUENCY RANGE

As mentioned in the introduction, blood composition is a good indicator of various pathologies. A change in one or another blood parameter indicates the appearance of various pathologies. In this section, we will consider how the optical response of blood changes depending on the variation in the concentration of bilirubin, creatine, uric acid, and triglycerides.

At first step, the preparation of the samples under study was carried out. The research involved 14 apparently healthy patients aged 24–42 years. Whole venous blood samples were taken after an 8- to13-hour fasting. In the serum concentrations, total bilirubin, creatinine, and triglycerides were measured by the colorimetric method, pseudo-kinetic method, and enzyme method, respectively. The glucose level was determined in the blood plasma by the enzymatic method. The studies were performed using a Cobas c311 analyzer (Roche, Switzerland). The values of the blood component concentrations and the reference intervals are presented in Table 10.3.

Each blood sample was placed in a special cell made of PMMA. There is a notch of 75 μm deep in the middle part of this cell for the blood sample (Figure 10.20).

To measure the optical parameters of blood, we used the method of terahertz time-domain spectroscopy in the transmission mode (Figure 10.1). The principle of the spectrometer operation is as follows: the femtosecond-laser generator (1040 nm, 200 fs, 70 MHz, 15 nJ, Solar Laser Systems, Belarus) produced a laser beam in a series of pulses. The beam was divided into two beams by a beam splitter; one beam passes through a time delay line and then hits the indium arsenide (InAs) semiconductor to generate THz radiation; then, the THz radiation penetrates through the sample and reaches the cadmium telluride (CdTe) detector. Meanwhile, another beam, known as a probe beam, having passed through a half-wave plate and a Glan prism, finally meets with the THz beam that transmitted through the sample on the CdTe detector surface. Then, the polarization of the probe beam is changed. The polarization of each probe beam is associated with a certain time. By detecting orthogonally polarized components of each probe beam using a balanced photodetector, the waveform of the THz pulse is finally recorded. In the experiment, an empty cell was measured first using THz TDS, and the corresponding signal is referred to as the "reference signal". Then, the cell with blood sample was measured, and the corresponding signal is referred to as the "sample signal". Using the method of THz time-domain spectroscopy, the waveforms of THz pulses passed

TABLE 10.3
Concentrations of Blood Components

Name of Blood Components	Concentration Range of Blood Components, μmol/l	Reference Value, μmol/l
Total bilirubin	3.4–18.3	3.4–20.5 μmol/l
Creatinine	59–86	53–106 μmol/l
Uric acid	150–690	140–340
Triglycerides	380–1300	Less than 1770
Glucose	4,290–5,550	3300–6100

Data from [42,60].

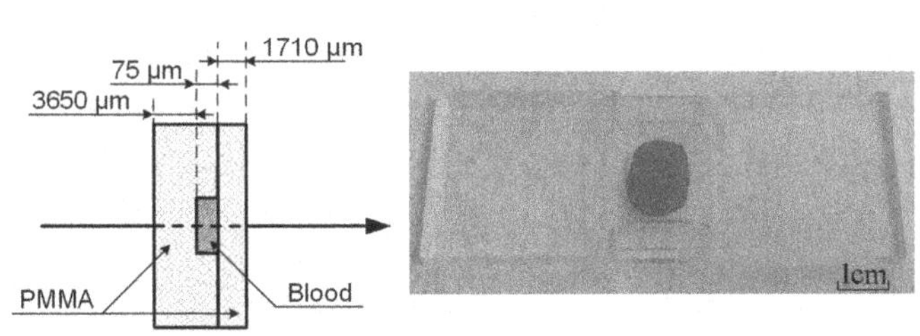

FIGURE 10.20 A cell with blood (a) and the layout of cell (b). (Reprinted with permission from [42].)

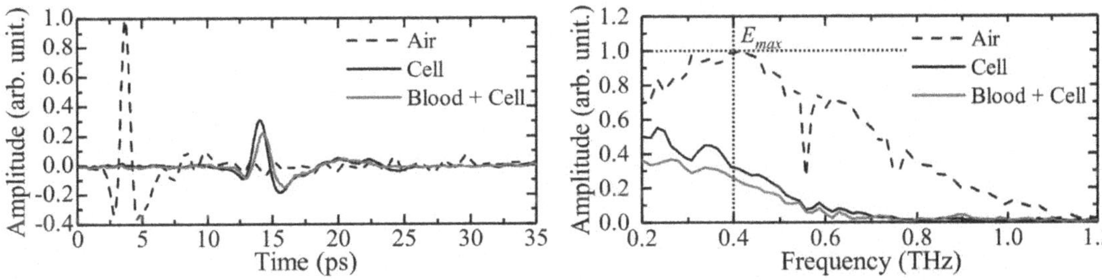

FIGURE 10.21 A sample of filtered THz waveforms (a) and their spectra map (b). (Reprinted with permission from [42].)

through an empty cell and a cell with blood were obtained. Fast Fourier transform (FFT) was applied to the obtained waveforms of THz pulses to extract the complex amplitude of the electric field strength of the transmitted signals (Figure 10.21b). To reduce the effect of noise on the result, wavelet filtering was used [59,61–64]. An example of two processed compared signals is shown in Figure 10.21a.

Knowing the amplitudes and phases of the signals passed through the empty cell and the blood cell, it is possible to calculate the refractive index and absorption coefficient of blood using the following formulas:

$$\alpha(f) = -2\ln\left[\frac{T(f)\,|\hat{E}_{\text{sample}}(f)|}{\left|\hat{E}_{\text{reference}}(f)\right|}\right] / d, \tag{10.15}$$

$$n(f) = 1 + c\frac{\phi_{\text{reference}}(f) - \phi_{\text{sample}}(f)}{2\pi f d}, \tag{10.16}$$

$$T(f) = 1 - \frac{\left(n(f)-1\right)^2}{\left(n(f)+1\right)^2}, \tag{10.17}$$

where f is the frequency, $n(f)$ and $\alpha(f)$ are the refractive index and the absorption coefficient of the blood sample correspondingly, $\hat{E}$ and ϕ are the amplitude and the phase of the signals correspondingly, d is the thickness of the blood sample, $T(f)$ is the transmission coefficient of the blood sample, and c is the speed of light.

Since the maximum intensity of the THz pulse signal of the THz TDS used is observed at a frequency of 0.4 THz (Figure 10.21b), this frequency was chosen for further study. It was shown that the glucose level affects the optical properties of blood in the THz frequency range [45,46]; therefore, in this experiment, blood samples were taken, the glucose level was varied from 5.0 to 5.5 mmol / l. The dependences of the optical properties of blood on the concentration of bilirubin (Figure10.22), creatinine (Figure 10.23), uric acid (Figure 10.24), and triglycerides (Figure 10.25) in blood samples are shown in Figures 10.22–10.25.

It should be noted that an important characteristic of absorption is its intensity. The absorption of THz radiation is observed only when the vibration leads to a change in the charge distribution inside the molecules [11–13]: the larger this change, the stronger the absorption and the higher the intensity of the absorption band. Consequently, the more polar group or bond, the higher the intensity of the corresponding absorption band, and vice versa – the absorption intensity of the non-polar bond is zero, i.e., this vibration in the THz region of the spectrum is inactive and does not manifest itself [11]. About 96% of the bilirubin in the blood is represented by non-polar insoluble bilirubin, which forms complexes with albumin. The remaining 4% of bilirubin binds to various polar molecules,

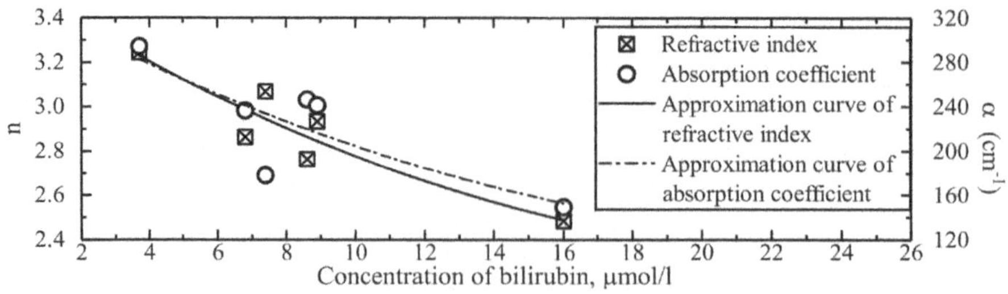

FIGURE 10.22 The relation between the optical properties of blood and the concentration of bilirubin. (Reprinted with permission from [42].)

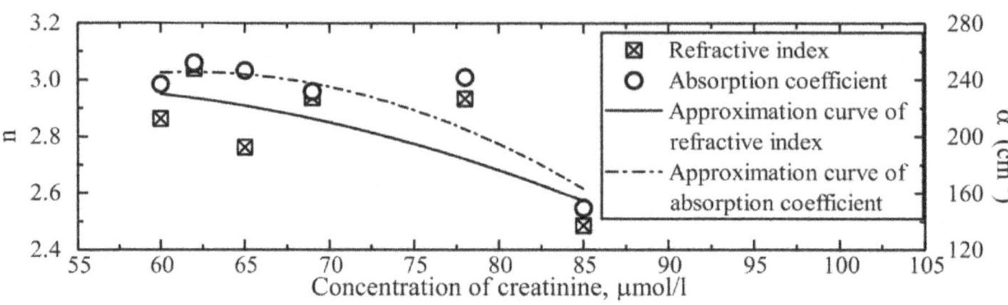

FIGURE 10.23 The relation between the optical properties of blood and the concentration of creatinine. (Reprinted with permission from [42].)

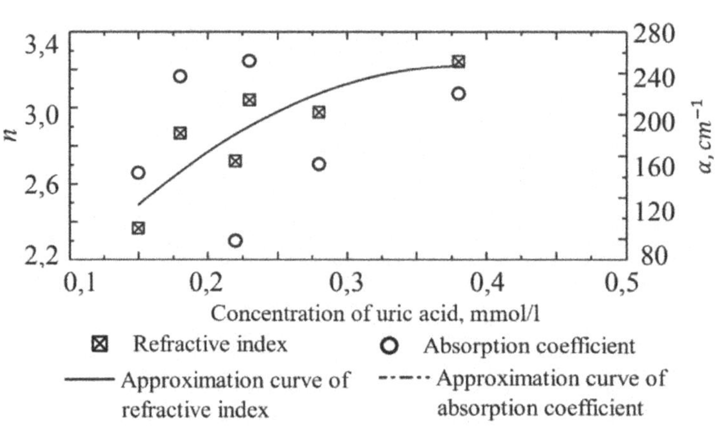

FIGURE 10.24 The relation between the optical properties of blood and the concentration of uric acid. (Reprinted with permission from [60].)

mainly glucuronic acid [64]. Because of this, with an increase in the concentration of bilirubin, the concentration of non-polar molecules in the blood increases, which reduces the absorption coefficient of blood (Figure 10.22). Creatinine is produced in muscle cells during the cleavage of energy-rich phosphate from creatine phosphoric acid and then released into the bloodstream [65]. Blood creatinine can be used to assess the renal function [65,66]. The level of creatinine increases with dehydration, muscle damage, and physical activity [67]. The molecule of water is the most important polar molecule in the blood. With dehydration of the body, the concentration of polar water molecules decreases, which also leads to a decrease in the absorption coefficient of blood (Figure 10.22) [11–13]. It should be noted that by themselves they do not harm the body and perform an

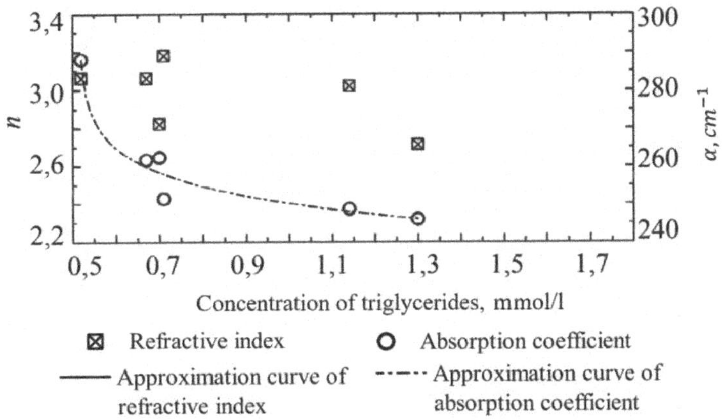

FIGURE 10.25 The relation between the optical properties of blood and the concentration of triglycerides. (Reprinted with permission from [60].)

TABLE 10.4
The Sensitivity of THz TDS

Name of Blood Components	Concentration Range of Sample Components, μmol/l	Sensitivity of Refractive Index, (μmol/l)$^{-1}$	Sensitivity of Absorption Coefficient, (cm · μmol/l)$^{-1}$
Total bilirubin	3.7–16	16.206	0.085
Creatinine	60–85	66.190	0.286
Uric acid	150–690	262	–
Triglycerides	520–1300	–	190

Data from [42,60].

important function in protein metabolism [68]. As the concentration of uric acid increases, the concentration of sodium salt increases accordingly [69]. It was shown in Ref. [70] that the increase of the sodium chloride concentration leads to growth of refractive index of an aqueous salt solution. Therefore, with an increase in the concentration of uric acid, the refractive index of the blood increases (Figure 10.24). Triglycerides are an organic fatty compound (lipid) formed from glycerol and three fatty acids [71]. Triglycerides are the main constituents of fat in humans and animals, as well as plant fat [72]. Triglycerides in their pure forms are poorly soluble in water and cannot mix with blood. Lipids absorb THz radiation weaker than water molecules [73]. Consequently, with an increase in the concentration of triglycerides in the blood, blood absorption decreases (Figure 10.24). The sensitivity of the refractive index and absorption coefficient of human blood to changes in the concentrations of blood constituents is shown in Table 10.4.

The effect of changes in the composition of blood on its optical properties at the frequency of 0.4 THz was studied by the method of time-domain terahertz spectroscopy. It was revealed that the refractive index and absorption coefficient of blood decrease with the increase in the level of bilirubin due to the increase in the concentration of non-polar bilirubin molecules, insoluble in water. It was found that the values of the refractive index and absorption coefficient of blood decrease with the increase in creatinine levels due to dehydration and the decrease in the concentration of polar water molecules in the blood. It is shown that with the increase in the level of uric acid in the blood, the increase in the refractive index of the blood was observed, associated with the increase in the concentration of sodium salt in the blood. It was demonstrated that the decrease in the absorption coefficient of blood with the increase in the concentration of triglycerides, caused by the increase in the concentration of lipid molecules in the blood, which weakly absorb terahertz radiation. These results can be used to develop a new optical method for analyzing the biochemical composition of blood based on time-domain terahertz spectroscopy.

10.5 SUMMARY

This chapter shows the application of THz time domain spectroscopy in the field of glucose sensing. Based on the refractive index dispersion for whole human blood, we demonstrate that different blood glucose levels have individual refractive index dispersions in frequencies ranging from 0.3 to 0.5 THz, and the relationships between the refractive index and glucose concentration at the single THz frequencies are described by the Gompertz function. Additionally, the dispersion of complex refractive index of human nails in the transmission mode was obtained. Based on these data, we proved the possibility of noninvasive glucose-level sensing in reflective mode using reflection of THz pulse from the nail plate/nail bed interface.

This chapter presents the study of metafilm application possibility as a biosensor for blood glucose concentration estimation purposes. The metafilm used in this research is a bandpass filter based on cross-shaped resonators for terahertz frequency range. Transmission spectrum of this filter is being affected by the changing optical properties of surrounding media. A change in glucose concentration leads to a change in blood optical properties; therefore, if a filter has a contact with blood, there is a shift of blood transmission spectrum observed. The shift is going to lower frequencies direction, and the shift value depends on glucose concentration. Three frequencies for more detailed study are chosen based on our previous research of blood permittivity dependence on glucose-level correlation. The general algorithm was developed for selection of filter metafilm geometrical parameters based on cross-shaped resonators for given resonance frequency and Q-factor. This algorithm was used for the calculation of three biosensor parameters with different spectral characteristics aimed at determination of the best properties of metafilm filter for biosensor purposes. The possibility of metafilm-based biosensor application was proved for the determination of blood glucose concentration. Modeling showed that a filter with higher resonance frequency (0.5 THz) has a higher sensitivity.

This chapter illustrates the study of blood biochemical composition (bilirubin, creatinine, triglycerides, uric acid) effect on its optical properties, refractive index, and absorption coefficient, in terahertz frequency range. It is shown that the refractive indices and the absorption coefficients of blood decrease with an increase in the concentration of bilirubin and creatinine. It has also been found that with an increase in the concentration of uric acid and triglycerides, the refractive index of blood increases and the absorption coefficient of blood decreases, respectively. The correlation between the refractive index and the concentration of triglycerides and the correlation between the blood absorption coefficient and the concentration of uric acid were not revealed.

This research may provide a quick, accurate, and continuous method to measure glucose concentration in human blood, which has a wide potential application in clinical practice.

ACKNOWLEDGMENT

This chapter was supported by the Government of Russian Federation, Grant 074-U01.

REFERENCES

1. S. K. Vashist, "Non-invasive glucose monitoring technology in diabetes management: A review," *Anal. Chim. Acta* **750**, 16–27 (2012). DOI: 10.1016/j.aca.2012.03.043.
2. M. Eadie and R. J. Steele, "Non-invasive blood glucose monitoring and data analytics," in *Proc. ACM Int. Conf. on Compute and Data Analysis*, 138–142, Lakeland, USA (2017). DOI: 10.1145/3093241.3093283.
3. R. O. Potts, J. A. Tamada, and M. J. Tierney, "Glucose monitoring by reverse iontophoresis," *Diabetes Metab. Res. Rev.* **18**, S49–S53 (2002). DOI: 10.1002/dmrr.210.

4. J. Yadav, A. Rani, V. Singh, and B. M. Burari, "Prospects and limitations of non-invasive blood glucose monitoring using near-infrared spectroscopy," *Biomed. Signal Process. Control* **18**, 214–227 (2015). DOI: 10.1016/j.bspc.2015.01.005.
5. R. Pandey, S. K. Paidi, T. A. Valdez, C. Zhang et al., "Noninvasive monitoring of blood glucose with Raman spectroscopy," *Acc. Chem. Res.* **50**(2), 264–272 (2017). DOI: 10.1021/acs.accounts.6b00472.
6. V. V. Tuchin, *Tissue Optics: Light Scattering Methods and Instruments for Medical Diagnosis*, 3rd ed., 882 p, SPIE PRESS, Bellingham, WA (2007).
7. Juvenile Diabetes Research Foundation Continuous Glucose Monitoring Study Group et al., "Variation of interstitial glucose measurements assessed by continuous glucose monitors in healthy, nondiabetic individuals," *Diabetes Care* **33**(6), 1297–1299 (2010).
8. N. A. Naser and S.A. Naser, *Clinical Chemistry Laboratory Manual*, Mosby, Maryland Heights, MO (1998).
9. M. Tonouchi, "Cutting-edge terahertz technology," *Nat. Photon.* **1**(2), 97 (2007).
10. A. Y. Pawar, D. D. Sonawane, K. B. Erande, and D. V. Derle, "Terahertz technology and its applications," *Drug Invent. Today* **5**(2), 157–163 (2013).
11. X.-C. Zhang and J. Xu, *Introduction to THz Wave Photonics*, Vol. 29, Springer, New York (2010).
12. J.-H. Son, *Terahertz Biomedical Science and Technology*, CRC Press, Boca Raton, FL (2014).
13. E. Pickwell and V. Wallace, "Biomedical applications of terahertz technology," *J. Phys. D: Appl. Phys.* **39**(17), R301 (2006).
14. K. Horiike, R. Miura, T. Ishida, and M. Nozaki, "Stoichiometry of the water molecules in glucose oxidation revisited: Inorganic phosphate plays a unique role as water in substrate-level phosphorylation," *Biochem. Educ.* **24**(1), 17–20 (1996).
15. J. Xu, K. W. Plaxco, and S. J. Allen, "Probing the collective vibrational dynamics of a protein in liquid water by terahertz absorption spectroscopy," *Protein Sci.* **15**(5), 1175–1181 (2006).
16. V. G. Sakai, C. Alba-Simionesco, and S. H. Chen, *Dynamics of Soft Matter: Neutron Applications*, Springer Science & Business Media, Belin (2011).
17. K. Bartik, "The role of water in the structure and function of biological macromolecules," http://www.exobiologie.fr/index.php/vulgarisation/chimie-vulgarisation/the-role-of-water-in-the-structure-and-function-of-biological-macromolecules/ (accessed: 08 August 2018).
18. M. Shur, "Terahertz technology: Devices and applications," *Proc. ESSDERC* **35**, 13–21 (2005).
19. L. Consolino, S. Bartalini, and P. De Natale, "Terahertz frequency metrology for spectroscopic applications: A review," *J. Infrared Millim. Terahertz Waves* **38**(11), 1289–1315 (2017).
20. I. Al-Naib and W. Withawat, "Recent progress in terahertz metasurfaces," *J. Infrared Millim. Terahertz Waves* **38**(9), 1067–1084 (2017).
21. C. B. Reid, et al., "Terahertz time-domain spectroscopy of human blood," *IEEE Trans. Terahertz Sci. Technol.* **3**(4), 363–367 (2013).
22. X. Yang, et al., "Biomedical applications of terahertz spectroscopy and imaging," *Trends Biotechnol.* **34**(10), 810–824 (2016).
23. L. Thrane, R. H. Jacobsen, P. U. Jepsen, and S. Keiding, "THz reflection spectroscopy of liquid water," *Chem. Phys. Lett.* **240**(4), 330–333 (1995).
24. A. J. Fitzgerald, et al., "An introduction to medical imaging with coherent terahertz frequency radiation," *Phys. Med. Biol.* **47**(7), R67 (2002).
25. K. Wang, S. Da-Wen, and P. Hongbin, "Emerging non-destructive terahertz spectroscopic imaging technique: Principle and applications in the agri-food industry," *Trends Food Sci. Technol.* **67**, 93–105 (2017).
26. I. M. Stratton, et al., "Association of glycaemia with macrovascular and microvascular complications of type 2 diabetes (UKPDS 35): prospective observational study," *Br. Med. J.* **321**(7258), 405–412 (2000).
27. S. Zoungas, et al., "Severe hypoglycemia and risks of vascular events and death," *N. Engl. J. Med.* **363**(15), 1410–1418 (2010).
28. G. Su, S. Mi, H. Tao, et al., "Association of glycemic variability and the presence and severity of coronary artery disease in patients with type 2 diabetes," *Cardiovasc. Diabetol.* **10** (1), 19 (2011).
29. M. Muggeo, et al., "Long-term instability of fasting plasma glucose, a novel predictor of cardiovascular mortality in elderly patients with noninsulin-dependent diabetes mellitus: The verona diabetes study," *Circulation* **96**(6), 1750–1754 (1997).
30. J. S. Krinsley, "Glycemic variability: A strong independent predictor of mortality in critically ill patients," *Crit. Care Med.* **36**(11), 3008–3013 (2008).
31. X. Wang, X. Zhao, T. Dorje, H. Yan, J. Qian, and J. Ge, "Glycemic variability predicts cardiovascular complications in acute myocardial infarction patients with type 2 diabetes mellitus," *Int. J. Cardiol.* **172**(2), 498–500 (2014).

32. M. J. Tierney, et al., "Clinical evaluation of the Glucowatch R Biographer: A continual, non-invasive glucose monitor for patients with diabetes," *Biosens. Bioelectron.* **16**(9–12), 621–629 (2001).
33. K. V. Larin, M. S. Eledrisi, M. Motamedi, and R. O. Esenaliev, "Noninvasive blood glucose monitoring with optical coherence tomography: A pilot study in human subjects," *Diabetes Care* **25**(12), 2263–2267 (2002).
34. O. P. Cherkasova, et al., "Application of time-domain THz spectroscopy for studying blood plasma of rats with experimental diabetes," *Phys. Wave Phenom.* **22**(3), 185–188 (2014).
35. Y. Segman, "New method for computing optical hemodynamic blood pressure," *J. Clin. Exp. Cardiol.* **7**(12), 1–7 (2016).
36. H. Chen, et al., "Quantify glucose level in freshly diabetics blood by terahertz time-domain spectroscopy," *J. Infrared Millim. Terahertz Waves* **39**(4), 399–408 (2018).
37. J. Smith, *The Pursuit of Noninvasive Glucose: Hunting the Deceitful*, 5th ed., NIVG Consulting, Turkey (2017). https://www.nivglucose.com/.
38. S. I. Gusev, et al., "Blood optical properties at various glucose level values in THz frequency range," *Proc. SPIE-OSA* **9537**, 95372 (2015).
39. S. I. Gusev, et al., "Influence of creatinine and triglycerides concentrations on blood optical properties of diabetics in THz frequency range," *J. Phys. Conf. Ser.* **735**(1), 012088 (2016).
40. V. A. Guseva, et al., "Optical properties of human nails in THz frequency range," *J. Biomed. Photon. Eng.* **2**(4), 040306 (2016).
41. V. G. Bespalov, et al., "Methods of generating super broadband terahertz pulses with femtosecond lasers," *J. Opt. Technol.* **75**(10), 636–642 (2008).
42. T. Zhang, et al., "The influence of bilirubin and creatinine on the refractive index of whole blood in terahertz frequency range: A qualitative analysis," *Optics in Health Care and Biomedical Optics VIII*, Vol. 10820, International Society for Optics and Photonics (2018).
43. X.-C. Zhang and X. Jingzhou, *Introduction to THz Wave Photonics*, 40–43, Springer, New York (2010).
44. S. I. Gusev, et al., "Study of glucose concentration influence on blood optical properties in THz frequency range," *Nanosyst. Phys. Chem. Math.* **9**(3), 389–400 (2018).
45. T. Biester, T. Danne, S. Blasig, K. Remus, et al., "Pharmacokinetic and prandial pharmacodynamic properties of insulin degludec/insulin as part in children, adolescents, and adults with type 1 diabetes," *Pediatr. Diabetes* **17**(8), 642–649 (2016).
46. M. Van Exter, C. Fattinger, and D. Grischkowsky, "Terahertz time-domain spectroscopy of water vapor," *Opt. Lett.* **14**(20), 1128–1130 (1989).
47. P. Bogner, et al., "Steady state volumes and metabolism-independent osmotic adaptation in mammalian erythrocytes," *Eur. Biophys. J.* **31** (2), 145–152 (2002).
48. S. K. Jain, "Hyperglycemia can cause membrane lipid peroxidation and osmotic fragility in human red blood cells," *J. Biol. Chem.* **264** (35), 21340–21345 (1989).
49. J.-H. Son, *Terahertz Biomedical Science and Technology*, 347–350, CRC Press, Boca Raton, FL (2014).
50. H. Nemec, et al., "Independent determination of the complex refractive index and wave impedance by time-domain terahertz spectroscopy," *Opt. Commun.* **260** (1), 175–183 (2006).
51. K. Hasegawa and B. P. Pereira, "The microvasculature of the nail bed, nail matrix, and nail fold of a normal human fingertip," *J. Hand Surg.* **26**(2), 283–290 (2001).
52. W. Xu, L. Xie, and Y. Ying, "Mechanisms, and applications of terahertz metamaterial sensing: A review," *Nanoscale* **9**(37), 13864–13878 (2017). DOI: 10.1039/C7NR03824K.
53. J. F. O'Hara, R. Singh, I. Brener et al., "Thin-film sensing with planar terahertz metamaterials: Sensitivity and limitations," *Opt. Express* **16**(3), 1786–1795 (2008). DOI: 10.1364/OE.16.001786.
54. V. Yu. Soboleva, S. I. Gusev, and M. K. Khodzitsky, "Metafilm-based biosensor for determination of glucose concentration in human blood," *Sci. Tech. J. Inform. Technol. Mech. Opt.* **18**(3), 377–383 (2018) (in Russian). DOI: 10.17586/2226-1494-2018-18-3-377-383.
55. A. M. Melo, A. L. Gobbi, M. H. O. Piazzetta, and A.M. P.A. da Silva, "Cross-shaped terahertz metal mesh filters: Historical review and results," *Adv. Opt. Technol.* Art. 530512 (2012). DOI: 10.1364/OE.16.001786.
56. A. Ferraro, D. C. Zografopoulos, R. Caputo, and R. Beccherelli, "Broad-and narrow-line terahertz filtering in frequency-selective surfaces patterned on thin low-loss polymer substrates," *IEEE J. Select. Topics Quant. Electron.* **23**(4), 1–8 (2017). DOI: 10.1109/JSTQE.2017.2665641.
57. S. I. Gusev, A. A. Simonova, P. S. Demchenko, M. K. Khodzitsky, and O. P. Cherkasova, "Blood glucose concentration sensing using biological molecules relaxation times determination," in *Proceedings of IEEE International Symposium on Medical Measurements and Applications*, Rochester, NY, 458–463 (2017).

58. R. V. Kuranov, V. V. Sapozhnikova, D. S. Prough, I. Cicenaite, and R. O. Esenaliev, "In vivo study of glucose-induced changes in skin properties assessed with optical coherence tomography," *Phys. Med. Biol.* **51**(16), 3885–3900 (2006). DOI: 10.1088/0031-9155/51/16/001.
59. S. Mallat, *A Wavelet Tour of Signal Processing*, 3rd ed., Academic Press, Cambridge, 832 p. (2009).
60. T. Zhang, Yu. A. Kononova, M. K. Khodzitsky, P. S. Demchenko, S. I. Gusev, A. Yu. Babenko, and E. N. Grineva, "Research of human blood optical properties with concentration changes of blood components in terahertz frequency range," *Sci. Tech. J. Inform. Technol. Mech. Opt.* **18**(5), 727–734 (2018) (in Russian). DOI: 10.17586/2226-1494-2018-18-5-727-734.
61. O. Rioul and M. Vetterli, "Wavelets and signal processing," *IEEE Signal Process. Mag.* **8**(4), 14–38 (1991). DOI: 10.1109/79.91217.
62. S. I. Gusev, M. A. Borovkova, M. A. Strepitov, and M. K. Khodzitsky, "Blood optical properties at various glucose level values in THz frequency range," in *European Conference on Biomedical Optics*, Munich, Germany, p. 95372A (2015). DOI: 10.1364/ECBO.2015.95372A.
63. S. I. Gusev, P. S. Demchenko, O. P. Cherkasova, V. I. Fedorov, and M. K. Khodzitsky, "Influence of glucose concentration on blood optical properties in THz frequency range," *Chinese Opt.* **11**(2), 182–189 (2018). DOI: 10.3788/CO.20181102.0182.
64. J. D. Ostrow, L. Pascolo, C. Tiribelli, "Mechanisms of bilirubin neurotoxicity," *Hepatology* **35**(5), 1277–1280 (2002). DOI: 10.1053/jhep.2002.33432.
65. I. V. Milyukova, *Handbook of Hypertension*. AST, Sova Publ., Moscow, 224 p. (2010) (in Russian).
66. A. Anvaer, *The Patient's Main Book*. AST, Moscow, 226 p. (2017) (in Russian).
67. J. T. Lumeij, "Plasma urea, creatinine and uric acid concentrations in response to dehydration in racing pigeons (Columba livia domestica)," *Avian Pathol.* **16**(3), 377–382 (1987). DOI: 10.1080/03079458708,436388.
68. D. H. Kang, S. K. Park, I. K. Lee, and R. J. Johnson, "Uric acid– induced C-reactive protein expression: Implication on cell proliferation and nitric oxide production of human vascular cells," *J. Am. Soc. Nephrol.* **16**(2), 3553–3562 (2005). DOI: 10.1681/ASN.2005050572.
69. L. Hou, M. Zhang, W. Han, Y. Tang, F. Xue, et al., "Influence of salt Intake on association of blood uric acid with hypertension and related cardiovascular risk," *PLoS One* **11**(4), Art. e0150451 (2016). DOI: 10.1371/journal.pone.0150451.
70. N. Q. Vinh, M. S. Sherwin, S. J. Allen, et al., "High-precision gigahertz-to-terahertz spectroscopy of aqueous salt solutions as a probe of the femtosecond-to-picosecond dynamics of liquid water," *J. Chem. Phys.* **142**(16), Art. 164502 (2015). DOI: 10.1063/1.4918708.
71. J. E. Vance and D. E. Vance, *Biochemistry of Lipids, Lipoproteins and Membranes*, Elsevier, Amsterdam, 639 p. (2008).
72. L. A. Belova, O. G. Ogloblina, A. A. Belov, and V. V. Kukharchuk, "Modification of lipoproteins. Physiological and pathogenetical role of modified lipiproteins: A review," *Biomeditsinskaya Khimiya* **46**(1), 8–21 (2000) (in Russian).
73. C. B. Reid, E. Pickwell-MacPherson, J. G. Laufer, et al., "Accuracy and resolution of THz reflection spectroscopy for medical imaging," *Phys. Med. Biol.* **55**(16), 4825–4838 (2010). DOI: 10.1088/0031-9155/55/16/013.

11 Noninvasive Photonic Sensing of Glucose in Bloodstream

Nisan Ozana and Zeev Zalevsky
Faculty of Engineering, Bar-Ilan University, Ramat-Gan 52900, Israel

CONTENTS

11.1 INTRODUCTION

Glucose plays a central role in metabolism: (1) it is the principal metabolic pathway for transporting carbohydrates and (2) it is an important fuel for spurring metabolic energy in all human cells. In many of the biochemical reactions that modulate the metabolic profile both physiologically and pathologically, glucose plays a main role as well.

Diabetes mellitus is a group of metabolic diseases in which a person has high level of glucose concentration in blood, which is due to the fact that the (1) pancreas does not produce enough insulin (type I diabetes), or due to the fact that (2) the cells do not respond to the insulin that is produced (type II diabetes). "Insulin-dependent diabetes mellitus" (IDDM) or "juvenile diabetes" is referred to type I diabetes. "Non-insulin-dependent diabetes mellitus" (NIDDM) or "adult-onset diabetes" is referred to type II diabetes, which is related to the modern lifestyle. The latter form of diabetes constitutes >95% of all diabetic patients and affects more than 26.8 million people of all ages in the U.S. (10.2% of the U.S population) [1] and more than 422 million people worldwide suffer from diabetes [2].

One of the most important goals of medicine is the development of a noninvasive method to measure the glucose concentration in blood. The main reason for the development of noninvasive sensors is due to the fact that diabetic patients must continuously monitor their blood glucose level in order to avoid the most harmful effects of the disease. The most common way to measure glucose is by pricking a finger and applying the blood droplet to a strip. The blood sample is inserted into a glucometer, and the flux of the glucose reaction generates an electrical signal. This electrical signal is converted to glucose concentration in blood readout. There are still many challenges related to the achievement of reliable glycemic monitoring, despite the impressive advances in glucose biosensors.

Currently, the most common way to measure glucose is by pricking a finger and applying the blood droplet to a strip that is inserted into a glucometer. The flux of the glucose reaction

DOI: 10.1201/9781003112099-11

generates an electrical signal, and the glucometer corresponds to the strength of the electrical current.

Despite the impressive advances in glucose biosensors, there are still many challenges related to the achievement of reliable glycemic monitoring. Desirable features of a biosensor system are accuracy, reliability, ultra-sensitivity, fast response, and low cost per test. In order for the procedure to be more comfortable and simpler, a few optical noninvasive methods have been developed.

As of 2014, there have been very few noninvasive glucose meters, which are being marketed in several countries. Measurement of glucose levels in the interstitial fluid is currently available in the form of continuous glucose monitors (CGMs); however, these are invasive devices that require having a sensor implanted below the surface of the skin.

The search for noninvasive glucose monitoring began after 1975 and has continued till present days without a clinically or commercially viable product. One method is photoacoustic spectroscopy, which is based on the conversion of optical energy into acoustic energy through a multi-stage energy conversion process that uses a piezoelectric detector. A change in the photoacoustic pulse was found when the glucose concentration in the blood was altered [3]. Another example is optical coherence tomography (OCT). By using an interferometer with a light that is aimed at the subject, the light that backscatters from structures within the tissue interferes with the light from the reference arm. This method is based on a phenomenon in which the increase in glucose concentration in the blood decreases the scattering coefficient [4, 5].

Another method for remote monitoring of glucose is based on measuring the amount of polarized light that is rotated by glucose in the front chamber of the eye (containing the "aqueous humor") [6]. Other examples include bio-impedance spectroscopy [7], electromagnetic sensing [8], fluorescence technology [9], optical polarimetry [10], Raman spectroscopy [11], and ultrasound technology [12]. Concluding with the observation, none of these had produced a well commercially, clinically reliable device, and therefore, much work remained to be done.

In this chapter, we present a few potential approaches for remote measuring of glucose. The first approach consists of illumination of the human skin, which is close to a blood artery, with a laser beam. The back-scattered light from the skin near the blood artery creates a secondary speckle pattern. These self-interference random patterns (i.e., speckle patterns) movements are due to the blood pulse stream changes that can be extracted [13–15]. Various bioparameters can be monitored from the blood flux pulsation. Via machine learning algorithms and trading process of the detected signals, the glucose concentration changes can be monitored [16]. The second effect includes the Faraday rotation effect, which is the rotation of the plane of vibration of linearly polarized light when passing through a medium exhibiting this effect [17]. Changing the polarization state of the wavefront will cause the speckle patterns to change as shown in Ref. [18].

Furthermore, we will present contactless measurement of acoustic excitations in a glucose solution [19]. To perform this measurement, we excited acoustic waves in a solution and measured the changes in the speckle pattern. The basic concept is that while the solution is acoustically excited, the acoustic waves modulate the density of the fluid under examination. This modulation will have two effects on the speckle pattern: the first is a spatial and time-varying modulation of the effective refractive index, and the second is a spatial and time-varying modulation of the optical rotation, which is induced by the presence of glucose. Both of these effects should change the speckle pattern, which if recorded with an exposure time, which is longer than the acoustic period, will be seen as a smearing of the pattern.

Finally, we will present a method for the diagnosis of diabetic foot. Diabetic foot usually affects patients suffering from peripheral arterial diseases (PAD). The autocorrelation function decay time was implemented to estimate the blood flow remotely. In this study, we present a clinical study with 15 subjects. Our findings show the difference between low perfused and a healthy foot via the presented approach in this chapter.

11.2 REMOTE SENSING OF ACOUSTICAL VIBRATIONS VIA TEMPORAL-SPATIAL ANALYSIS OF SPECKLE PATTERNS

Via the analysis of the generated secondary speckle patterns, we extract the tilting movement of a tissue. Mathematically, the light distribution can be expressed as follows:

$$A(x_0, y_0) = \left| \int\int \exp[i\phi(x,y)] \exp\left[i(\beta_x x + \beta_y y)\right] \exp\left[\frac{\pi i}{\lambda Z}\left((x - x_0)^2 + (y - y_0)^2\right)\right] dxdy \right|, \quad (11.1)$$

where φ is the random phase generated by the tissue roughness, λ is the illuminated wavelength, Z is the axial distance to the imaging plane, and β expresses the tissue tilting movement:

$$\beta = \frac{4\pi \tan\alpha}{\lambda}, \quad (11.2)$$

where α is the tissue tilting angle. To sense the tissue tilting movement by using a simple spatial correlation calculation, the image captured by the camera was strongly defocused. Thus, the imaging plane was moved to the far-field regime; therefore, the tilting movement can be expressed as follows:

$$A(x_0, y_0) = \left| \int \int \exp[i\phi(x,y)] \exp\left[i(\beta_x x + \beta_y y)\right] \exp\left[\frac{-2\pi i}{\lambda Z}(xx_0 + yy_0)\right] dxdy \right|. \quad (11.3)$$

An example of simple system that senses the optical phonocardiogram (OPG) signal is shown in Figure 11.1a. The optical setup consisted of (a) an eye-safe 780-nm laser and (b) a camera (Basler acA1920–25um, monochrome). The camera captures the speckle images at 300 frames per second (fps). The distance from the laser to the subject's leg was approximately 90 cm. First, a big spot pattern was illuminated on the subject's leg. Later, the big spot was divided into 25 spatial subspots to perform the numerical analysis. Using a simple correlation-based algorithm, the 2-D movement of the blood vessels can be extracted. Temporal movement of the reflecting surface causes changes to the random speckle pattern over time due to the temporal change in its tilting angle. In the first step, a set of images as a function of time was captured. In the second step, the sequential 2-D row data is correlated. The relative movement of patterns can be extracted using a 2-D correlation. The position of the correlation peak over time expresses this relative tilting movement. The temporal movement of the speckle is caused by the blood flow changes. This signal presents an OPG signal as shown in Figure 11.1b.

An example of glucose relative measurement after 12 hours of fast is shown in Figure 11.2. In this measurement, the optical sensor consists of a green laser (at 532 nm) and a camera attached to the

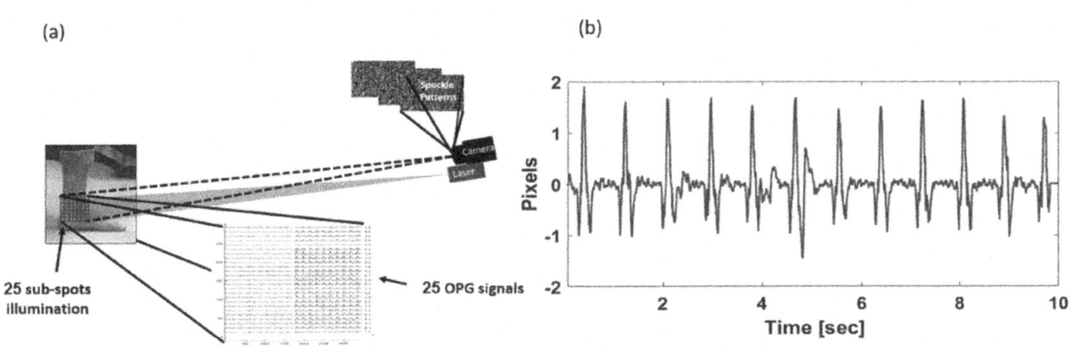

FIGURE 11.1 (a) Schematic sketch of one of the presented configurations. (b) An example for an OPG signal from the leg. (This figure has been adapted from Ref. [20].)

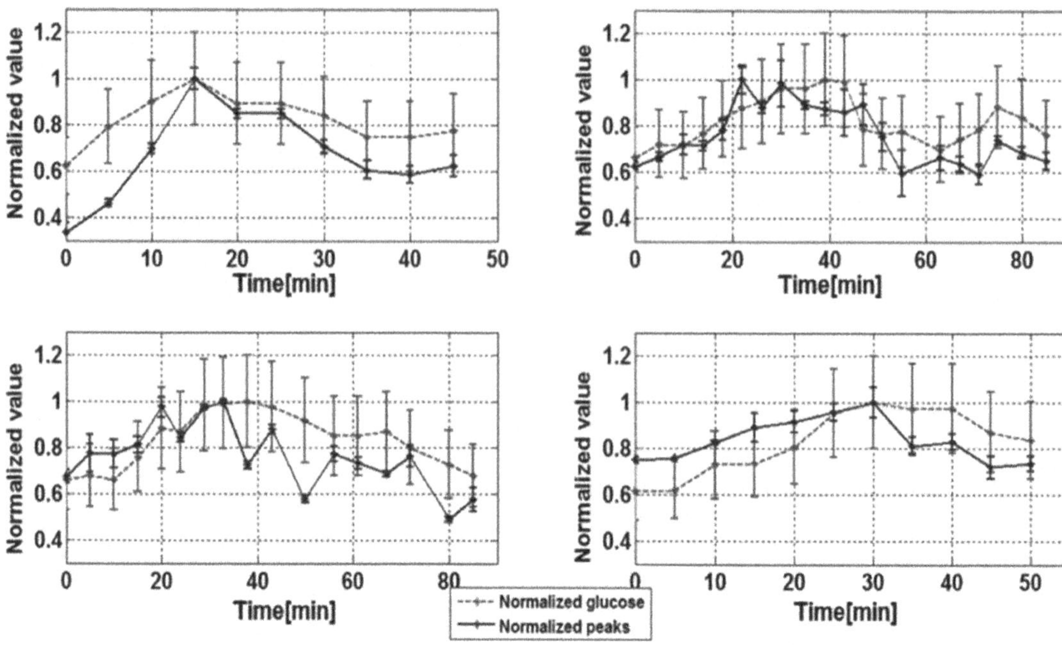

FIGURE 11.2 Normalized glucose level in blood and the normalized peak amplitude of the same subject. Glucose level is denoted with a dotted line, and the optically measured parameter is denoted with a solid line. (This figure has been adapted from Ref. [21].)

TABLE 11.1
Subject Information

#	Gender	Age [Years]	Weight [kg]
1	Male	22	95
2	Male	28	75
3	Male	47	84
4	Male	37	88

bracelet. The camera (PixelLink PL-E531) captures images of the secondary speckle pattern reflected from the subject's hand at the rate of 545 frames per second (fps). The distance from the laser to the subject's hand is approximately 5 cm. The laser output power is approximately 5mW. The subject came to the laboratory after 12 hours of fasting. First, both the optical parameter and a blood sample were taken for a calibration purpose. Immediately afterward, the subject drank 500 ml of grape soda. Here, the glucose level of four other subjects was measured, whose characteristics are summarized in Table 11.1.

The protocol according to which those subjects were tested is described above. The duration of the experiment was between 45 and 85 minutes until the glucose readings returned to baseline levels and remained stable. Note that in order to achieve good temporal resolution, we want to increase our sampling rate. For example, each pulse is about 0.2 seconds; we want to sample at least 20 samples in this period. On the other hand, if the frame rate will be too high, the measured temporal peaks will be too low (less energy per sample). In the experiment, we choose the sampling rate to be 545 frames per second as a result of this tradeoff.

The observed parameters from the video files were analyzed by a specific MATLAB program that was written for this purpose. Each file contained approximately 5 seconds of video samples at a rate of 545 fps. From the OPG graph, the maximum pulse amplitude that refers to the highest amplitude during one heartbeat was analyzed. In each measurement, 4–7 peaks were taken and then averaged. Our aim is to demonstrate that the average amplitude in each frame is correlated with the two biomedical parameters measured in this paper as demonstrated in the following sections. The results are presented in Figure 11.2.

11.3 MACHINE LEARNING-BASED ANALYSIS

In this experiment, the sensor was measuring the back-reflected patterns from the subject's leg main blood artery area. To retain only signals of good quality, preprocessing recordings were preprocessed. Again, first, the glucose levels were taken from the subject after 12 hours of fasting. Each 10 minutes, a blood sample from a finger was taken to measure the glucose concentration with a glucometer (FreeStyle Lite Blood Glucose Monitoring System). First, the subject was measured with the optical method as well as with the reference device before drinking a sweetened drink. Afterward, the subject drank 500 ml of a sweetened drink. The ingredients of this drink are shown in Table 11.2.

During the tests, five different levels of glucose were measured. During the first step of the machine learning processing using random forest algorithm, a training of the algorithm was performed. During the tests, 10% of the samples from each glucose level was randomly chosen for training using random forest classifier with 50 trees. Later, using the training process, the signals were tested to predict the glucose concentration value. One can see in Figure 11.3 the predication of each illuminating laser spot separately.

Summary of two glucose tests is shown in Figure 11.4. These tests include all the test samples from all of the spots. The percentage of the signals that passed the quality test is also shown in Figure 11.4. One can see the good predication using the presented method.

During the second part (Figure 11.5) of these tests, the subject drank the same amount of water without glucose. It is shown that during the first test (i.e., glucose test), there is an increase in the measured distribution corresponding to the real value of the glucose concentration in the bloodstream. However, as expected, during the second experiment (i.e., the water test), there is no change of the predicted values. The aim of the water test is to demonstrate that the technique is not affected by the changes in the volume of the blood, due to the addition of the drinking water, but rather the variations in the measured values are indeed due to the modification in the glucose level in the bloodstream.

TABLE 11.2
Nutritional Ingredients of the Sweetened Drink

	Quantity	Units
Energy	195	Calories
Carbohydrates	50	G
Sodium	50	Mg
Vitamin C	30	Mg

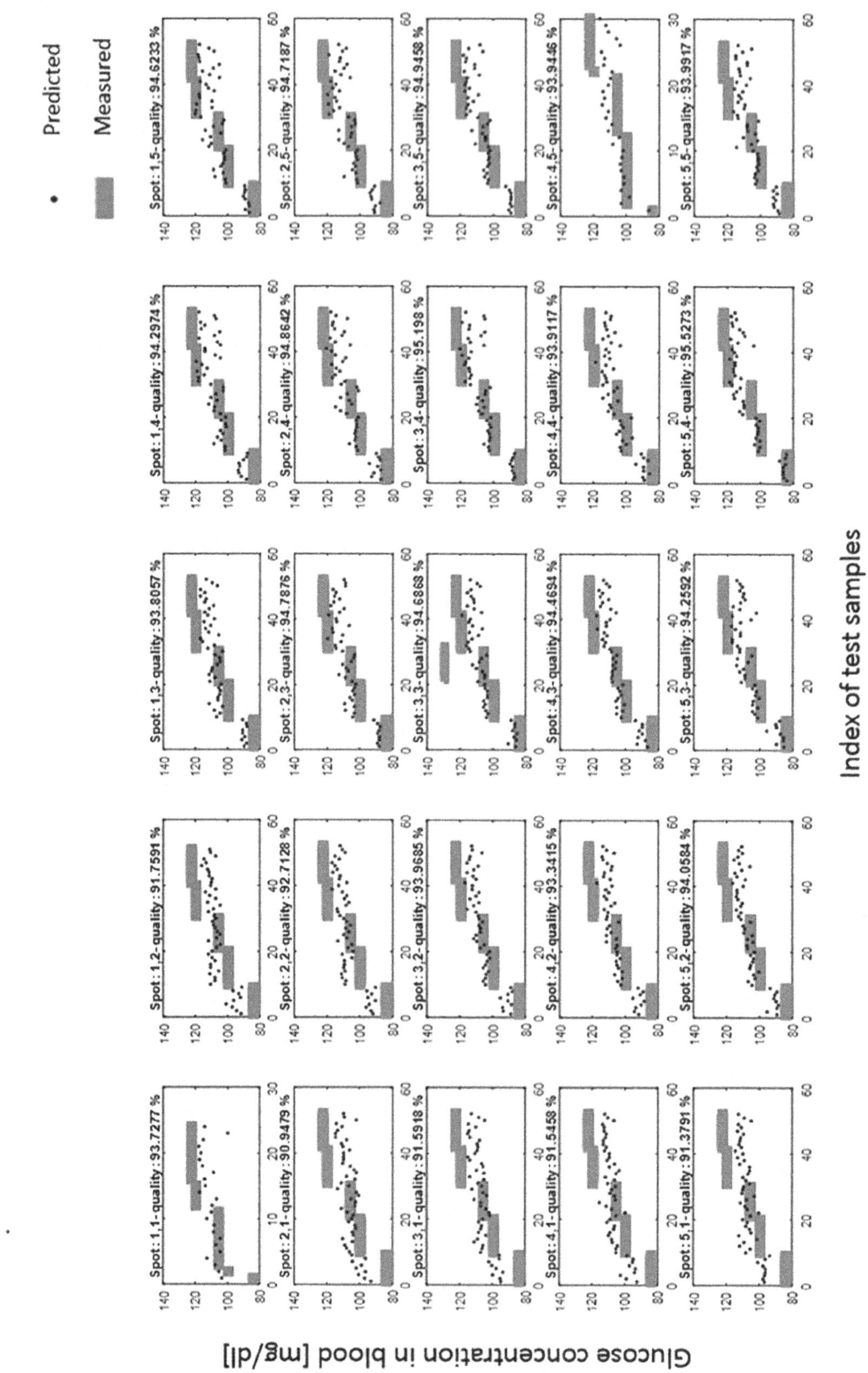

FIGURE 11.3 A prediction map of each spatial illuminating laser spot. Each spot presents a different location next to the subject's leg main blood artery. The graphs show the predicted values denoted by dots with respect to real glucose values denoted by grey bars. (This figure has been adapted from Ref. [20].)

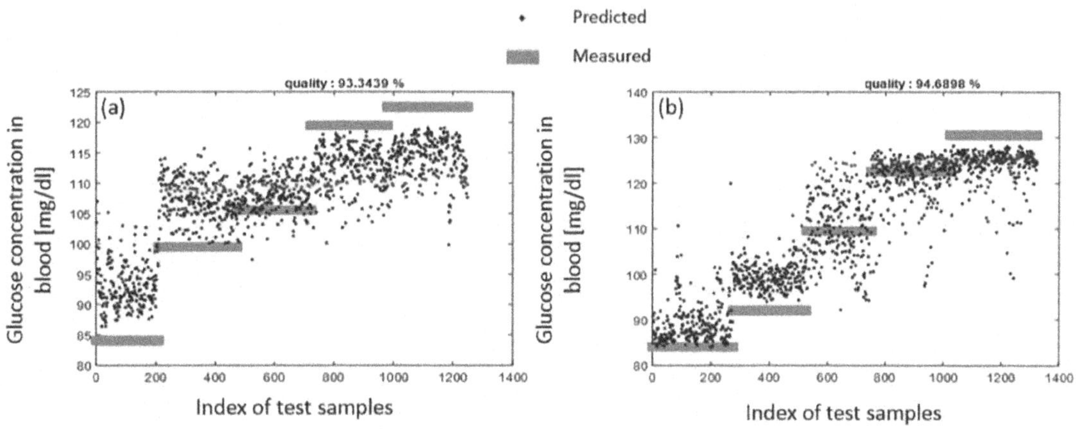

FIGURE 11.4 Two tests of glucose prediction using the presented method. The tests are denoted by (a) and (b). The graphs show the predicted values denoted by dots with respect to real glucose values denoted by grey bars. (This figure has been adapted from Ref. [20].)

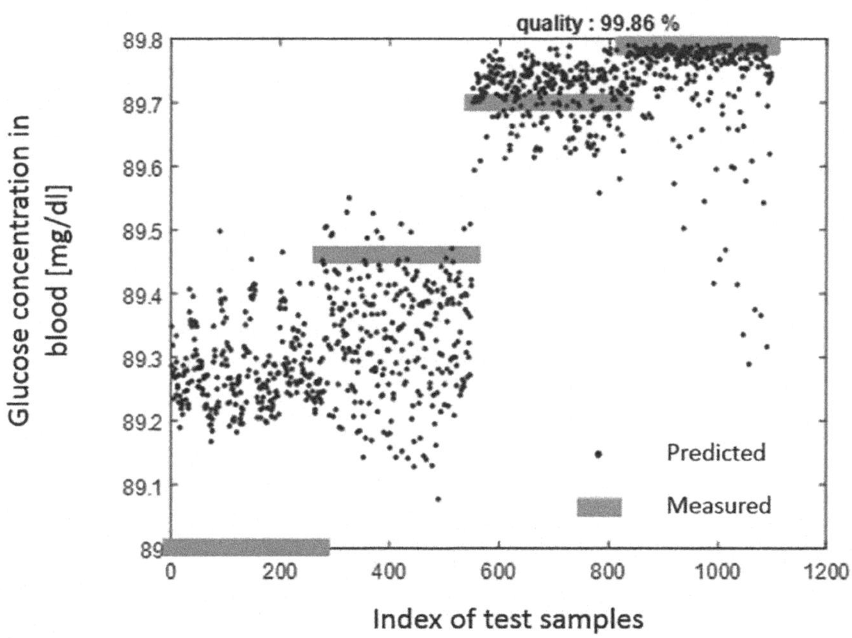

FIGURE 11.5 A graph of a water test. During this test, the subject drank only water without glucose. The graph shows the predicted values denoted by dots with respect to real glucose values denoted by grey bars. (This figure has been adapted from Ref. [20].)

11.4 MAGNETO-OPTIC EFFECT-BASED MEASUREMENTS

As shown in Ref. [17], a change of a wavefront polarization state can be caused by the glucose concentration changes. In magneto-optic materials, polarization will be rotated according to the following expression:

$$\theta = \vartheta BL = {}^{\pi L \Delta n(B)}\!\big/_{\lambda}, \tag{11.4}$$

where ϑ is a Verdet constant, B is the magnetic field, L is the interaction length, λ is the optical wavelength, and Δn is the difference in the index of refraction between two circularly polarized states leading to the rotation. Verdet constant is defined as:

$$V = \frac{\alpha}{l \cdot H \cdot \cos(\varphi)}, \tag{11.5}$$

where α is the angular rotation, l is the length path through the substance, H is the intensity of the magnetic field, and φ is the angle between the magnetic field and the path of the light. As proven in Ref. [18], the minimal magnetic field $B_{\min}$ that will de-correlate the speckle field is proportional to:

$$B_{\min} \propto \pi L \theta R, \tag{11.6}$$

where R is the radius of the illuminating beam and L is the interaction length. It was shown in Ref. [17] that sensitivity of the glucose measurement can also be improved with an AC magnetic field due to the lock-in amplification, while magnetic square wave at specific frequency was generated. However, the main disadvantage of this method is the vibration noise at the same frequency of the magnetic field. The experimental optical remote configuration is shown in Figure 11.6. The configuration consists of a camera (PixelLink PL-E531), which captures images of the time-varied speckle patterns at 2000 fps, an eye-safe 532-nm laser, a polarizer, and a filter. The coil generated magnetic short pulses of 1 ms at 120 Hz, and the detected AC magnetic field was at a strength of 100 Gauss (measured by Gaussmeter, AlphaLab, GM2). Each glucose sample consisted of 1% of intralipid (IL), 1% of agarose, and different concentrations of glucose. Each glucose sample was made in a cuvette that was inserted inside the coil.

One can see in Figure 11.7a–c three different magnetic field change indicators: (1) the intensity change of the image, (2) the speckle pattern correlation coefficient temporal change, and (3) the maximum peak location of the temporal correlation between one speckle pattern and the next. The correlation coefficient was normalized due to the energy of the image. The goal is to prove that the change of the magnetic field, while polarized light passes through a glucose solution, results in the change of the speckle pattern. In addition, change of the wavefront due to change of the polarization angle will affect the intensity. However, the intensity parameter can also be affected by extraneous factors such as changes of external illumination conditions, whereas normalized correlation coefficient will not be affected by such changes.

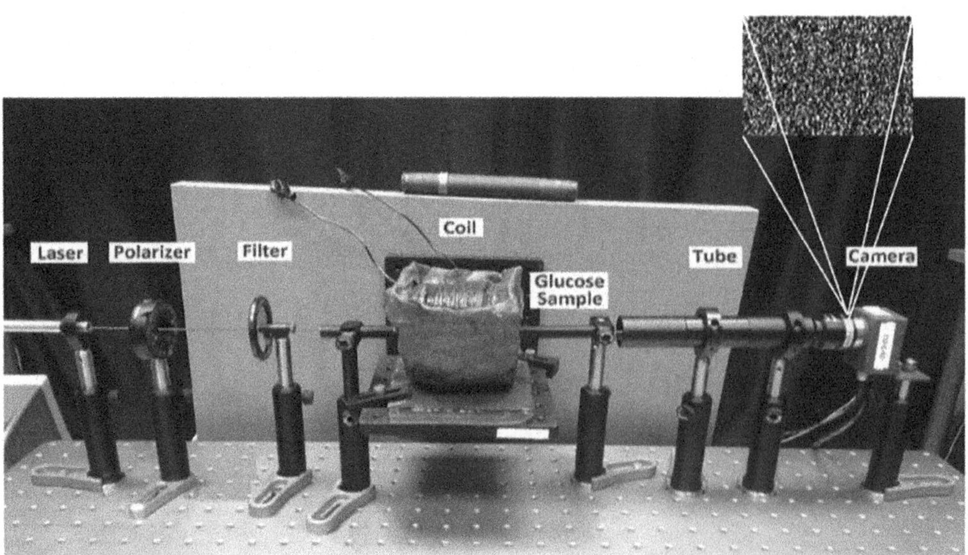

FIGURE 11.6 The remote configuration for the magneto-optic effect. (This figure has been adapted from Ref. [20].)

The main reason for calculating the correlation coefficients is that the reflected light complex amplitude changes with polarization and hence changes the resulting speckle field. Change in the magnetic medium that is generated by time-varied glucose concentration caused temporal changes of the speckle field, which is evaluated by the correlation coefficient of the speckle images row data.

In order to examine the magnet effect while the magneto-optic medium material is glucose substance, we applied three different dc magnetic fields per each trial. During this experiment, the glucose concentration was 0.2% (200 mg/dl). First, a series of reference speckle patterns (without any magnetic field) were recorded for 60 seconds at 2 Hz frame rate (fps). When the magnetic field was applied, the plane of polarization of the beam rotated. A correlation coefficient between the current

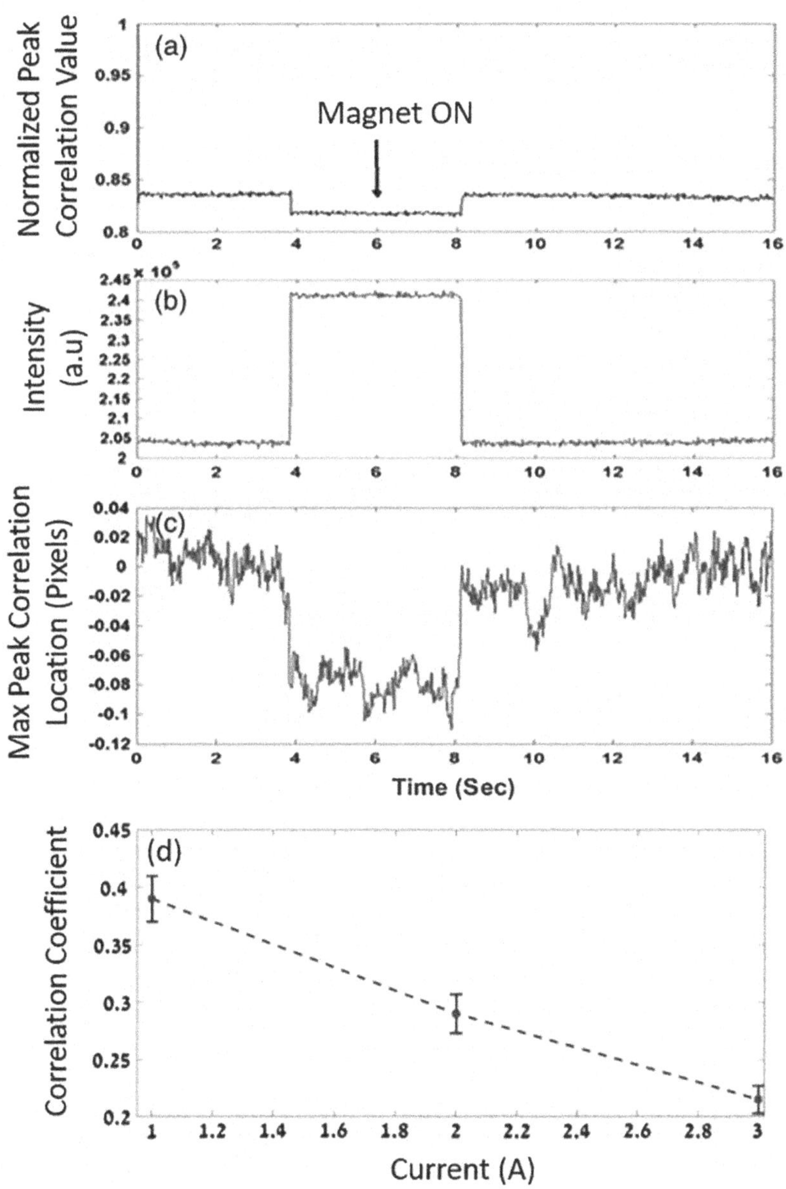

FIGURE 11.7 (a) The value of the maximum peak of correlation between each image and the first image, (b) the intensity of each image, and (c) the location of the maximum peak of correlation between each image and the first image. (d) Change in time-averaged speckle correlation coefficient versus the change in the strength of the applied current. (This figure has been adapted from Ref. [17].)

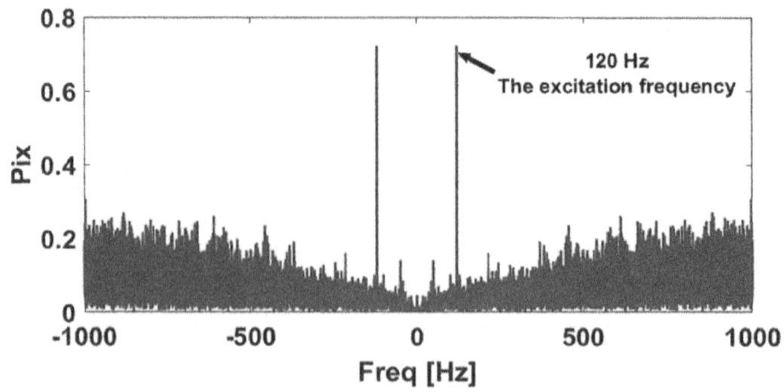

FIGURE 11.8 The frequency response of 1ms magnetic pulses at a repetition rate of 120Hz. (This figure has been adapted from Ref. [20].)

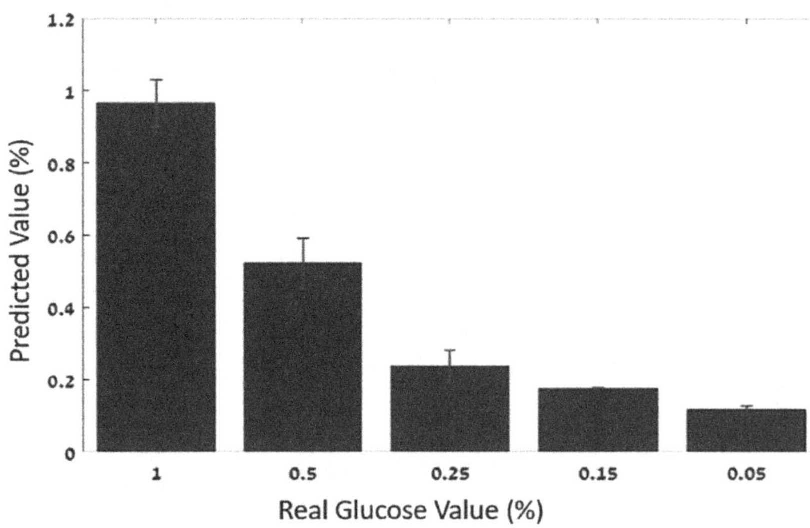

FIGURE 11.9 The glucose prediction values according to multiple linear regression process. (This figure has been adapted from Ref. [20].)

frame and the reference frame, for each of these videos, was computed and was time-averaged. Figure 11.7d shows the difference between the correlation coefficients of different magnetic fields that were applied.

In order to enhance the effect, an ac magnetic field was applied. After extracting the speckle pattern in each frame, the correlation was calculated, revealing a change in the two-dimensional position of the correlation peak and in the value of the peak as a function of time. The algorithm is primarily based upon correlation between the images of the speckle patterns in temporally adjacent frames and the movement in the position and the value of the obtained correlation peak. One can see in Figure 11.8 an example for the frequency response corresponding to 1 ms sec magnetic field pulses at repetition rate of 120 Hz. The frequency response is presented in the Y-axis in pixels units (shifts of the speckle pattern in cameras pixels units).

To extract the glucose concentration value, a multiple linear regression processing was used. This process was calculated according to the equations presented in Ref. [22]. During these measurements, five different glucose concentrations were examined: 1%, 0.5%, 0.25%, 0.15%, and 0.05%. Each sample was measured three times. For this calculation, the frequency responses of X-axis and Y-axis at the excitation frequency were calculated during this process. The results are shown in Figure 11.9. One can see the predicted value according to the suggested regression process.

11.5 SPECKLE-BASED SENSING OF CHEMICALS BY AN ACOUSTIC EXCITATION IN AQUEOUS SOLUTIONS

A schematic illustration of the setup is presented in Figure 11.10. A 780-nm laser source was polarized and diffused into a glass cuvette filled with a solution. The speckle pattern from the diffused light was recorded using a CCD (Basler) or a high-speed camera (Fastcam, Photron). The fluid in the glass cuvette was excited acoustically with a 45×45 mm plate (Steminc).

To drive the PZT, a function generator amplified with an RF power amplifier was used. Since the resonance frequency of the PZT was found to be extremely sensitive to temperature variation, the PZT temperature was stabilized using a thermoelectric cooler (TEC). The TEC was driven by 12V applied through an H-bridge (L298N). The average current to the TEC was tuned by switching the H-bridge on and off with another function generator and controlling its duty cycle. The PZT temperature was measured with a 10K thermistor, sampled with a 24-bit analog-to-digital converter (LTC2400).

The results of the experiment described above are summarized in Figure 11.11. The average speckle size was calculated from the recoded images at various frequencies of the acoustic excitation for deionized (DI) water, salt solution (1g/DL), and glucose solution (1g/DL). Between each swap of the solution, the cuvette was rinsed with DI water. As a control experiment, the cuvette was filled with DI water once again, and as can be seen, a very similar curve was obtained for responsivity of the speckle size to the acoustic excitation.

We discover that the responsivity curve is extremely sensitive to the temperature of the PZT and to avoid this effect the PZT. The temperature of the PZT was stabilized to better until the temperature fluctuations were smaller than 0.01°C. To verify that the temperature is sufficiently stable, we performed several consecutive measurements at time intervals of about 5 minutes. As can be seen in Figure 11.11 (for water and for glucose), the responsivity curves overlap with one another.

It is clear from the results presented here that the responsivity curve of the speckle size to an acoustic excitation which was recorded through a cuvette filled with liquid is sensitive to the presence of substances in that liquid. There is a basis to think that the presence of these substances changes the mechanical properties of the fluid and shifts the resonance frequency of the PZT while it is embedded in them and changes the efficiency at which it excites acoustic waves in the fluid at each given frequency.

While the method presented here is not fully contactless, we clearly showed that it is possible to remotely sense these minute changes in the composition of solutions. Based on these observations,

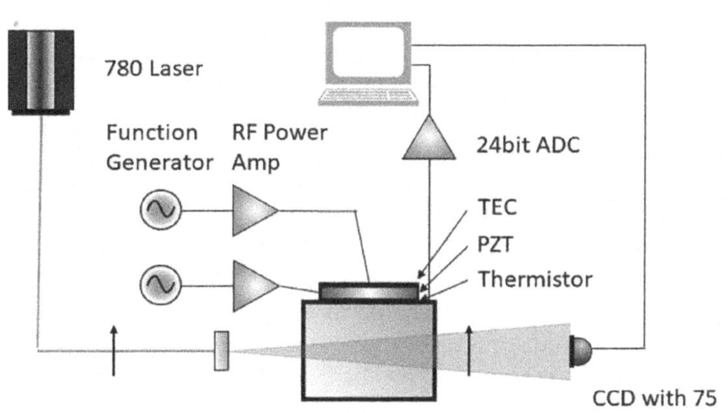

FIGURE 11.10 The schematic system for measuring different chemicals via different acoustic pulses. (This figure has been adapted from Ref. [19].)

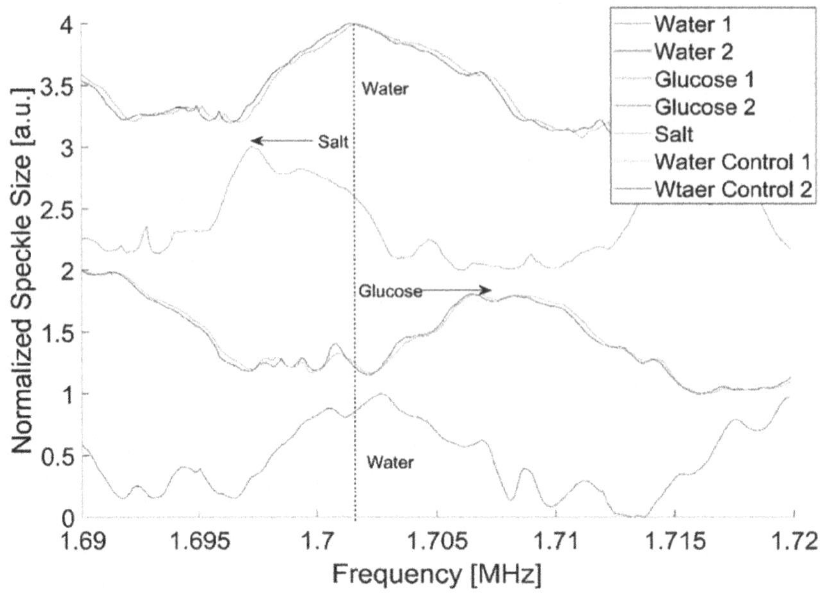

FIGURE 11.11 Summary of the results. Each of the graphs indicates the variation in the speckle size with respect to the frequency of the acoustic excitation. The graphs were vertically shifted for clarity. (This figure has been adapted from Ref. [20].)

it would be interesting to investigate the variation in the resonance frequency of acoustic sources, which can excite acoustic waves in the fluid from a distance.

11.6 REMOTE SENSING OF TISSUE PERFUSION IN THE LOWER LIMBS

In order to control the hemodynamic and thermal state of the patient and to regulate the metabolism, sensing continuously the blood microcirculation is critical, even though it accounts for approximately only 5% of the total blood volume. Fast and convenient monitoring with minimal interference to the patient is an important tool for the clinicians to be able to monitor the state of flow and its distribution in the compartments of the microvasculature.

Microcirculation plays an important role in several vital functions, such as regulation of blood pressure, metabolic exchanges, homeostasis of interstitial fluids, and thermoregulation, to ensure proper tissue nutrition. Therefore, monitoring the distribution of blood between nutritional and non-nutritional skin microvessels remotely and continuously may be an early indication of diabetic foot.

LDF is common noninvasive and remote approach for evaluating the wound depth in patients suffering from PAD injuries. Another method is transcutaneous oximetry (TcPO2) to measure local oxygen. This method is based on measuring the transfer of oxygen molecules to the skin surface. A local perfusion index can be extracted by calibrating TcPO2 with a TcPO2 value measured from the patient's chest. Laser speckle imaging (LSI), known also as laser speckle contrast analysis (LASCA), measures the temporal contrast of back-scattered light. In this method, the tissue is illuminated by a long coherence length near-infrared laser, and the back-scattered pattern formed by moving scatterers, mostly red blood cells. The decay of the measured temporal intensity autocorrelation function originated by the speckle fluctuations is proportional to blood flow.

Here, we describe LASCA-based method for the evaluation of perfusion in lower limbs of PAD patients. We will show a simple experimental setup for the detection of PAD feet. About 23 feet were enrolled: 15 healthy and 8 sick.

The optical sensor consists of a high-speed camera (Basler, Aca 800-510uc) and a diode laser at 780 nm (50 mW, illuminates a ~5 mm spot diameter (Figure 11.12a). The camera frame rate was set

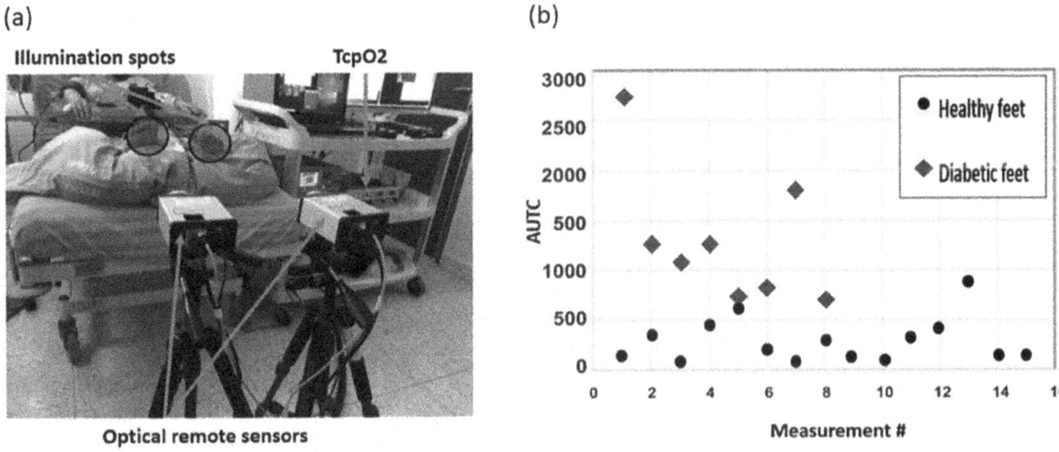

FIGURE 11.12 (a) Remote sensing of diabetic foot perfusion state. (b) AUTC-based comparison. (This figure has been adapted from Ref. [23].)

to 800 frames per second. The laser illuminates the inspected tissue and the spatial temporal back-reflected light is analyzed and the temporal autocorrelation function is calculated.

To distinguish between PAD patients suffering from low perfusion in their feet, and between healthy control groups, an occlusion test was demonstrated. The protocol consists of the following steps: (1) 3 minutes of baseline following by 3 minutes of lower limbs occlusion (via dedicated sphygmomanometer). (2) 9 minutes of post-occlusion time. Later, the area under the decay curve (AUTC) was calculated in order to distinguish between healthy and diabetic feet as shown in Figure 11.12b.

11.7 SUMMARY

There are three main ways to diagnose diabetes: (1) fasting plasma glucose (FPG): the blood glucose concentration after fasting 8 hours is greater than 126 mg/dl. (2) Oral glucose tolerance test (OGTT): 2-hour test that monitors the subject's glucose blood levels before and 2 hours after the subject drank a sweet liquid. Diabetes is diagnosed in this test as greater than or equal to 200 mg/dl. (3) Casual plasma glucose test: blood check at any time of the day in a case of severe diabetic symptoms. Diabetes is diagnosed in this test as greater than or equal to 200 mg/dl. There might be a big usefulness of our technique in the medical practice by providing the trend of the glucose concentration in their bloodstream via the change of the OPG together with machine learning techniques that can improve the sensitivity of the measurement.

Speckle-based Faraday effect sensing with an AC magnetic field has still several technological challenges to be overcome in order to make the proposed sensor an applicable device. The main challenge is to add a mechanical and acoustic cancelation mechanism since at the moment; acoustic and mechanical noises are still evident during the measurement, while the AC magnetic field is applied. Several methods can enhance the measurement resolution such as acoustic cage or using a rotating magnetic field for the locking amplifier method. Further maturing of a more robust, generic, and automatic calibration process that will translate the optical readout into the exact value of the estimated biomedical parameter is also required.

In the clinical study of the tissue perfusion in the lower limbs (i.e., "diabetic foot"), we presented one possible application of remote optical measurement configuration based on the analysis of back-reflected speckle patterns generated when illuminating the inspected tissue with a laser light. In the measurements, we were able to transfer the dynamics of the speckle patterns to the real-time estimation of physiological quantity in the real world, being in this case the amount of blood perfusion. We used commercial-off-the-shelf components to demonstrate the robustness and ease of use of the sensor.

ACKNOWLEDGMENTS

The authors would like to acknowledge the following collaborators that contributed to the scientific work on which this chapter is based: Y. Beiderman, A. Anand, B. Javidi, S. Polani, A. Schwarz, A. Shemer, J. Garcia, H. Genish, M. Golberg, R. Califa, O. Goldstein, A. Zailer, M. Niska, R. Talman, J. Ruiz-Rivas, N. Arbel, V. Mico, and M. Sanz.

REFERENCES

1. National Diabetes Information Clearinghouse (NDIC), http://diabetes.niddk.nih.gov/dm/pubs/statistics/.
2. G. Danaei, M. M. Finucane, Y. Lu, G. M. Singh, M. J. Cowan, and C. J. Paciorek, "National, regional, and global trends in fasting plasma glucose and diabetes prevalence since 1980," *Lancet* **378**, 31–40.
3. H. A. MacKenzie, H. S. Ashton, S. Spiers, Y. Shen, S. E. Freeborn, J. Hannigan, J. Lindberg, and P. Rae, "Advances in photoacoustic noninvasive glucose testing," *Clin. Chem.* **45**, 1587–1595 (1999).
4. R. O. Esenaliev, K. V. Larin, I. V. Larina, and M. Motamedi, "Noninvasive monitoring of glucose concentration with optical coherence tomography," *Opt. Lett.* **26**, 992–994 (2001).
5. B. D. Cameron and Y. Li, "Polarization-based diffuse reflectance imaging for noninvasive measurement of glucose," *J. Diabetes Sci. Technol.* **1**, 873–878 (2007).
6. A. Tura, A. Maran, and G. Pacini, "Non-invasive glucose monitoring: Assessment of technologies and devices according to quantitative criteria," *Diabetes Res. Clin. Pract.* **77**(1), 16–40 (2007).
7. J. Chung, C. So, K. Choi, and T. K. S. Wong, "Recent advances in noninvasive glucose monitoring," *Med.Devices* **5**, 45 (2012).
8. M. Hofmann, M. Bloss, R. Weigel, G. Fischer, and D. Kissinger, "Non-invasive glucose monitoring using open electromagnetic waveguides," in *European Microwave Week 2012: "Space for Microwaves", EuMW 2012, Conference Proceedings - 42nd European Microwave Conference (EuMC 2012)*, pp. 546–549, IEEE Computer Society (2012).
9. D. C. Klonoff, "Overview of fluorescence glucose sensing: A technology with a bright future," *J. Diabetes Sci. Technol.* **6**(6), 1242–1250 (2012).
10. A. M. Winkler, G. T. Bonnema, and J. K. Barton, "Optical polarimetry for noninvasive glucose sensing enabled by Sagnac interferometry," *Appl. Opt.* **50**(17), 2719–2731 (2011).
11. C. C. Pelletier, J. L. Lambert, and M. Borchert, "Determination of glucose in human aqueous humor using Raman spectroscopy and designed-solution calibration," *Appl. Spectrosc.* **59**(8), 1024–1031 (2005).
12. L. Zhu, J. Lin, B. Lin, and H. Li, "Noninvasive blood glucose measurement by ultrasound-modulated optical technique," *Chinese Opt. Lett.* **11**(2), 021701–021705 (2013).
13. N. Ozana, Y. Bishitz, Y. Beiderman, J. Garcia, Z. Zalevsky, and A. Schwarz, "Remote optical configuration of pigmented lesion detection and diagnosis of bone fractures," in *Photonic Therapeutics and Diagnostics XII*, B. Choi, N. Kollias, H. Zeng, H. W. Kang, B. J. F. Wong, J. F. Ilgner, G. J. Tearney, K. W. Gregory, L. Marcu, M. C. Skala, P. J. Campagnola, A. Mandelis, and M. D. Morris, Eds., p. 968916, SPIE, 9689 (2016).
14. N. Ozana, J. A. Noah, X. Zhang, Y. Ono, J. Hirsch, and Z. Zalevsky, "Remote photonic sensing of cerebral hemodynamic changes via temporal spatial analysis of acoustic vibrations," *J. Biophoton.* **13**(2), e201900201 (2020).
15. Y. Tzabari Kelman, S. Asraf, N. Ozana, N. Shabairou, and Z. Zalevsky, "Optical tissue probing: Human skin hydration detection by speckle patterns analysis," *Biomed. Opt Express* **10**(9), 4874 (2019).
16. G. Janatsch, H. M. Heise, J. D. Kruse-Jarres, P. Bhandare, R. Marbach, R. A. Peura, and Y. Mendelson, "Multivariate determination of glucose in whole blood using partial least-squares and artificial neural networks based on mid-infrared spectroscopy," *Appl. Spectrosc.* **47**(8), 1214–1221 (1993).
17. N. Ozana, Y. Beiderman, A. Anand, B. Javidi, S. Polani, A. Schwarz, A. Shemer, J. Garcia, and Z. Zalevsky, "Noncontact speckle-based optical sensor for detection of glucose concentration using magneto-optic effect," *J. Biomed. Opt.* **21**(6), 065001 (2016).
18. A. Anand, V. Trivedi, S. Mahajan, V. K. Chhaniwal, Z. Zalevsky, and B. Javidi, "Speckle-based optical sensor for low field faraday rotation measurement," *IEEE Sens. J.* **13**(2), 723–727 (2013).
19. H. Genish, N. Ozana, and Z. Zalevsky, "Speckle based sensing of chemicals by an acoustic excitation in aqueous solutions," *Proceedings of SPIE Conference 10895, Frontiers in Biological Detection: From Nanosensors to Systems XI*, SPIE Photonics West, San Francisco, CA (February 2019).

20. N. Ozana, R. Talman, A. Shemer, A. Schwartz, S. Polani, R. Califa, Y. Beiderman, J. Ruiz-Rivas, J. García, and Z. Zalevsky, "Remote photonic sensing of glucose concentration via analysis of time varied speckle patterns," *Adv. Mater. Lett.* **9**(9), 624–628 (2018).
21. N. Ozana, N. Arbel, Y. Beiderman, V. Mico, M. Sanz, J. Garcia, A. Anand, B. Javidi, Y. Epstein, and Z. Zalevsky, "Improved noncontact optical sensor for detection of glucose concentration and indication of dehydration level," *Biomed. Opt. Express* **5**(6), 1926 (2014).
22. S. H. Brown, "Multiple linear regression analysis: A matrix approach with MATLAB," *Alabama J. Math.* (2009).
23. M. Golberg, R. Califa, S. Polani, O. Goldstein, A. Zailer, M. Niska, and Z. Zalevsky, "Assessment of tissue perfusion in the lower limbs using laser speckle analysis," *Proceeding of SPIE Conference 11641, Dynamics and Fluctuations in Biomedical Photonics XVIII*, SPIE Photonics West, San Francisco, CA (March 2021).

Index

Note: **Bold** page numbers refer to tables and *italic* page numbers refer to figures.